T0077246

Praise for
Grace in All Simplicity

"In *Grace in All Simplicity*, Cahn and Quigg share with us their ring-side view of the discoveries that led to our current picture of our most fundamental physical laws. Ironically, the path taken that led to our emerging view of the Universe is neither simple or graceful, but spell-binding and charming in all its human complexities."

—**Professor Steven Chu, Nobel laureate, Professor of Physics, Molecular and Cellular Physiology and Environmental Science and Engineering, Stanford University**

"Engaging and captivating, *Grace in All Simplicity* is filled with riveting stories of the ups and downs of the remarkable scientists who made key experimental and theoretical advances in the search for a fundamental theory that might explain it all. The last chapter on 'The Best of All Possible Worlds?' is itself worth getting this book."

—**E. William Colglazier, former science and technology adviser to the U.S. Secretary of State, former executive officer of the U.S. National Academy of Sciences**

"Cahn and Quigg regale us with non-stop fascinating stories about the quirky characters and inventive experiments that have built our understanding of the physical world and how it works. And those physicists, their experiments—and the rules of our universe that they discovered—really are quirky! A rare insider view into the dramatic developments of particle physics and cosmology."

—**Saul Perlmutter, Nobel Laureate in Physics, Professor of Physics at University of California, Berkeley**

"*Grace in All Simplicity* is an accessible and comprehensive narrative that reveals the history, personalities, experiments, accomplishments, failures, and metaphysical speculations involved in the rise of modern science. Cahn and Quigg deliver a realization of amazing grace in this extraordinarily satisfying and extremely educational read."

—**S. James Gates, Jr., Clark Leadership Chair in Physics and Public Policy, recipient of the National Medal of Science**

"A grand history of the brilliant physicists who answered the question of all questions: *What is the whole universe made from?*"

—**Edward Tufte, statistician, visualizer, artist, and professor**

"In this insightful and accessible book, Quigg and Cahn celebrate the people behind these discoveries, following their paths around the globe to explore the extraordinary experiments that probe the fundamental properties of the universe. The vast collective of minds that collaborate in this building, across space and time, is an astonishing feat itself—one driven by the most essential curiosity: that of our existence. A captivating book that reveals the interconnectedness of science's most profound advances."

—*Kirkus Reviews*

"Cahn and Quigg's marvelous book gracefully leads us on a promenade through the many wonders that physicists and cosmologists, like themselves, have uncovered in the past hundred plus years. In doing so, their writing is a model of lucidity. I remain in awe of their achievement."

—**Gino Segrè, Professor of physics, University of Pennsylvania,**
author of *The Pope of Physics*

"Robert Cahn and Chris Quigg share an uncanny ability to open up complex and bedrock scientific concepts to ordinary people in a way that clarifies the mind, makes what seemed impenetrable clear, and most of all sparks curiosity. After any given page of this book, a reader wants to know more."

—**Greil Marcus, author of *Lipstick Traces***

"At heart, this book is a travelogue that takes readers on a journey through conceptual landscapes so exotic they challenge the imagination. It's a highly engaging account of how successive generations of scientists broke through to the strange reality that underpins the familiar world we inhabit, beautifully capturing the wonder and fun of it all."

—**J. Madeleine Nash, former *TIME* senior science correspondent and**
author of *El Niño: Unlocking the Secrets of the Master Weather-Maker*

Erudite yet accessible, *Grace in All Simplicity* interweaves scientific adventure with remarkable human stories. *Grace* is also personal, as the authors Cahn and Quigg themselves participated in—and were affected by—many of these great discoveries. This makes the book a fine embarkation for those just beginning their journey, for those already on a scientific path, or for anyone who has followed this great story."

—**David Saltzberg, Professor of physics and astronomy,**
University of California Los Angeles, Science Consultant for
***The Big Bang Theory, Young Sheldon*, and *Oppenheimer*.**

GRACE IN ALL SIMPLICITY

BEAUTY, TRUTH, AND WONDERS ON THE PATH TO
THE HIGGS BOSON AND NEW LAWS OF NATURE

ROBERT N. CAHN
AND CHRIS QUIGG

PEGASUS BOOKS

NEW YORK LONDON

GRACE IN ALL SIMPLICITY

Pegasus Books, Ltd.
148 West 37th Street, 13th Floor
New York, NY 10018

Copyright © 2023 by Robert N. Cahn and Chris Quigg

First Pegasus Books cloth edition November 2023

Interior design by Maria Fernandez

All rights reserved. No part of this book may be reproduced in whole or in part
without written permission from the publisher, except by reviewers who may quote
brief excerpts in connection with a review in a newspaper, magazine, or electronic
publication; nor may any part of this book be reproduced, stored in a retrieval
system, or transmitted in any form or by any means electronic, mechanical,
photocopying, recording, or other, without written permission from the publisher.

Library of Congress Cataloging-in-Publication Data is available.

ISBN: 978-1-63936-481-7

10 9 8 7 6 5 4 3 2

Printed in the United States of America
Distributed by Simon & Schuster
www.pegasusbooks.com

To the colleagues, teachers, and students who enrich our lives in science, and to all who view the natural world with a sense of wonder.

CONTENTS

INVITATION

*G*race in All Simplicity celebrates and illuminates science in the making, featuring the people who make the discoveries and reveal the insights that redefine our place in the universe. Many of its stories tell how new truths were found, while others illustrate the puzzles for which we have as yet no answer. To provide answers, today physicists and astronomers are exploring distances from a billionth of a billionth of the human scale to the entire cosmos, and contemplating time intervals that range from less than a trillionth of a trillionth of a second out to far longer than the age of the universe. To explore—to leave home in a metaphorical way—requires devising new instruments that spectacularly expand our senses and conceiving new ways of thinking that expand our minds.

We invite you to join us as we follow the path of extraordinary people who uncovered new laws of nature. The trail leads to mountaintops, to caverns, to the coldest place on Earth, and to the farthest reaches of the cosmos. We will be traveling light, with no equations, no diagrams, no tables, no pictures. The best part of many journeys is often the people you meet along the way, so we will introduce you to some of our friends and colleagues who made fundamental discoveries over the past fifty years and to some of our heroes of the past. Among these are professional scientists and amateurs, eccentrics and normal folk, performers and introverts.

Grace in All Simplicity is your guidebook. The journey we've laid out is not chronological but our recommended route. Together, we will travel the path to the Higgs boson, weigh the evidence for subliminal dark matter, and learn what makes scientists invoke a mysterious agent named "dark energy." We will behold the emergence of a compelling picture of matter and forces, simple in its structure, graceful in the interplay of its parts, but still a tantalizing work in progress. We thrill to the prospect that, at any moment, vast new vistas could open before us. This trail isn't designed for experts. You don't need to be a scientist to appreciate the tour any more than you need to be a musician to appreciate a symphony. All you need to bring is your curiosity.

Our first stop will be along the border between Italy and Switzerland, where three young men are stringing cables across deep chasms in the hope of harnessing the might of lightning to split the atom. Would they be the first to succeed in tearing apart the nucleus on a grand scale, or would makers of the first particle accelerators win the prize?

We next step back in time to 18th-century London, where Stephen Gray, a pensioner living in the former monastery Charterhouse, devotes himself to playing with static electricity, distinguishing conductors from insulators, and making the first primitive telegraph.

A bit before the outbreak of World War I, we visit the laboratory of Ernest Rutherford, the first to surmise that atoms contain a compact nucleus surrounded by electrons. Young Harry Moseley finds a way to measure the positive charge on each nucleus, refining Mendeleev's periodic table and predicting the existence of four missing elements.

We climb through the mists that enshroud the top of Ben Nevis, the highest peak in the Scottish Highlands, to join C. T. R. Wilson in 1911 as he wonders whether he can re-create the scene in the laboratory. When he does, his cloud chamber reveals contrails left by visitors from outer space, the cosmic rays, and opens a new world of varieties of matter never before seen. Cloud chambers in the Alps and the Andes then reveal a zoo of bizarre particles, and more appear once particle accelerators replace the random showers of cosmic rays with tamed beams of charged particles. Then a pattern emerges: everything can be explained if there are three building blocks called quarks. But no one had ever seen an isolated quark. Still no one has, and yet . . .

You will walk with us, your humble tour guides, on November 11, 1974, when everything changes, when the quarks become so real that we can *almost* touch them (but can't). What made quarks real was a new particle made of a fourth quark, the charmed quark, and its antimatter counterpart.

In pursuit of theoretical understanding of this new discovery, we step back to 1928 to join Paul Dirac who has set out to write the simplest equation that might describe an electron. When he deciphers what he himself has written, it is a single line that describes the electron, its interactions with a light quantum, and—to make the picture hang together, an antimatter electron, the positron. His equation has become a script for a multi-act play, a theory called quantum electrodynamics. After six acts, theory matches experiment at the level of one part per trillion.

Emmy Noether proves that symmetries in the laws of nature lead to laws like energy conservation, and that symmetries can dictate forces. Hermann Weyl finds that a symmetry inherent in quantum mechanics implies the electromagnetic force. Their reasoning has led us to a profound new understanding of the forces of nature.

The idea of quarks seemed too simple to be true, and no one could explain why they combined to make the observed particles. For the story to make sense, quarks needed to have a new attribute, called color. Following Noether's example, symmetry among the colors implies a force that ties the quarks together so tightly that no one has ever found an isolated quark. It is a new law of nature—quantum chromodynamics.

Wolfgang Pauli confronts a startling result: careful measurements show that when an element emits an electron to become a neighboring element in the periodic table, some energy seems to disappear, which would mean that energy is not conserved. Pauli proposes a desperate remedy: the energy is carried away by a particle that is not seen, and that he imagines will never be seen. Enrico Fermi joins Pauli's conjecture with Dirac's methods to make a theory whose influence is still felt.

We take a break from alien particles and recondite theories to visit the Dutch university town of Leiden, the coldest place on Earth in 1911. Heike Kamerlingh Onnes, the first to liquefy helium at 4.2°C above absolute zero, could survey how substances behave in the ultracold. There the electrical resistance of mercury simply vanished, meaning that a current could circulate forever, undiminished—superconductivity! At Bell Labs in the 1950s, Ted Geballe and Bernd Matthias conjure up new recipes for superconductors that will make electromagnets of unprecedented strength.

Fermi's theory of weak interactions endured for three decades with one simple amendment. But unlike the theory of electromagnetic interactions, Fermi's script ended at the first act. The insight leading to an improved theory came from superconductivity. The successor to Fermi's theory, formulated by Steven Weinberg and others, adds the insight that many symmetries in nature are hidden to the principle that symmetries dictate interactions. The electroweak theory predicts a new weak interaction and a new massive particle, the Higgs boson.

What innovations would be required to test the theory? Today's premier accelerator, the Large Hadron Collider at CERN, takes a tamer path to high energies than the trio had tried on Monte Generoso. A ring twenty-seven kilometers around, the LHC uses many gentle pushes to accelerate protons through 6.8 *trillion* volts.

Inventions that extend our senses drive fundamental discovery. The cloud chamber and bubble chamber allowed us to see the interactions of subatomic particles with our eyes. But our eyes are no match for electronics, which can both observe and record. The magic of silicon, which has defined our digital age, has also transformed particle physics. But the legacy of the cloud chamber and the bubble chamber has not been lost. Dave Nygren, a latter-day wizard of particle detectors, invented the Time Projection Chamber, which gave electronic detectors the capability to follow tracks in three dimensions.

Inspiration and innovation can come from unexpected places. Scores of American children in our generation dreamed of playing baseball in the major leagues. For most, this dream was out of reach. But we had *All-Star Baseball*, a board game in which a turn of the spinner would reenact with proper probabilities the batting prowess of the great Babe Ruth and the heroes of our youth. This game of chance, invented by major leaguer Ethan Allen, could simulate with satisfying verisimilitude what could have happened. We were using the Monte Carlo method, essential to contemporary particle physics and finance, long before we even knew its name.

By the summer of 2012, forty years of evidence argued that symmetries and hidden symmetries lie behind the strong, weak, and electromagnetic interactions. One by one, the predictions of that paradigm had been borne out in the interactions of neutrinos and electrons and in the highest energy collisions. But the Higgs boson, the keystone to the electroweak theory, was still an object of desire. Then on the Fourth of July, two experiments at the Large Hadron Collider revealed a new particle that looked for all the world like the Higgs boson of our dreams. Was it, indeed? And what might we learn from it? What would come next?

The picture of fundamental particles and the forces that rule them looks complete. This might tempt us to ask, as every trail walker sometimes does, "Are we there yet?" We may have accounted for the microcosm, but the macrocosm tells us this isn't at all the end of the trail. By cosmic accounting, we have achieved an excellent description of just less than 5 percent of what must actually be present in the universe.

The birth-cry of the universe reaches us in the form of a steady radio hum, very nearly uniform across the sky. Three decades after the discovery of this cosmic microwave background, George Smoot and his team detected exquisitely subtle stippling in the baby picture of the universe when it was, in human terms, just a day old. Variations in the temperature of patches in the heavens of perhaps one part in 100,000 represent the seeds of all the large-scale structure in the universe. Cosmology becomes a quantitative science.

Would the universe expand forever, gradually slowed by the pull of gravity? Would the expansion slow so much that the universe would begin to contract, as a ball thrown in the air returns to Earth? When Saul Perlmutter's team and their rivals produced growth histories of the universe, the astonishing answer was "Neither!" More than half the energy in the universe is a mysterious something that accelerates the expansion. It is not matter of any sort, so we give it the name dark energy. To add to the puzzle, about four-fifths of the matter around us is not the stuff we see, but a nearly imperceptible something that we call dark matter.

Our trail ends, for now, at one of the deepest questions of all, explored more than half a century ago by Andrei Sakharov. How can we understand why there is anything tangible at all in the universe? Why didn't all the matter and antimatter annihilate at the beginning? Sakharov set down three criteria for the persistence of matter that guide our search for an explanation today.

These are just some of the highlights of the path that *Grace in All Simplicity* will follow. As one puzzle after another has been resolved by experiment or insight, new ones appear. The stakes grow higher and higher. How can we have a perfect picture of quarks and the like and still be missing most of the matter in the universe? What is that missing energy, the dark energy?

We have made enormous progress toward making sense of the physical world. But the consequences of our best theory, the provisionally named "standard model of particle physics," depend on some two dozen parameters—among them, the masses of the elementary particles and the strengths of their interactions. These dictate the nature of our quotidian world. How they got their values we cannot yet say. Changing these parameters, even a little, would lead to a very different world. Are we just lucky? Is ours the best of all possible worlds? Here physics runs up against metaphysics, but many remaining questions are unmistakably scientific. Answers can only come from new explorations and the new discoveries they will bring. Remarkably, the more we have learned and the more we have tried to explain, the simpler and more satisfying the full explanation has become. In that simplicity lies not just grace, but beauty and truth.

1

SPLITTING THE ATOM: 1927

Thunderstorms visit most of Europe only eleven days a year, but during the summer season on Monte Generoso in the Italian-Swiss Alps, they give a breathtaking performance every other day. Afternoon clouds sweep in, welling up deep and rich as bruises, purpling the waters of Lake Lugano nearly 5,000 feet below. Lightning slips down the throat of the sky like a shot of grappa.

Thunder echoes majestically through mountain corridors. The next morning, the newly washed air reveals the great canine of the Matterhorn, sixty miles away. Now and then, Monte Viso appears, a faint obelisk of snow more than a hundred miles to the southwest.

Visitors to Monte Generoso in the summer of 1927 found new reason to be conscious of the might of thunderstorms. A net woven of iron cable, nearly half a mile long, stretched between the summit and the end of a ridge 500 feet below. At night, a corona of bluish-green light, sometimes many yards across, shimmered around the net, as mesmerizing as the aurora borealis. On several occasions during storms, loud reports sounded once a second from the vicinity of the net. According to a dispatch published in the far-off *New York*

Times, "the hills rang as if a giant machine gun were laying a barrage on the enveloping clouds."

One can picture summer residents exchanging reports and speculations over *aperitivi* about the construction on the ridge and the three young men from Berlin who were leading the work. Did the displaced northern lights and the rapid-fire thunder mean that scientists were able to control the weather? Or had engineers devised a means to collect energy from thunderstorms? A persistent theory, endlessly embroidered, was that the young Germans were trying to harness the power of the atom, so that the world could dispense with oil and coal.

Arno Brasch, Fritz Lange, and Kurt Urban welcomed their celebrity, and took some pleasure in the talk of atomic energy sources. That possibility had lived in popular imagination since the very beginning of the 20th century, when Ernest Rutherford and Frederick Soddy, a twenty-three-year-old chemist, grasped that in the glow of radium they were seeing a process far more powerful than any chemical reaction: the transmutation of one element into another.

Liberating atomic energy was not the immediate goal that drew scientists to Monte Generoso, but the tales invented to account for the strange goings-on nevertheless contained more than a grain of truth. By recreating Benjamin Franklin's mythic kite experiment on an epic scale, the three young physicists hoped to summon from nature enough energy to dismantle the atomic nucleus. What drove them to labor with their ungainly contraption was a passion to know what the world is made of, following in Rutherford's footsteps.

Ernest Rutherford came from a family of New Zealand homesteaders. A stellar student, he was already doing impressive research on magnetism by the age of eighteen, but life in colonial New Zealand demanded more prosaic pursuits. Twenty-three-year-old Ernest was toiling in the family garden when his mother brought the telegram announcing that he had won an 1851 Exhibition Fellowship that would take him from the antipodes to the University of Cambridge. "That's the last potato I'll dig!" he crowed.

It was a propitious moment to launch a scientific career. In 1895, the year Rutherford came to Cambridge, Wilhelm Conrad Roentgen discovered X rays. The next year Henri Becquerel found that uranium emitted unseen radiation that would fog photographic plates. Maria Skłodowska-Curie and her husband, Pierre, isolated polonium and radium, new elements even more radioactive than uranium. In 1897, Rutherford's mentor, J. J. Thomson, discovered that electrons, electrically charged particles much lighter than an atom, were ubiquitous constituents of matter. Thomson hypothesized that an atom was composed of thousands of electrons, whose negative charge was balanced by a positive charge spread uniformly throughout the volume containing the electrons. In Thomson's atom, electrons were scattered about like raisins in a plum pudding.

In Thomson's laboratory, Rutherford joined the pursuit of the mysterious uranium rays. He found that radium emitted two distinct kinds of radiation. One, which he called alpha, was blocked by thin aluminum foils, or even by sheets of cardboard, and was deviated only slightly by a magnetic field. The other, which he called beta, penetrated such barriers easily but was influenced more strongly by a magnetic field. Later, he gave the name gamma to a still more penetrating kind of radiation that was unaffected by a magnetic field.

Already by the fall of 1898, Rutherford's research into radioactivity had gained such recognition that he was offered a professorship at McGill University in Montreal. Superb equipment compensated for Canada's remoteness from the great European centers, and new discoveries followed quickly.

Working with Soddy, who had come to Canada from Oxford, Rutherford established that radioactivity actually changed one element into another. "Rutherford, this is transmutation," Soddy declared one day as they were isolating radium that had begun as thorium. "For Mike's sake, Soddy," Rutherford burst in, "don't call it transmutation. They'll have our heads off as alchemists!" (In 1937, a decidedly less timorous Rutherford published a thin volume titled *The Newer Alchemy: The transmutation of elements, how it has been accomplished, and what it means.*) When radium breaks apart, it expels an alpha particle moving at about 10,000 miles per second—roughly a twentieth of the speed of light—and leaves behind an atom of a new element, the radioactive "radium emanation" now familiar as radon, the insidious infiltrator of homes. When Rutherford and Soddy calculated the energy carried away by alpha particles in the disintegration of radium, they glimpsed a reservoir of energy within the atom that "may be a million times as great as the energy of any molecular change," like burning wood or exploding dynamite.

In the enthusiastic final chapter of his popular account of their discoveries, *The Interpretation of Radium* (1909), Soddy tried to imagine the implications of the newly recognized stores of energy. He compared his moment in history to the time when primitive man first noticed, but had not yet mastered, the energy liberated by fire, and imagined how the world could be changed if humans could learn to kindle the radioactive flame. "Radium," Soddy wrote, "has taught us that there is no limit to the amount of energy in the world available to support life, save only the limit imposed by the boundaries of knowledge." But, he cautioned, a single mistake had the potential for infinitely disastrous consequences. He even ventured that a past calamity in which knowledge of the atom was imprudently used might possibly account for the biblical tale of the Fall of Man.

Soddy's prophetic vision of a civilization with limitless possibilities poised on the brink of catastrophe resonated with both the optimistic faith in technology and the reformist zeal of H. G. Wells. In *The World Set Free*, written a year before the start of the Great War and dedicated to *The Interpretation of Radium*, Soddy's conditionals became Wells's statements of historical fact: *coulds* were replaced by *dids*. Wells put 1933 as the date when mankind would learn how to speed up the natural process of radioactivity and tap the energy within the atom to light a great city, to drive the wheels of industry, to power aircraft, and, inevitably, to explode devastating weapons. The threat that a world war fought with atomic bombs would end civilization was an ideal vehicle for Wells's campaign for a world government.

The atom was far from the pages of popular fiction when Rutherford came on the scene.

Although we can trace the notion of fundamental constituents of matter—minimal parts—through the 1st century B.C.E. Roman poet Lucretius to the ancients, the experimental reality of the atom is a profoundly modern achievement, established during the lifetime of our grandparents. Through the end of the 19th century, controversy seethed over whether atoms were real material bodies or merely convenient computational fictions. The simple proportions of chemical compounds and the indivisibility of the elements supported the notion of real atoms, but a reasonable person could resist because no one had ever seen an atom. Wilhelm Ostwald, one of the founders of physical chemistry, wrote influential chemistry textbooks that had no use for atoms. In a passionate 1895 lecture, he declared, "We must . . . definitively renounce any hope of interpreting the physical world by describing real phenomena with evocative images of the mechanics of atoms." Then, channeling Moses, Ostwald exhorted, "Thou shalt not make unto thee any graven image, or any likeness!" The physicist, philosopher, and psychologist Ernst Mach likened "artificial and hypothetical atoms and molecules" to algebraic symbols, tokens devoid of physical reality that could be manipulated to answer questions about nature.

In the end, the atomists won not because they could see atoms—atoms are far too small to see—but because they learned to determine the size and weight of a single atom. Even under conditions of macroscopic tranquility—perfect equilibrium and constant temperature—particles suspended in a liquid are in perpetual erratic motion. At the Sorbonne in 1908, Jean Perrin established that the wild-wandering "Brownian" movement results from the relentless buffeting of suspended pollen grains by agitated molecules of the surrounding medium. Verifying a relation for the rate of wandering derived by Albert Einstein, he deduced the mass of an individual molecule. In demonstrating the mechanical effects of tiny atoms and molecules, Perrin effectively ended skepticism about their physical reality. Ostwald announced his conversion in 1909, the year he won the Nobel Prize; Mach went to his grave in 1916, doggedly fighting a futile rearguard action. But scientists of the old school could not hold back the idea that each element had its own smallest building block, its own indivisible and immutable atom. To Rutherford's generation, the reality of atoms was self-evident. Ironically, the atom became real just as it came apart.

Radioactivity arrived just in time to rock the new wave's own comfortable order. Whether you dared utter the word "transmutation" or not, radioactivity caused one element to turn into another when an alpha or beta particle was emitted. How could scientists continue to think of atoms like unbreakable bricks? To Rutherford and Soddy, the conclusion was irresistible: that if one atom could become another, it must do so by rearranging its parts, and so it must have constituents and structure. Their corollary—that the properties of the elements could be understood from the structure of the atoms—was revolutionary.

In 1907, Rutherford returned to England and a new position at the University of Manchester, ready to use his alpha particles as a scalpel to cut to the heart of the atom. By now, he had learned much more about alpha particles. Rutherford's alpha particles emanated from radon sealed inside a glass tube. The captive radon threw off alpha particles, some of which escaped through a thin window at the end of the tube. Rutherford tracked the alphas as they passed through sheets of mica. To "see" the alphas, he used an early version of the phosphor-coated screen that later produced television images. He and his disciples spent long hours in a darkened room counting scintillations—brief flashes that appeared on the screen when an alpha particle struck. Some contemporaries resorted to belladonna drops as a performance-enhancing drug, to dilate their pupils and gain sensitivity. The flashes showed that the alphas changed course by a degree or so on their way through the mica—a very subtle bend, but too large to be explained by the plum-pudding atom. Because atoms are only a few billionths of an inch across, Rutherford concluded that very intense electrical forces operating within the atom must cause the deflections.

We can draw an instructive metaphor from the game of golf, which became Rutherford's Sunday pastime. The atoms in the mica sheet could be viewed as small mounds steering alpha particles like golf balls putted across a green. The speeding alphas deviated only slightly from their original path as they passed over the atomic contours, but the tiny deviations gave evidence of an atomic topography.

At Manchester, Rutherford passed the question of atomic structure on to his postdoctoral assistant, Hans Geiger, and a young student, Ernest Marsden, suggesting they look for larger deflections. In June 1909, an excited Geiger reported to Rutherford that about one alpha particle in 8,000 was reflected from a thin gold foil—that the fast, massive projectiles were bouncing back-ward. The great man was flabbergasted: "It was quite the most incredible event that has ever happened to me in my life," he recalled. "It was almost as if you fired a fifteen-inch shell at a piece of tissue paper and it came back and hit you." The fraction of alpha particles thus reflected was much greater for targets made of heavy elements than for light elements.

In 1911 Rutherford announced his explanation of the phenomenon. Golf balls rolling across gently sloping manicured greens do not suddenly bounce back to the spot from which they were putted. The atomic links must be a freakish miniature golf course with a towering peak rising up from the center of a mostly featureless green. If the atomic golfer putted straight for the peak, the ball would climb until it could rise no higher, then roll back down toward the golfer. A ball aimed just off-center would veer wildly from its initial course. In an entire session of putting practice, only a few balls would encounter the narrow spire, but their trajectories would be astonishing.

Rutherford calculated that a backward recoil must be the result of a head-on collision of an alpha with a minute central body—a nucleus—in which most of the mass and positive charge of the atom were concentrated. The new Rutherford atom resembled a miniature solar system, in which electrons orbited like distant planets around a small, massive core that was the seat of the properties of an element. Rutherford's atom was mostly empty space: if the nucleus were the size of a small pearl, the nearest electron would be fifty yards away.

The pace of revolutionary discoveries in Rutherford's laboratory slowed when the Great War called his young researchers into battle. Rutherford himself was occupied with anti-submarine warfare, but kept a small research program going with the help of a single technician. He used a radium C (Bismuth-214) source, which emitted more energetic alphas than radon, to irradiate nitrogen gas. What emerged occasionally from the collisions were not deflected alpha particles, but protons—hydrogen atoms stripped of their lone electron. Rutherford painstakingly demonstrated that the protons were not the result of contamination nor were they emitted directly by the radioactive source, and concluded that "The nitrogen atom is disintegrated under the intense forces developed in a close collision with a swift alpha particle."

The more energetic alphas produced by radium C were able to climb further up the hill toward the very center of the nucleus, close enough to allow the nitrogen nucleus to swallow the alpha and spit out a single proton, becoming an oxygen nucleus. In radioactive decay, elements transmuted themselves spontaneously. Rutherford had shown that he could induce transmutation artificially. He foresaw that "if alpha particles—or similar projectiles—of still greater energy were available for experiment, we might expect to break down the nucleus structure of many of the lighter atoms."

Now it was open season on the inner workings of the nucleus. What was inside? How were the pieces assembled? What forces held the parts together? And, always in the background, the question that most interested the prophets: Could the incredibly vast reserves of energy within the atom be put to work? The insight that the elements were compound made it possible to consider questions that had seemed beyond the reach of science: Why do these particular elements occur in nature? Why do they have the properties they do? It allowed scientists to see a path to understanding why the world is the way it is, instead of merely describing the world they observed.

Radium had provided both motive and means for exploring the structure of the atom, but the questions scientists wanted to answer exposed its limitations as a probe. Radium is so rare that in most laboratories in the 1920s, only a fraction of a gram was available for use as a radioactive source. Half the radium atoms in any sample will fly apart in the course of its half-life of 1,620 years. Half the remainder will disintegrate in the next 1,620 years, and so on. That time is a blink of an eye compared with the Earth's age of about four and a half billion years, so whatever radium came as part of the Earth's original equipment has long since decayed away. The only reason any radium exists on Earth today is that it is continuously being resupplied in small amounts by the disintegration of uranium, whose half-life is about the same as the age of the Earth. Extracting that radium from even the richest ore required a heroic effort. The Curies, working in their "miserable shed" in Paris, showed the magnitude of the challenge: from five tons of pitchblende ore they culled but a single gram of radium. A 1930 British Pathé newsreel on the extraction process called radium "the rarest substance known." The price was correspondingly high: a gram could be had in 1927 for about $60,000. The total supply in all the world's laboratories was no more than a pound.

Radium has other shortcomings. Even with the highest-energy source, radium C, the particles emitted move at a mere one-sixteenth of the speed of light. More energetic alpha particles would open new possibilities: to break apart nuclei that proved resistant to radium's missiles, to disturb familiar nuclei in novel ways. Perhaps other kinds of fast particles—protons and electrons—would penetrate more readily the electric field that guards the nucleus against intruding alphas. To understand the nucleus, physicists needed more energy, more intense beams, and a greater variety of projectiles. They had to free themselves from their reliance on radium.

The three young physicists who went to Monte Generoso—Arno Brasch, Fritz Lange, and Kurt Urban—were captivated by the problem of taking apart the atomic nucleus to learn its structure and composition. Frustrated by the limitations of the tools at hand, they conceived a way to extend the experiments in which Rutherford had shattered atoms with alpha rays: they would create faster particle beams containing as many atom-smashing missiles as would be emitted by several hundred pounds of radium.

Electrical forces can accelerate alphas and other charged particles. Hitch up the positive end of a battery to one metal plate and the negative end to another plate. Place the positive alpha particle near the positive plate and it will rush toward the negative plate. To challenge Rutherford and his coworkers, the intrepid trio would have to generate millions of volts and harness those high voltages in a practical way to accelerate particles. Brasch, Lange, and Urban came up with the audacious idea of using the electrical energy in thunderstorms to develop the high voltages they required. If one of nature's gifts, radium, was not up to the task of laying bare the nucleus, perhaps another gift, atmospheric electricity, was.

When Ben Franklin reported drawing "electrical fire" from the string of his kite during a thunderstorm in 1752, he showed the world that lightning is an electrical discharge. Franklin's successors found that thunderstorm activity around the globe leaves the whole atmosphere electrified, even in fair weather. Over the oceans, or above the wide-open spaces of flat desert country, it is as if little batteries were stacked in towers from the surface of the Earth up to the skies, each battery delivering about four times the voltage for its size as the ones that power our flashlights.

The best time to extract energy from the atmosphere is during thunderstorms, when the atmospheric batteries near the ground under a storm can give 10,000 volts for each inch of their length. Although kites and balloons can easily ascend hundreds of feet, where the potential reaches several million volts, they cannot survive the high winds of thunderstorms. A collector stretched far above the ground between separated summits would serve as one plate and the ground below as the other. The atmosphere would supply a battery of millions of volts. This was the energy source that Brasch, Lange, and Urban sought to develop. Choosing a site was troublesome because they needed an easy-to-reach spot with frequent thunderstorms. Meteorologists advised the three Berliners that the best sites for thunderstorms were in South Africa or the Andes, but those were dismissed because of high transportation costs. In Europe, the Swiss canton of Ticino was the place to go. Ticino's advantages as a spawning ground for thunderstorms include the frequent invasion of masses of warm, moist air that drift up from the Adriatic Sea, steep valleys to capture it, mountain slopes to push it upward, and, for good measure, a supply of surrounding cool air chilled by alpine glaciers. Monte Generoso won out over other peaks because of the rack railway that winds from the village of Capolago on the shore of Lake Lugano nearly to the summit, providing ready access of people and matériel.

The Berliners first sketched a mile-long metal hammock to collect electric charge, suspended between the summit of Monte Generoso and the neighboring peak of Monte Sant'Agata. The average height of the collector above the valley floor would have been 2,300 feet, as high as a 200-story building. At that altitude, the potential during thunderstorms could be expected routinely to reach twenty million volts—more than enough for their experiments. Despite their eagerness, the physicists concluded that it did not seem wise to start with a full-scale apparatus right away.

Instead, they decided first to erect a more modest prototype, a mere half-mile long, between two peaks of the Generoso ridge. Two tasks had to be accomplished: building an efficient collector of charge, and suspending it in midair, electrically isolated from the ground, lest the charge flow ineffectually into the mountainside. Building a large collector—a coarsely woven wire net, to which they attached thousands of spikes that would assist in gathering charges—turned out to be the easy part of the task. Franklin's silk-handkerchief kite was succeeded by a two-ton iron hammock that covered an area as large as a tennis court.

Insulating the collector raised problems that dogged the experimenters throughout their work. Brasch, Lange, and Urban had known from the start that the net had to be isolated from the anchors. In the support cable high above the rocks they placed special ceramic insulators that would protect against large voltages and could bear the weight of the apparatus in the swirling winds of thunderstorms.

The corona that played around the net on stormy nights, a larger-scale version of the hissing and glowing you can observe around high-voltage transmission lines on sultry evenings, was an unforeseen problem. The high voltage ripped apart surrounding air molecules just the way the mountaineers hoped to rip apart the nucleus. The ionized air provided an alternate conduction path that would bleed away the accumulated charge. Because the effect is more pronounced around small conductors or sharp points, the corona can be reduced by increasing the size of the conductor, but weight and cost make thick, solid conductors impractical. The boys strung hollow metal beads seven feet long on the 500 feet of cable closest to the anchorage like pearls on a string. The diameter of the beads increased steadily from three inches near the net, where the corona had been small, to three feet where the cable met the insulators, where the corona was largest.

The experimenters' ambition was not simply to accumulate high voltage, but to measure it and use it to accelerate beams of particles that could penetrate the nucleus. They built a voltage-measuring device near the lower anchorage, where the terrain was comparatively gentle. On a tower under the last hollow cylinder of the collector, they mounted a thick wire, connected to the rocks below, that they could raise or lower with a lever. The experimenters operated the lever from a lightning-proof corrugated-metal hut that resembled a suburban backyard garden shed, placed out of harm's way some 200 yards from the tower. For the young scientists to venture outside this fortress during a storm, reported the *New York Times* in the spring of 1928, "means to throw themselves into a bristling mass of unknown current waves which would make hair stand on end and cause the body to tingle."

When the collector was sufficiently charged up during storms, fat sparks jumped like small lightning strokes from the hollow dome at the end of the cylinder to the wire. The higher the voltage they had accumulated, the longer a gap the spark could bridge. A spark that leaped fifteen feet, the maximum distance between the two poles of the spark gap, meant that the collector had reached a potential of more than two million volts. Preparations for experiments in 1927 took much longer than planned because of the ruggedness of the site and numerous mishaps during construction. Solving the problem of corona discharges had diverted considerable effort. The capabilities of the equipment could only be tested in a single late-season thunderstorm. The maximum distance of the spark gap was easily surpassed, but, as on several occasions earlier in the summer, sparks repeatedly jumped across the insulators, showing that even larger voltages were to be had if only the experimenters could learn to contain them.

After a summer's work on the mountain, Brasch, Lange, and Urban had confirmed that immense electrical energy was theirs for the taking. They returned to their Berlin laboratory determined to solve the insulator problems that had limited the voltage they could gather, and to learn how best to use high voltages to disrupt the atom.

The daring young Berliners were not the only ones devoted to the quest for higher energies. In his presidential address to the Royal Society of London in November 1927, Ernest Rutherford, now a knighted elder statesman of fifty-six, reviewed some of the other work in progress around the world. So great had been Rutherford's success with modest equipment and elegantly conceived experiments that he had disdained complicated or large-scale apparatus. For five years, he had shown little interest in appeals from his Cambridge University colleagues to establish work in the Cavendish Laboratory on the acceleration of particles by high voltages. But now, at last, he was prepared to embrace the goal.

Though the Berlin physicists were not alone in the pursuit of high voltages or in their desire to dismantle the atom, their spiritual identification with Franklin's heroic deeds made them charismatic figures within the romantic fraternity of atomic physicists. To a watching world, they combined an adventurer's disregard for personal safety with the awe inspired by the unknown consequences of disrupting the atom. The *Times* report on their plans for the summer of 1928 bore the subhead, GREAT RISK IS INVOLVED. In a breathless tone, the *Times* explained, "They are working with the comparatively unknown and the danger is all the greater because they are unacquainted with what is safe and what is perilous. They estimate that they will get power equal to Alpha rays from 220 pounds of radium. What may happen if this force gets beyond their control after it is released no one is willing to try to estimate."

The *Times* editors showed the same excitement and optimism in a Sunday editorial. It concluded, "Clearly, the Germans are dealing with forces compared with which those that blotted out whole villages in the recent war were puny indeed. In this effort to shatter atoms all the resources of the inventor, physicist and engineer must be taxed to the utmost. But what a triumph if the experiment succeeds! A new tube of tremendous potentialities. Rutherford's list of disrupted atoms extended by the addition of the more complex. The boundaries of physical chemistry pushed out still further into the unknown. Man given a little more control over the stuff of which the universe is made."

Over the winter, while back in their Berlin laboratory, Brasch, Lange, and Urban devised a way to accelerate particles using sudden bursts of voltage like those they had observed in the collector when lightning struck nearby. Accordingly, in the summer of 1928, the experimenters concentrated on exploiting the influence of lightning bolts. Charge would build up so quickly that the corona discharge would have little effect and they could dispense with the metal beads installed at such great cost in time and effort in 1927.

As a first step, they discarded the collector mesh, saving weight and enabling them to add insulators to prepare for larger voltages. Extremely high voltages were induced in the new device—a horizontal antenna—by lightning strikes within a mile, but even the enlarged insulator chains could not keep the charge from escaping before it could be used. Unwanted discharges made their way part way along the insulator chains with loud bangs, then leaped down to the ground below. Nature was providing more voltage than the scientists were able to use: another long summer would be spent struggling with insulators.

Doubling the chains of insulators made the antenna so heavy that it could not be raised high enough to gather the desired high voltages. The physicists tried replacing the chains of ceramic insulators with lighter 300-foot-long insulating ropes two inches in diameter, to raise the high-voltage section of the apparatus farther from the ground. But what if the insulation failed and a spark set the rope afire? To eliminate this hazard, they inserted more insulators at the ends of the ropes. It was now possible to raise the antenna higher than before, because the sleeker apparatus weighed only half as much as in 1927. Streamlining brought another benefit: without the mesh, the ceramic insulators, and the metal beads to prevent corona discharges, the antenna was no longer buffeted by winds.

After going to such lengths to increase the voltage their antenna could gather and hold, the Berliners had to find a way to measure larger voltages. The solution they invented was simple and ingenious. They hung a wire from the antenna and tied its free end to an insulating rope that ran over pulleys on the tower into their metal shed. By pulling in the rope or playing it out, they could raise or lower the end of the high-voltage wire. A second wire, connected to the mountainside, was attached a little higher on the tower and supported by an insulating rope hung from the antenna.

During a storm, one of the physicists would gradually reel in the rope—hoping that it truly was an insulator—to raise the slack high-voltage wire until it came close enough to the ground wire to induce a spark. A dazzling flash, a sizzling crackle, and a loud boom gave the signal to record the distance between the wires, from which the antenna's voltage could be figured. Adjusting the spark gap was not the only aspect of life on the mountain that required iron nerves. Although the apparatus had been designed to exploit the influence of nearby lightning strokes, not to serve as a lightning rod, its exposed location made it an inviting target. During one storm, the physicists claimed that the antenna had taken thirty direct hits—so much for the folk wisdom that lightning never strikes twice. *Popular Science Monthly* named Kurt Urban the "'most shocked man in the world' from being knocked unconscious by sky currents." The Monte Generoso experiment was not for the faint of heart.

During thunderstorms, sparks regularly jumped a sixty-foot gap with apparent ease, signaling a potential of at least eight million volts. The prototype had provided all the energy that Brasch, Lange, and Urban would need for their atomic explorations. The grandiose apparatus linking Monte Generoso with the summit of Monte Sant'Agata would not be required. The first goal of the Monte Generoso installation, collecting nature's high voltages, had been met. The late summer of 1928 was devoted to making detailed measurements under different conditions, to learn better how to use the apparatus, and perhaps how to improve it. But the prime question was now the one that had inspired Brasch, Lange, and Urban to come to the Alps in the first place. Could the three comrades put to work the high voltages they had collected to accelerate beams of protons, electrons, and alpha particles that would disrupt the nucleus and unlock its secrets?

They would never find out. About seven o'clock in the evening of Monday, August 20, twenty-four-year-old Kurt Urban fell from the antenna onto the rocks 150 feet below and was killed instantly. Newspaper accounts of his death are brief and confused, but scientific legend has it that he was struck by lightning. Urban did not receive the same attention in death that his efforts had attracted in life. The *Times*, which had called attention to the heroic and perilous nature of the undertaking, took no notice at all.

Much greater notoriety had been accorded the first martyr to the study of atmospheric electricity. A year after Franklin's kite experiment, Georg Wilhelm Richmann, professor of experimental physics at the Imperial Academy of Sciences that Peter the Great had established in St. Petersburg, attempted to bring atmospheric electricity into his home laboratory for systematic study. Richmann fixed an iron rod on the roof of his house and connected the rod by a chain to his measuring device. He met his maker in the company of a master engraver named Ivan Alexeevich Sokolow, who had come to illustrate a report of the experiments. (Perhaps there is a message for modern scientists who invite film crews to record their moments of discovery for posterity.)

"Richmann was describing his apparatus," according to a 19th-century popular history, "when a terrific clap of thunder alarmed the whole city, and from the rod a ball of fire leapt to the head of the unfortunate professor, who was standing at a distance of about a foot. He instantly fell backward, dead. Sokolow was stupefied for a few minutes but was not otherwise injured."

Richmann's death made a strong impression on scientists and the public alike, for it showed that—even in the hands of a skilled experimenter—the power of lightning was fearsome, unlike the parlor tricks with static electricity that had delighted Europe for twenty years. But Richmann was universally admired for having made the supreme sacrifice in the name of science. In his 1767 treatise, *The History and Present State of Electricity*, the English chemist and investigator of electrical phenomena Joseph Priestley declared, "It is not given to every electrician to die in so glorious a manner as the justly envied Richmann."

Urban's tragic accident ended the work on Monte Generoso. Brasch and Lange returned to Berlin, where they pursued by other means the race to disintegrate the atom.

In 1929, the course of that race was changed decisively by separate events in Berkeley, California, and Cambridge, England. One evening early that year, Ernest Orlando Lawrence, a new faculty member at the University of California, was glancing through scientific journals in the university library. Scanning a German electrical engineering journal, he came across an article by the Norwegian Rolf Wideröe on the acceleration of protons or alpha particles. Born in South Dakota into a Norwegian-immigrant community, Lawrence earned his doctorate at Yale and served on the faculty there before being lured to California, where he emerged as a charismatic leader. According to Lawrence, his command of German was so primitive that he had to puzzle out Wideröe's general approach from the accompanying drawings and photographs. He saw that Wideröe had exploited the technique familiar to anyone who has ever pushed a child on a playground swing: many gentle pushes, delivered with the right rhythm, will eventually lift the swing far higher than a single, uncontrollably forceful push. Likewise, charged particles could be raised to very high energies with a succession of gentle accelerations, requiring only modest voltages, delivered at the right moment. Lawrence went further: he used a magnetic field to guide the particles in a circular path past the same accelerating gap many times.

The first truly successful cyclotron, as Lawrence called his invention, was constructed by M. Stanley Livingston, a Berkeley student, in the fall of 1930. Protons were pushed to 80,000 volts by the successive application of less than 1,000 volts on the accelerating electrodes. The cyclotron principle established, Lawrence and Livingston set out to produce 1,000,000-volt protons to use in experiments. Livingston constructed the first practical cyclotron during 1931 and 1932, using a specially designed ten-inch magnet. Lawrence and Livingston achieved protons of 1.2 million volts early in 1932, but were still seeking more intense beams at higher voltages when they heard that John Cockcroft and Ernest Walton at Rutherford's Cavendish Laboratory in

England had beaten them to the finish. A nuclear transmutation had been produced in Cambridge by means entirely under human control.

The theoretical insights that guided Lawrence and his collaborators to the idea of changing the caliber of their ammunition from alpha particles to protons—and had helped the Cantabrigians to redefine the technical problem in the way that prevailed in the short run—sprang from the brow of the Russian theorist, George Gamow. Like many of his contemporaries, Gamow was puzzled by the phenomenon of radioactivity. Some nuclei disintegrated in the blink of an eye, others at a leisurely pace, and still others—most of the substances around us—seemed eternal. An observation that Rutherford had made in 1927 sharpened the puzzle for Gamow. Uranium, everyone knew, turned itself into thorium by ejecting an alpha particle moving at about 8,000 miles a second. Yet Rutherford found that the most energetic alpha particles at his disposal, moving at about 12,000 miles a second, were deflected harmlessly from uranium atoms, held at bay by the strong electric field of the uranium nucleus. How could alphas moving at 8,000 miles a second escape the clutches of the uranium nucleus when a barrier kept alphas one and a half times as fast from getting anywhere near the inner sanctum of the nucleus?

In Göttingen, where he arrived on tour with his doctorate from Leningrad, Gamow found an explanation in the new science of quantum mechanics: no physical barrier is perfectly impenetrable. Run at a wall enough times and eventually—if you happen to be an atomic particle—you will run through it. This surprising phenomenon is called quantum tunneling. Gamow calculated that an alpha particle within the uranium nucleus had one chance in 100 trillion-trillion-trillion (10^{38}, or one followed by thirty-eight zeroes) of penetrating the barrier and emerging. Although this is a very slim chance, alpha particles inside a nucleus have nothing else to do with themselves, so they run into the nuclear wall about a billion-trillion (10^{21}) times a second. After only a few billion years, on average, an alpha will succeed in slipping through the nuclear wall, leaving behind a thorium nucleus that in turn decays to radium. So, Gamow reasoned, to disrupt the nucleus, it was not necessary to supply enough energy to surmount the wall the nucleus erects around itself.

Instead, if enough missiles are hurled at the wall from the outside, eventually one will tunnel through. In this instance, beam intensity could substitute for beam energy.

John Cockcroft, then a thirty-one-year-old fresh PhD, wasted no time in noticing that Gamow's picture enabled him to calculate the probability of disintegrating light atoms by bombarding them with artificially accelerated protons. He estimated that an intense beam of protons with an energy of only a few hundred thousand volts could produce several million disintegrations per minute of the light metallic element boron. Suddenly—thanks to quantum tunneling—the technological problem seemed far more tractable than before. With Rutherford's blessing, Cockcroft and twenty-five-year-old research student Ernest Walton, already a veteran of unsuccessful attempts to produce fast particles, joined forces.

Their first attempts yielded erratic 280,000-volt beams with the intensity Cockcroft had calculated they would require. But when Cockcroft and Walton bombarded both heavy and light elements, they saw no sign of the gamma rays—high-energy X rays—they expected to signal nuclear disintegrations. By the middle of 1931, having lost their modest laboratory to physical chemists in a space war, they moved into larger quarters in a disused lecture theater and started anew. Using a new voltage multiplier of Cockcroft's design, they worked their way gradually to steady potentials of 500,000 to 600,000 volts, battling vacuum leaks and insulation failures all the way. By December, the improved apparatus was behaving itself. For more than three months, Cockcroft and Walton measured the range of their fast protons in air and checked their velocity by using magnetic fields to steer them.

Rutherford, known around the Cavendish as the Crocodile because he always pressed forward and was—like the reptile—incapable of looking backward, did not conceal his impatience with these preliminaries. (A second origin myth for the sobriquet holds that his penetrating voice and heavy footfall alerted the staff to their master's approach, as the ticking of a swallowed alarm clock warned that the *Peter Pan* character was on the prowl.) The issue, the newly created First Baron Rutherford of Nelson reminded his junior colleagues, was not technique; what mattered was whether the beam could produce any nuclear effects. Folklore has it that Lord Rutherford gently reasoned with his boys with the words, "If you don't put in a scintillation screen and look for alpha particles by the end of the week, I'll sack the lot of you!"

Under implacable pressure from the boss, Cockcroft and Walton installed a lithium target and a scintillation screen inside the vacuum chamber of their accelerating tube. Lithium is the third element in the periodic table, after hydrogen and helium, and the lightest that is a solid under ordinary conditions. On Thursday, April 13, 1932, Walton gradually brought the voltage and current of protons up to a reasonably high value. As a beginning research student, Walton had never observed alpha particle scintillations, but Cockcroft had been called to another lab on an errand. Walton crawled over to the shrouded packing crate that served as an observing hut. Immediately he saw blips on the screen that looked just like the alpha particle scintillations he had read about. When he switched off the power, the scintillations ceased. After repeating the procedure a few times to convince himself that the effect was real, he phoned Cockcroft. Once Cockcroft had seen the scintillations for himself, the two summoned Rutherford. With no little difficulty, Cockcroft and Walton maneuvered their bulky chief into the cramped observing hut, from which he took command, barking out orders with the air of a beached walrus in a waistcoat until he was satisfied that alpha particles had been produced by nuclear disintegration.

By Saturday evening, Cockcroft and Walton had made a complete set of measurements and were ready to present their discovery to the world in a one-column letter to the editor of *Nature*. Their observations indicated that a proton combined with a lithium nucleus to form an unstable nucleus of beryllium—the fourth element—that divided into two alpha particles, each carrying an energy of about eight million volts. Remarkably, scintillations were observed with an accelerating potential as low as 125,000 volts. Had Cockcroft and Walton only looked for alpha particles instead of gamma rays with their first machine, their triumph would have come a year sooner! To be sure, the rate of disintegrations rose rapidly with energy. At 250,000 volts, approximately one proton in a billion provoked a disintegration; at twice that energy, the rate had increased tenfold. The race to disrupt the nucleus was won. The competition to understand the physics of the nucleus was about to begin.

∞

Scant weeks after Cockcroft and Walton's dramatic observations, Lange announced that he and Brasch had surpassed the British achievement by using vast power to pulverize atoms by the millions. What the Berlin press termed "Dr. Fritz Lange's sensational report" to a regional association of German chemists explained that, in experiments at the Oberschöne-weide transformer works of the Allgemeine Elektricitäts-Gesellschaft, the German General Electric Company, the collaborators had produced beams far exceeding the destructive power of the Cockcroft-Walton apparatus. Atoms were smashed at such a tremendous rate by his fast proton bullets, proclaimed Lange, that the standard methods of detection were completely inadequate to measure the wreckage.

While the initial motivation for their research lay in understanding the nucleus, Brasch and Lange had always been attuned to novel applications of the technology they had developed. In August 1932, Brasch announced their first results on treating cancer with electron beams from their five-million-volt accelerator. X-ray treatments of the day subjected patients to hours of irradiation, but the new electron-beam therapy could be administered in only a ten-thousandth of a second.

Citing evidence that tumorous growths of various animals were held in check or cured without damage to the fur, Brasch offered the hope of easier and more effective treatments, if electron-beam therapy could be properly controlled. However, there was much to learn before clinical trials could be imagined: one mouse irradiated for a ten-thousandth of a second died only a few days later. On January 27, 1933, Brasch told a popular audience that harnessing atomic energy would require new ways of disturbing the nucleus. He likened bombarding atomic nuclei with energetic electrons from the most intense existing sources to "shooting at an ocean very sparsely dotted with a few islands. All too many of our shots land in the sea." For therapeutic applications, on the other hand, accelerated beams and the unstable nuclei they produced could soon replace costly radium sources.

Brasch and Lange would not complete their programs of research in nuclear physics and nuclear medicine on German soil. Only three nights after Brasch's Berlin lecture, Hitler's storm troopers streamed through the Brandenburg Gate, bearing Nazi flags and firebrands, to celebrate their führer's installation as chancellor. Political and economic chaos gave way to Hitler's absolute rule. Berlin's reputation for cultural brilliance and intellectual ferment was extinguished within a few months. Like so many others, the two young physicists would find the situation intolerable and flee, Lange to the Soviet Union and Brasch to the United States.

An active member of the German Communist Party, Fritz Lange emigrated in 1935 to the Ukrainian Physical-Technical Institute in Kharkov, then the leading research center of the USSR. Installed as head of the laboratory of high-voltage studies, he constructed powerful X-ray sources used for factory inspection of crankshafts and turbines. In 1940, he coauthored "The Use of Uranium as an Explosive," which the defense commissar shelved after receiving negative reviews from leading Soviet scientists. When the USSR established an atomic bomb project later in the Great Patriotic War, Lange developed centrifuges to separate fissionable uranium-235. He was permitted to return to Berlin in 1959, becoming director of the Biophysics Institute of the East German Academy of Sciences.

Having found refuge on American soil, Arno Brasch attracted the attention—and soon, the patronage—of the businessman and philanthropist Lewis Strauss, who would later chair the Atomic Energy Commission. Strauss foresaw the possibility that Brasch's technique could produce vast amounts of the radioactive isotope cobalt-60 at minimal cost, to be distributed to hospitals for cancer therapy—a memorial to his parents, both of whom had succumbed to the disease. Drawing on his remarkable network of influential friends, and on his personal fortune, Strauss enabled Brasch to begin work on a new "surge generator" in the Kellogg Radiation Laboratory at Caltech. But in 1940, after the federal government became interested in the possibility of nuclear weapons and began to support projects in the universities, "it became necessary," Strauss wrote in his autobiography, "to exclude aliens from the laboratory at Caltech and to defer the surge generator program." Brasch eventually settled in Brooklyn, where, ever the original thinker, he developed and commercialized a powerful electron gun to sterilize pharmaceuticals and preserve food without cooking it or altering its flavor. His method for improving the properties of polymerized materials by electron-beam processing is the technology by which all forms of heat-shrinkable materials are produced today.

2

ENCOUNTERING RESISTANCE

Chris's friendship with electricity began with a battery, a No. 6 dry cell from the Sears, Roebuck store, a weighty cylinder two-and-a-half inches in diameter and six inches high, with two brass terminals on top. A dry cell was power, stored energy; it wore its stiff sleeve of coated blue paper as an emblem of readiness. It was heavy with potential.

I was about eleven years old and I had just found out from a book that you could make an electromagnet out of an iron spike and some insulated doorbell wire connected to a dry cell. A magnet that turned on and off sounded like fun, especially because my grandfather's crane had one.

My mother's father came to the steel mills of eastern Pennsylvania in 1911 from a peasant village near Kraków. After winning a competition staged by his foreman, he had been chosen to operate an overhead crane at the Bethlehem Steel plant, shuttling back and forth day or night on important errands. He could be trusted with raw materials and finished products—and with other men's lives. One of my unfulfilled childhood fantasies was to ride along in his crane. It never occurred to me that it might be tedious to spend eight hours perched above the world in a little cab, maneuvering a huge electromagnet to lift a load here and put it down there according to some divine plan. It just seemed clearly important. I thought that if my grandfather didn't show up for work, the plant would shut down and the whole country, maybe the entire world, would come to a halt.

I had found a spike as black as a seasoned frying pan and two bottle-green glass insulators for telegraph wires next to a railroad track while on a twenty-mile hike at Boy Scout camp the summer before. I didn't have any use for them at the time, but they were exotic objects, of pleasing shape and texture like interesting shells on a beach, so I put them in my knapsack and carried them home as trophies. I displayed the insulators on my windowsill where they caught the afternoon light. The spike was waiting on my desk when I read about making an electromagnet.

With an ancient pocketknife, I scraped the insulation gingerly from the ends of a yard of wire, fighting its unwillingness to let go while trying not to nick either the wire or the blade. At last, half an inch of untarnished copper lay revealed—reddish-brown, cool and smooth to the touch, with a satiny luster that was almost soft on my finger and on the groove above my upper lip to which I instinctively pressed it. I bent the wire around the dry cell's outer terminal and screwed the knurled nut down on it. It mattered which way I bent the bare end around the post. Clockwise, the nut drew the wire closer around the threaded post, but counterclockwise, the descending nut unbent the wire and pushed it away. My magnet wasn't finished, but already I felt that I had learned an important truth. Then I wrapped a dozen turns of wire around my spike in a careful helix. Holding the flat head of the spike above a pile of finishing nails, I touched the free end of the wire to the terminal in the middle of the dry cell. A small, friendly spark greeted the wire as the nails hopped up to meet the spike. When I lifted the wire from the terminal, evoking another sparkle, the nails fell back to the floor. I had a crane of my own that I could use to move nails from one significant pile to another around the concrete cellar floor.

While I transported loads of nails and paper clips, I found that the book hadn't told me everything—which may be why I liked it. I noticed that the wire became warm when I kept the circuit closed for several seconds. The insulation stretched and loosened beneath my fingers. The longer I kept the electromagnet on, the hotter the wire became—until I had to let go. That's how I discovered electrical resistance: not by reading about it, but by feeling the wire heat up. Electricity and magnetism suddenly seemed much more interesting than cranes—though neither could compete with an after-school ball game.

Even when the wire didn't overheat my fingers, it was clumsy to hold the bare end against the terminal, so I decided to invest in a switch. My parents had bought me the dry cell and a little reel of eighteen-gauge wire, but my need for a switch was urgent. I had a small bank made of shiny red metal in the shape of a cash register. When you put a nickel or dime or quarter in the slot and pulled a sturdy lever, the proper amount was added to a running total. The bank enforced the discipline of saving for a worthy goal by opening only when the total reached exactly $10.00. This feature must have ensured the peace of mind of many parents, including mine, but my sister and I had learned how to break our matching banks within a few days of receiving them. Our safecracking enabled me to make a pilgrimage—with a pocketful of coins—to the electrical supply store on the west side of town. There I discovered a wire stripper—an elegantly simple device—the first tool I ever bought for myself. It is like a lightweight pair of pruning shears whose short jaws are kept from closing all the way by an adjustable bar that can be set for different gauges of wire. Matched notches in the opposing jaws meet to form a square opening that closes around the wire.

Walking home, I practiced scoring the insulation with a quick twist of the wrist, rocking the heel of my right hand down and up, then sliding the severed tube off without damaging the wire. I slipped the first tube I removed into the gap between my front teeth, then savored the smooth coolness of the bare wire on the sensitive flap of skin that hung above it. I practiced all the way home, until the motion became completely natural. The trail of red plastic bucatini that I left behind was part of learning the art of experimentation. Stripping insulation cleanly from wire is one of life's great tactile pleasures, like crushing a garlic clove under the flat of a sturdy blade.

I also bought a switch. I chose a knife switch because the contacts were exposed, so I could see the little spark when I made or broke the connection. I started changing things to see what mattered. I substituted a screwdriver, a pencil, an iron wrench, even my finger, for the spike. I tried to pick up different kinds of nails, aluminum foil, a lump of coal, pieces of wire. I made bigger and smaller coils out of different kinds of wire. Solid copper was my favorite: it felt good, it was agreeably stiff, it held its shape better than stranded wire. Aluminum wire didn't suit me. Its color was colder, it was not as shiny or as smooth as copper, it seemed insubstantial—too compliant to be serious. Before my first dry cell was exhausted, I discovered that a coil wrapped around thin air would not lift any nails, but did disrupt a compass—an electric current has a magnetic influence.

Edward O. Wilson celebrated this style of learning in *Naturalist*: "Hands-on experience at the critical time, not systematic knowledge, is what counts in the making of a naturalist. Better to be an untutored savage for a while . . . Better to spend long stretches of time just searching and dreaming."

I talked interchangeably about "doing experiments" and "playing with electricity." Only much later did I learn how much my play had in common with the researches of the 18th-century founders of electrical science, who were among the first to put their trust in orderly experimentation. The Nobel laureate I. I. Rabi said of his fellow physicists, "They never grow up, and they keep their curiosity." How joyously true of the early electricians is this description!

∞

On a soft spring day in 1746, the Abbé Jean-Antoine Nollet, electrician to the court of Louis XV, arrayed an entire community of Carthusian monks in a serpentine loop that snaked for more than a mile through their convent grounds, now the Luxembourg Gardens in Paris. The brethren were linked hand to hand by short lengths of iron wire. With appropriate ceremony, Nollet placed the free hands of the monks at the ends of the human chain on a small jug and transmitted an electric shock through the community. To the delight of

spectators, the 700 Carthusians sprang into the air as one, all experiencing the same convulsive sensation. There is a poverty of detail in historical accounts, but we are much taken with the idea that, as the charge passed through the brothers' torsos, the hair of their shirts stood on end.

Nollet's small jug was a Leiden jar, an early vessel for the storage and transportation of an electric charge. Only a few months before, electricians in the Dutch city of Leiden had discovered that an ordinary nail suspended in a jar of water could be electrified so strongly that it stunned the experimenter who dared touch it while holding the jar. This invention brought electricity out of the laboratory and into the streets. Fervid reports of terrifying shocks and their physical effects caused every electrician in Europe to try the Leiden experiment for himself. Once many repetitions had established that the subject would survive to tell the tale, persons of every rank and description flocked to laboratories to see and feel the new sensation.

The electric shock and the production of long, crackling sparks quickly became fashionable sport, which people were willing to travel—and pay—to see. Electricians sought practical applications in every realm, trying electric-shock therapy to reverse paralysis and devising schemes to kill and tenderize poultry at a single stroke. Lectures and demonstrations of electrical phenomena became available to an eager public on demand, in cities great and small. Itinerant professors, electrical demonstrators, and mountebanks took the new developments to every part of Europe and to the American colonies.

In fact, the public was not far behind the scientists in attending to the wonders of electrical phenomena. Electricity had been born as a science only seventeen years before, largely through the labors of the devoted English scientific amateur Stephen Gray. Gray was in his sixty-third year when he undertook the series of observations that we can take as the beginning of electrical science.

Gray was born in 1666 into a family of Canterbury artisans. He worked at the dyer's trade that he learned from his father, but his passion was natural philosophy—observing and deciphering natural phenomena. At the age of thirty, having built a water-drop microscope of his own design, he

communicated "several microscopal observations and measurements" for the first time to the Royal Society of London. The report was deemed worthy of inclusion in the Society's *Philosophical Transactions*, a journal "Giving Some Accompt of the Present Undertakings, Studies, and Labours of the Ingenious in Many Considerable Parts of the World." Over the next three decades, Gray sent the Royal Society other ideas for instruments and won a reputation as a conscientious and accurate observer of astronomical ephemera: eclipses of the sun and moon, sunspots, and the motions of Jupiter's four largest moons. Some of his reports were of practical value as well. In the days before reliable chronometers, tables of the times—once or twice a day—that the moons of Jupiter passed into and out of the planet's face gave a sea captain blessed with clear skies, a good telescope, and smooth sailing an accurate time standard to aid his navigation.

Gray's landmark accomplishments in electricity began with the same puttering about and attentiveness to interesting detail or unexpected effect that had filled his adult life. In Gray's time, an "electrified body" was one that had the power of attracting lightweight bodies. (The word *electricity* comes from the Greek name for *amber, elektron*, which the ancient Greeks discovered could attract bits of chaff when rubbed, just as a comb drawn rapidly through dry hair on a winter day can pick up small pieces of paper.) The best "electric" of Gray's day was a large flint-glass tube, whose electricity could be excited by rubbing. Gray set out to discover whether this electricity could be communicated to other bodies, measuring a body's electrification by its ability to attract lightweight objects, such as down or bits of leaf-brass, a cheap imitation of gold leaf that the English disparaged as Dutch gold.

One of Stephen Gray's prized possessions was his "great Glass Tube," more than three feet long and over an inch in diameter. Concerned that dust attracted to the inside of the electrified tube might diminish its ability to electrify other things, Gray fitted a cork to each end, to keep dust out when the tube was not in use. "The first Experiment I made," wrote Gray to the Royal Society, "was to see if I could find any Difference in its Attraction, when the Tube was stopped at both Ends by the Corks, or when left open, but could perceive no sensible Difference; but upon holding a Down-Feather over against the upper End of the Tube, I found that it would go to the Cork, being attracted and repelled by it, as by the Tube when it had been excited by rubbing. I then held the Feather over against the flat End of the Cork, which attracted and repelled many times together; at which I was much surprized and concluded that there was certainly an attractive Vertue communicated to the Cork by the excited Tube."

By taking care to check that a cosmetic change in his equipment would not interfere with his intended experiment, Gray managed to answer his question almost before asking it: he discovered that electricity—the power to attract—was communicated from the electrified glass to the cork stoppers. The report in *Philosophical Transactions* reveals the swiftness and enthusiasm with which Gray pursued his discovery. Since electricity could be communicated, he asked himself how far, under what conditions, and to what substances? Gray attacked these questions with a controlled frenzy, marshaling every piece of equipment he could lay hands on.

He fixed a small ivory ball to the cork with as many different substances as his surroundings could provide him: wooden sticks of four, then eight, then twenty-four inches in length, iron wire, and brass wire. The ball attracted and repelled the feather "with more Vigour than the Cork had done." When wires two or three feet long were stuck through the cork, the vibrations excited by rubbing the tube made the ball jump about wildly, so he tried hanging the ball from the tube by a loop of packthread, a stout twine made of hemp. It worked.

Having found that electricity could be communicated from the tube by a variety of materials, Gray then went on to see what happened when he substituted different bodies for the ivory ball. He experimented with metals, "first in small Pieces, as with a Guinea, a Shilling, a Half-penny, a Piece of Block-Tin, a Piece of Lead; then with larger Quantities of Metal, suspending them on the Tube by Packthread. Here I made use of a Fire-Shovel, Tongs, and Iron Poker, a Copper Tea-Kettle, which succeeded the same, whether empty, or full of either cold or hot water; a Silver Pint Pot; all of which were strongly Electrical, attracting the Leaf-Brass to the Hight of several Inches.

After I had found that the Metals were thus Electrical, I went on to make trials on other Bodies, as Flint-Stone, Sand-Stone, Load-Stone, Bricks, Tiles, Chalk; and then on several vegetable Substances, as well green as dry, and found that they had all of them an Electrick Vertue communicated to them." Who can doubt that, had a kitchen sink been close at hand, Gray would have tested it?

Gray's methods look somewhat random and bumbling next to modern accounts of scientific investigations that depict only orderly progress. Sometimes, when he refuses to draw a conclusion in the face of what appears to be overwhelming evidence, he seems a bit thick. But the care and coherence of his investigations shine through. The tediously detailed descriptions ensured that his experiments could be repeated by others. By systematically changing conditions, he learned what factors controlled the outcome of an experiment. His reluctance to leap to conclusions, or to generalize prematurely, was a mark of discipline shared by few of his contemporaries.

Gray next proceeded to ask how far he might carry the "Electrick Vertue." He succeeded with walking canes and with a fourteen-foot fishing rod of wood, cane, and whalebone, topped with a cork ball. Then he lashed several cane and wooden sticks into a rod that, together with the tube, was more than eighteen feet long, "the greatest Length," Gray explained, "I could conveniently use in my Chamber."

Unlike most of his scientific contemporaries, Stephen Gray was not a gentleman of independent means. Nearly all his life, he had to fit his observations in whenever the demands of his trade permitted. He concluded a report to the first astronomer royal, the Reverend John Flamsteed, with the gentle complaint, "This time of the year we are in the Greatest Hurry of our business soe that I have very little time and am soe fatiged that I Can Make but few astronomical observations." In 1711, Gray appealed for relief from his mean circumstances in the form of a warrant for admission to the Charterhouse of London. Nine years later Gray entered the Charterhouse as a pensioner by the nomination of Prince George, the future King George II. There the "hoary

Sage," as he was immortalized by an adoring poet, lived out his years, free at last to devote all his energies to science.

The Charterhouse had been established in 1611 as a home for elderly gentlemen and a school for boys. The founder, one Thomas Sutton, was a Lincolnshire patrician who found great earthly reward as a public servant to Queen Elizabeth—so much that his own purse was said to be fuller than the Queen's exchequer. Toward the end of his life, he began to consider how to share his bounty in a way that would burnish his reputation. A timely letter from his bishop steered him toward a pure act of charity. In the year of his death, Sutton endowed a chapel, almshouse, and school on the site of a former Carthusian monastery that had been dissolved in 1535, after Henry VIII rejected papal authority. Sutton provided a home for single, elderly gentlemen of professional status—civil servants, military officers, clergy, schoolmasters, artists, or musicians—who had fallen on hard times. These poor brothers had rooms of their own and were looked after completely.

We cannot be sure whether Thomas Sutton's bequest bought him a place in heaven, but it did help him achieve a kind of immortality. William Makepeace Thackeray, an old Charterhouse boy, writes in *The Newcomes*, "The custom of the school is, that on the 12th of December, the Founder's Day, the head gown-boy shall recite a Latin oration, in praise *Fundatoris Nostri*, and upon other subjects; and a goodly company of old Cistercians is generally brought together to attend this oration: after which we go to chapel and hear a sermon, after which we adjourn to a great dinner, where old condisciples meet, old toasts are given, and speeches are made. Before marching from the oration hall to chapel, the stewards of the day's dinner, according to old-fashioned rite, have wands put into their hands, walk to church at the head of the procession, and sit there in places of honour. The boys are already in their seats, with smug fresh faces, and shining white collars; the old black-gowned pensioners are on their benches, the chapel is lighted, and Founder's Tomb, with its grotesque carvings, monsters, [and] heraldries, darkles and shines with the most wonderful shadows and lights. There he lies, *Fundator Noster*, in his ruff and gown, awaiting the great Examination Day."

Besides the Charterhouse, the other important institution in Stephen Gray's life was, to give its full name, the Royal Society of London for Improving Natural Knowledge. The Royal Society was founded in 1662 under the patronage of Charles II, eager to rival his cousin Louis XIV as a patron of the arts and sciences. The Society's motto, *Nullius in verba* ("Don't take anybody's word for it"), marked a declaration of independence from the set orthodoxies of the ancient masters, particularly Aristotle, and a proclamation of reliance on experiment. Contributors to *Philosophical Transactions*, who were encouraged to write in the vulgar tongue, English—as opposed to the Latin favored earlier by scholars—were requested to use "a close, naked, natural way of Speaking; positive expressions, clear senses; a native easiness." *Phil. Trans.*, as it came to be known, carried news of scientific progress outside the small circle of the elect (the Fellows of the Society) and encouraged an extended, even transnational, community of natural philosophers.

Gray's pursuit of the discovery that electricity could be communicated to other bodies—the discovery he had set out to make—led eventually to an unexpected result of far greater significance. In May 1729, Gray went to the country, taking along several glass canes and other materials that he would need to continue his experiments. He went first to the home of Flamsteed's cousin, John Godfrey, Esq., at Norton Court, in Kent. Godfrey, a country gentleman whose hobby was astronomy, had become a fellow of the Royal Society in 1715, about the time he met Stephen Gray. The experiments Gray and Godfrey made were straightforward extensions of the work Gray had done in his Charterhouse chamber. In these more spacious surroundings, Gray fabricated a pole that extended the glass tube to thirty-two feet. When the tube was rubbed, the cork ball that crowned the pole vigorously attracted the leaf-brass, but "the pole bending so much, and vibrating by rubbing the tube, made it . . . troublesome to manage the experiment."

The communication of electricity by rigid bodies having reached a technological limit, Gray and Godfrey moved on to packthread, to take advantage of Norton Court's balcony. A twenty-six-foot length of packthread carried the Electrick Vertue from the glass tube to Gray's ivory ball. Next, the two experimenters devised a hybrid instrument, tying a thirty-four-foot line to an eighteen-foot pole. Gray stood in the balcony with tube and pole, while his collaborator held a tray of leaf-brass in the courtyard below. When Gray rubbed the tube, electricity "passed from the Tube up the Pole, and down the Line to the Ivory Ball, which attracted the Leaf-Brass, and as the Ball passed over it in its vibrations, the Leaf-Brass would follow it, till it was carried off the Board: But these Experiments are difficult to make in the open Air, the least Wind that is stirring, carrying away the Leaf-Brass." The world record for the communication of electricity stood at fifty-two feet.

The final investigations in this series were unsuccessful—but not insignificant—attempts to transmit a charge horizontally. Gray hung a length of packthread from a nail in a ceiling beam to support a line joining the glass tube to his ivory ball at waist level. When he rubbed the glass, the leaf-brass under the ball showed "not the least Sign of Attraction." Gray concluded that "when the Electrick Vertue came to the Loop that was suspended on the Beam, it went up the same to the Beam; so that none, or very little of it at least, came down to the Ball." The failure, while it discouraged Gray from further attempts to carry electricity horizontally, made him dream of trying the experiment from the cupola of St. Paul's Cathedral, to increase the vertical distance still more.

Fortunately, Gray's failure at Norton Court was not the last word on the horizontal propagation of electricity. At the end of June, Gray went to Otterden Place, near Faversham in Kent, to visit another wealthy fellow of the Royal Society, Granvil Wheler. Wheler's country estate was an ideal setting for Gray's experiments, and Wheler—with a curiosity as acute as Gray's and a knack for asking the right question—proved to be the perfect collaborator.

Gray traveled light to Otterden Place, carrying just one small glass cane, "designing only to give Mr. Wheler a Specimen of my Experiments." After seeing the "vertue" communicated from the highest accessible point, the clock tower, to leaf-brass thirty-four feet below, Wheler asked to see electricity communicated horizontally. Gray described his failed attempt at Norton Court. Wheler suggested that the packthread that was to carry the "Electrick Vertue" be supported with silk thread, which Gray thought "might do better [than the packthread he had used before] on account of its Smallness; so that there would be less Vertue carried from the Line of Communication." The more ambitious experiments that ensued required technical support, ably supplied by Wheler's servants. The team of Gray, Wheler, and company undertook investigations that none of them could have done alone.

Their first joint experiment was made in the largest room in the house, a long matted gallery with high ceilings. When a packthread eighty feet long was suspended by a silk thread at the end of the gallery, it communicated to the ever-present ivory ball the power to attract and hold leaf-brass. Wheler then hung another silk thread at the far end of the gallery, so that the packthread could run back and forth, a length of 147 feet. "Though both Ends of the Line were at the same End of the Gallery, yet Care was taken that the Tube was far enough off from having any Influence upon the Leaf-Brass, except what passed by the Line of Communication: Then the Cane being rubbed . . . the Electrick Vertue passed by the Line of Communication to the other End of the Gallery, and returned back again to the Ivory Ball, which attracted the Leaf-Brass, and suspended it as before." To investigate whether a line doubled back on itself communicated electricity less effectively than a single line of the same length, the experimenters moved into larger laboratory space in Wheler's barn, where they stretched packthread 124 feet long in a single run.

There Fortune intervened. Gray and Wheler added a return leg that doubled their line to 293 feet, without diminishing the attraction of the ivory ball for the leaf-brass. This success encouraged them to add yet another return line. But when Gray began to rub the tube, the silk threads severed under the strain. Iron wire of similar size also gave way. When they substituted a slightly larger brass wire, which was strong enough to support the line of communication, something unexpected happened: "[T]hough the Tube was well rubbed, yet there was not the least Motion or Attraction given by the Ball, neither with the great Tube, which we made use of when we found the small solid Cane to be ineffectual: By which we were now convinced, that the Success we had before, depended upon the Lines that supported the Line of Communication, being Silk, and not upon their being small."

Thus the discovery of two kinds of materials, conductors and insulators, emerged unexpectedly from the efforts of Gray and Wheler to see how far they could transmit electricity. Their new insight enabled the two Englishmen, and much later, an electrical industry, to carry electricity as far as they wished. It also gave them new questions to ponder.

A few days later, Gray and Wheler placed their leaf-brass detector in the garret of the house and suspended the end of the line holding the ivory ball on a silk thread across the garret window, about forty feet above the ground. About a hundred feet away, in the Great Garden, they erected the first telegraph poles: two rods about ten feet long, joined by a silk thread to support the line of communication. "Beyond these, in the great Field, that is separated from the Garden by a deep Foss, about the same Distance from the first, were another pair of poles fixed; then four others at a like Distance." The length of the line was 650 feet. After Gray had rubbed the tube for some time, his collaborators called to him to report that what we might call the first electrical telegraph was a success. Then Gray and Wheler changed places, so Gray might see the effect for himself. A subsequent experiment carried electricity 765 feet, without any noticeable attenuation.

For the next two years, Gray, Godfrey, and Wheler worked to classify substances into conductors and nonconductors, trying to learn what distinguished the two groups. Asking whether living matter behaved differently from inanimate objects led the group to a new series of experiments.

Stephen Gray's most original contribution to instrumentation—which took advantage of the natural resources of the Charterhouse—addressed this issue, and also helped introduce electrical phenomena into popular culture. Picture the aged electrician, clad in his Poor Brother's black livery gown, at work in his chamber in April 1730. "I made the following Experiment on a Boy between eight and nine Years of Age. His Weight, with his Cloaths on, was forty-seven Pounds ten Ounces. I suspended him in a horizontal Position, by two Hair-Lines, such as Cloaths are dried on . . . with his Face downward, one of the Lines being put under his Breast, the other under his Thighs: Then the Leaf-Brass was laid on a Stand. . . . Upon the Tube's being rubbed, and held near his Feet, without touching them, the Leaf-Brass was attracted by the Boy's Face with much Vigour, so as to rise to the Hight of eight, and sometimes ten Inches."

The suspended-boy experiment, which soon became widely known, awakened a passionate enthusiasm for electrical investigations. Charles François Dufay, superintendent of the King's Gardens in Paris and a member of the Academy of Sciences, was an assiduous follower of Gray's experiments. Reasoning that a real scientist might perceive subtleties that would escape the attention of an untrained child, Dufay had himself suspended on silk lines and electrified. When his assistant, who was none other than the Abbé Nollet, brought his hand within about an inch of Dufay's person, there immediately issued from his body "one or more prickling shoots, attended with a crackling noise," causing pain to both master and pupil. In a darkened room, the snappings appeared as little sparks of fire.

News of Dufay's "important luciferous" experiment inspired Gray to repeat it for himself, on a visit to Otterden Place in July 1734. "Mr. Wheler, soon after my coming to him, procured silk Lines strong enough to bear the Weight of his Footboy, a good stout Lad; then having suspended him upon the Lines, the Tube being applied to his Feet or Hands, and the Finger of any one that stood by held near his Hands or Face, he found himself pricked or burnt, as it were by a Spark of Fire, as Mr. Dufay had related, and the snapping Noise was heard at the same Time; but it did not succeed with us, when we applied our Hands to any part of his Body through his Cloaths, except upon his Legs, upon which he felt the Pain through his Stockings, although they were very thick ones." This confirmation of Dufay's results stimulated Gray to begin thinking about the possibility of a connection between electricity and lightning. In recognition of "his many curious Experiments and Observations laid before the Society," Stephen Gray was made a fellow of the Royal Society on March 15, 1732, at the age of 65, accepting a fellow's extraordinarily civilized obligation "to be present at the Meetings of the Society, as often as Conveniently we can."

Perhaps the Continental variations on the suspended charity-boy are revealing of national character. In Germany, a glass globe, rotated rapidly in contact with the soles of a boy's feet, was used instead of the glass tube, making the boy an element in an electrical machine. The French genius for improvements in instrumentation was revealed by Nollet, who replaced the charity boy with a fair damsel. In this refined form the experiments of Gray and Dufay were brought to the court of Louis XV and the smart salons of Paris. Thus was public interest whetted for the sensation of the Leiden jar.

Gray's enthusiastic dedication to science ended only with his life. His last ideas were "taken from his Mouth" by Cromwell Mortimer, MD, secretary of the Royal Society, the day before he died in 1736. Gray's determined curiosity about the natural world, his meticulous attention to detail, and his assiduous devotion to figuring things out by experiment made him the personification of the Royal Society's injunction not to take anybody's word for anything.

William Thomson, the future Lord Kelvin, whose catholic knowledge of natural phenomena made him a dominant force in 19th-century science, introduced a Friday Evening Discourse, "On Atmospheric Electricity," delivered May 18, 1860, at the Royal Institution in London, by recalling Gray's conjecture that a way might be found "to collect a greater quantity of the electrical fire, and consequently to increase the force of that power, which by several of these experiments . . . seems to be of the same nature with that of thunder and lightning."

"The inventions of the electrical machine and the Leiden phial immediately fulfilled these expectations as to collecting greater quantities of electric fire," Thomson continued, "and the surprise and delight which they elicited by their mimic lightnings and thunders, and above all by the terrible electric shock, had scarcely subsided when Franklin sent his kite messenger to the clouds, and demonstrated that the imagination had been a true guide to this great scientific discovery—the identity of the natural agent in the thunderstorm with the mysterious influence produced by the simple operation of rubbing a piece of amber, which 2,000 years before had attracted the attention of those philosophers among the ancients who did not despise the small things of nature."

3

NUMBERING THE ATOMS

A fleck of gold is made of one single kind of atom. Grind it, boil it, react it with what chemicals you will, gold is gold. In ordinary experience, gold can never be reduced to components. Gold is an element.

Laying aside the radioactive elements, atoms are practically eternal. Time works change and decay on rocks and trees and us, but the atoms survive to be recycled again and again. What yesterday was oxygen, or carbon, or gold, will be so tomorrow, though the atoms may be rearranged and differently combined. Atoms are not only immutable, they are innumerable. It's nearly certain that you have breathed an atom that once passed through Cleopatra's nose. Far more of her exhaled atoms are in circulation than relics of the True Cross during the Crusades or people of our age who claim to have been at Woodstock.

Ten of our modern elements were already known to the ancients. By the middle of the 19th century, chemists knew about five dozen indivisible substances as diverse as the oxygen of the air, the carbon of soot, the gold of a monarch's crown. The chemists had not only catalogued many elements, they had learned how the elements behave and in what proportions they combine.

Through careful measurements of the amounts of different elements that joined in various compounds, they determined the relative weights of the atoms. With hydrogen, the lightest element, assigned weight one, the elements stretched out to uranium, near 240. The atomic weights advanced fairly regularly among the lighter elements, but by jumps and starts among the heavier ones. At first, this wealth of descriptive information gave no clue to how many elements there might be, no indication whether the catalogue was complete.

In 1862, Alexandre-Émile Béguyer de Chancourtois, a professor of geology in the École des Mines in Paris, noticed that elements whose weights differed by sixteen had similar chemical properties and frequently occurred together in the Earth's crust. He wrapped a list of the elements around a cylinder in a "telluric spiral," so that chemically similar elements appeared one above the other. Perhaps because Béguyer's account in the *Comptes Rendus* of the French Academy of Sciences did not include an image of his chart, his insight did not immediately take hold.

Two years later, the British chemist John Newlands formulated a similar pattern of repeating properties that he called the "law of octaves." As the eighth note in a musical scale sounds in perfect consonance with the keynote, the eighth element in sequence by weight was for Newlands a kind of harmonic repetition of the first. The musical analogy was too much for the Royal Chemical Society, which gave Newlands and his octaves a decidedly frosty reception.

If Béguyer and Newlands confounded their colleagues with romantic attempts to account for cycles of chemical properties, the cycles themselves were there for the perceptive eye to see. In 1869 the Russian chemist Dmitri Ivanovich Mendeleev noted that when he listed the sixty-three elements he knew in order of increasing atomic weight, their chemical properties had a tendency to repeat again and again. Now, both sodium (weight 23) and potassium (weight 39) are soft, silvery-white metals that tarnish rapidly in air and react vigorously with water, so it is not so hard to guess that they are siblings. It requires more intimate knowledge of the elements' habits to link the strong-smelling, greenish-yellow gas of chlorine with the stinking, fuming, ruby-red liquid of bromine and the dark, violet-wreathed crystals of iodine. Chemists of Mendeleev's day had learned to look beyond appearances to behavior.

Mendeleev constructed an array of rows and columns that displayed the repeating waves of chemical properties clearly. To make the patterns regular, he needed to tinker with some atomic weights that seemed out of order. More audaciously, he claimed that some elements were missing, still to be discovered. Mendeleev predicted the properties of the missing elements from what he knew of the surrounding elements in his chart. Between 1875 and 1886, the anticipated partners of silicon, aluminum, and boron turned up, testaments to the power of Mendeleev's insight. They were named germanium, gallium, and scandium, to honor their discoverers' homelands.

Over the years, Mendeleev's periodic table expanded with the addition of other elements that he had no way to foresee. In the 1890s came the noble gases like helium and neon, so listless and unsociable that they had never called attention to themselves by forming compounds. And then there were the rare-earth elements—ytterbium, samarium, thulium, gadolinium, praseodymium, neodymium, dysprosium, lutetium, holmium, and the rest—so similar to some known elements that no one had noticed they were missing. The rare-earth elements behave so nearly alike that they were difficult to place within the periodic table.

They all joined lanthanum, their chemical act-alike, in a single crowded pigeonhole. Obscure no more, the rare earths are now of high strategic value for their applications to liquid-crystal displays, fiber-optic cables, electric-vehicle batteries, permanent magnets in wind turbines, luminescent biological markers, and light-emitting diodes.

A list of elements too long for most of us to remember may not seem like a great breakthrough, but the discovery of elements and Mendeleev's classification represented both progress and promise for understanding nature. Progress because the endless variety of nature's substances results from the combination of a far smaller number of elements. Before chemistry became a science, every substance was a special case; now there were links of common understanding. Promise because the repeating properties of the elements suggested an underlying order that might be understood in terms of the structure of the atoms, a structure that Ernest Rutherford and his colleagues were the first to perceive.

"In order to explain the results of experiments on scattering of alpha rays by matter Prof. Rutherford has given a theory of the structure of atoms. According to this theory, the atoms consist of a positively charged nucleus surrounded by a system of electrons kept together by attractive forces from the nucleus." Thus wrote Niels Bohr in *The Philosophical Magazine* of July 1913. "In an attempt to explain some of the properties of matter on the basis of this atom-model we meet, however, with difficulties of a serious nature."

When the young Dane came to Manchester to work under Rutherford in the spring of 1912, Bohr was a shy new PhD of twenty-six; Rutherford was an ebullient, established star of thirty-nine. Bohr was deferential and polite, his speech indistinct and nearly inaudible. Our teachers who knew Bohr joke (we think!) that he was equally unintelligible in all languages. Rutherford was plain-spoken and direct, with a booming voice that was legendary for its disruptive effect on sensitive laboratory equipment. A famous photograph shows Rutherford visiting a colleague's experimental setup, standing beneath an illuminated sign that reads, TALK SOFTLY PLEASE, in stenciled letters.

More important than these differences were the affinities between the two. Theorist Bohr and experimenter Rutherford both emphasized physical reasoning over mathematical formalities. They shared a passion for exploring promising new ideas and attacking the fundamental problems. During his visit to Manchester, Bohr found a calling; he returned to Copenhagen full of enthusiasm for the nuclear atom.

In Rutherford's picture, the electrons whirled about the nucleus like planets around a sun. The established laws of electricity and magnetism insisted that the orbiting electrons must emit light, give up energy, and approach the nucleus in narrowing gyres. In a fraction of a second, the electrons should spiral toward the nucleus like space junk burning up in Earth's atmosphere. But that plainly doesn't happen. In Bohr's measured words, "the actual atoms in their permanent state seem to have absolutely fixed dimensions . . ."

Something had to give: either Rutherford's atom was wrong, or the intuition gained from the everyday world was not a trustworthy guide to the atomic landscape.

A visitor to a foreign land may find the local customs and language baffling. Often, there is no way to derive the internal logic of another language or culture; we must accept it without seeking to justify its rules in terms of our mother tongue or our upbringing. Could the subatomic world be such a foreign land, with rules that seem bizarre and at odds with our common sense? Radioactivity had already given a first hint that the atom's behavior was unconventional.

Since the time of Isaac Newton, physicists, philosophers, and much of the public at large had come to view the physical universe as a great and intricate machine that functioned like clockwork according to precise, reliable, and discoverable rules. If only we mortals could solve the innumerable equations that described the motion of the seas and stars and planets, we would be able to predict their minutest behavior. Atoms, too, were the handiwork of the Divine Watchmaker. Once the rules that governed their behavior were known, an atom's every move could be calculated.

Then came radioactivity. If a radioactive atom has a half-life of an hour, the odds are fifty-fifty that it will decay during any one-hour interval. If it does not decay during one hour, then it has a fifty-fifty chance of decaying during the next. Though half of a lump of radioactive material decays away during every half-life, the atoms that remain are always starting anew. A radioactive atom does not age.

Radioactivity is a random process, like gambling on a roulette wheel. Even if red has come up ten times in a row, the odds that black will come up the next time are the same, on an honest wheel, as the odds that red will come up yet again. The roulette wheel doesn't remember those ten consecutive reds and the atom doesn't remember that it has survived ten half-lives.

It is not that we do not have enough information to predict when the sudden and spontaneous transmutation of a radioactive atom into a new element will occur. There is simply no information that can reveal it. Radioactive transmutation is an aleatory event about which we can say only that, on average, it takes a certain amount of time to occur.

The puzzling stability of the atom was another suggestion that the atomic world was unlike the everyday world. Niels Bohr had the humility to realize that human experience doesn't necessarily extend to other realms. He had the boldness to postulate new laws that do not follow from the Newtonian tradition. Bohr did not require that atoms play by the rules of everyday life, but only that they play by a self-consistent set of rules. He resolved the problem of Rutherford's atomic model by fiat. In three articles "On the Constitution of Atoms and Molecules" communicated to *The Philosophical Magazine* by Rutherford, Bohr declared that an atom has a state of lowest energy that does not radiate; he simply forbade the death-spiral into the nucleus. Moreover, he said, the electrons in an atom may follow only certain prescribed orbits. Anywhere in between is off limits, out of bounds, no-electron's-land. And he offered a test for his ideas.

An atom heated or excited in some way will emit light. Pass a spark through a tube of neon gas and you will be dazzled by the jazz-age tangerine of advertising signs. Sprinkle salt on an open flame and you will see the goldenrod signature of sodium. Dispersed through a prism, an atom's light does not paint all the colors of the rainbow, but only narrow stripes of distinct hue. Each stripe is the light of a particular wavelength, like a pure musical tone. Together, an atom's colors are a characteristic chord that distinguishes one kind of atom from another.

Chord is perhaps too anodyne a term for the thicket of colored lines that make up an atomic spectrum. Better to have in mind the thick unresolved dissonances of Stravinsky's *Rite of Spring*, which had its debut a few weeks before the July 1913 publication of Bohr's first paper. To produce such a spectrum, it was said at the time, an atom must be at least as complicated as a grand piano. Who could imagine deducing the structure of a piano by listening to the noise it makes when thrown down a flight of stairs? The simplicity of Bohr's picture enabled him to understand the entire spectrum of hydrogen, line by line. In Bohr's picture, the hydrogen atom has a single electron, but the electron has its choice of many orbits. Electrons can change orbits only in definite steps—the now-famous quantum leap—by emitting or absorbing light of a definite color, or energy: the energy difference between the two orbits. Bohr's rules for allowed orbits explained the discrete spectrum of light emitted by hydrogen quantitatively, predicting the wavelength of every spectral line. He had taken the first step toward revealing the law of the nanoworld—a billion times smaller than the everyday world—which would be codified a decade later as quantum mechanics.

The atoms beyond hydrogen have more than a single positive charge in the nucleus, and correspondingly many orbiting electrons. Bohr wasn't prepared to explain the entire complicated spectra of these atoms, but he could say something about the light emitted by the electron nearest the nucleus. The innermost electron would be little affected by the others, so its permitted orbits should closely resemble those of the hydrogen atom. The orbits in a heavy atom are smaller in scale because the multiply charged nucleus has a stronger pull than hydrogen's lone proton. A quantum jump in a heavy atom would produce light with more energy than its counterpart in hydrogen. Indeed, for heavy atoms, this light would not be visible, would not lie in the ultraviolet, but would actually be X rays. According to Bohr, these X rays would produce a spectrographic fingerprint of the atom that would resemble the pattern for hydrogen, if only we had X-ray vision.

Our eyes don't, of course—but photographic film does. It was only scant weeks after discovering X rays in 1895 that William Conrad Roentgen made his iconic image of the bones of his wife's hand. So it is no problem to "see" X rays. But how to tell their "color," or wavelength?

When visible light passes through a prism, it fans out into different colors, ordered according to their wavelengths. Different wavelengths of light can also be separated—and measured—when light is reflected from a diffraction grating, a mirror with fine parallel lines scratched into it at regular intervals—thousands of lines to the inch. To measure the wavelengths of X rays requires a grating with a hundred million lines to the inch—lines spaced as closely as the atoms in the mirror themselves. In that statement of the challenge, Max Laue, a university privatdozent in Munich, found the solution: the regular lines of atoms in a crystal form a diffraction grating for X rays. Here was a new tool that could be used in two ways. Crystals with known dimensions can be used to measure the wavelengths of X rays. On the other hand, X rays with known wavelengths can be used to determine the arrangement of atoms in a crystal.

The Manchester laboratory had worked exclusively on radioactivity, but Harry Moseley and Charles Galton Darwin persuaded Rutherford that they should follow up Max Laue's insight. C. G. Darwin was the grandson and namesake of the author of *On the Origin of Species*. Henry Gwyn Jeffreys Moseley's lineage was scarcely less distinguished. His father and both his grandfathers were fellows of the Royal Society. His maternal grandfather, John Gwyn Jeffreys, was a conchologist; his paternal grandfather, the Reverend Henry Moseley, a mathematical physicist. Henry Nottidge Moseley, his father, was chief naturalist on the three-and-one-half-year circumnavigation of the globe by HMS *Challenger*, which included the first crossing of the Antarctic Circle by steamship.

Together, Moseley and Darwin used Laue's insight, the new technique of X-ray diffraction, to map the spectrum of the metal platinum. The stage was set for a test of Bohr's theory. When Harry Moseley developed his mastery of X-ray diffraction, Mendeleev's periodic chart of the elements was still beset with inconsistencies. It was clearly not quite right to order the elements by their atomic weights. As the noble gas neon preceded the metal sodium, so argon's inertness should have placed it ahead of the reactive metal potassium in the periodic table, but an atom of argon weighs nearly a unit more than an atom of potassium. Moseley and Bohr discussed another puzzle: cobalt preceded nickel in Mendeleev's classification, but its atomic weight was greater. In Mendeleev's time, it was possible to blame such anomalies on uncertain measurements, but now it was clear that the problems were real. Moseley resolved to determine the true order of the elements.

Other experimental studies had begun to give form to the idea of an atomic number: not just the sequence number in the periodic table, but a real property of the atom. Rutherford's "boys," Hans Geiger and Ernest Marsden, refined the scattering experiments that had inspired the planetary atom to determine the electric charge carried by the nucleus. They found that gold's nucleus carried a positive charge of roughly a hundred units, about half its atomic weight. Within the uncertainties of their measurement, 100 was consistent with gold's apparent order as Number 76 in the list of the elements. But it did not seem inconceivable that unrecognized gaps in Mendeleev's periodic table might push gold's number in line closer to 100.

Attacking the elements individually as he and Darwin had done with platinum was inadequate for Moseley's new challenge. He now constructed a small trolley inside a glass cylinder about a yard long and a foot in diameter. He loaded the trolley with samples of several elements. Using a system of handles and pulleys and fishing line, he moved the trolley to and fro, positioning one sample after another at the spot where he had trained an electron gun. When bombarded with electrons, the samples emitted X rays that passed through a crystal, where they were dispersed according to their wavelengths, and onto a photographic plate, where they were registered. In this lineup of the elements, Moseley scrutinized the spectrographic fingerprint of each sample.

Moseley was an obsessive worker, often toiling through the night, constantly tearing his apparatus apart and rebuilding it to make real or imagined improvements. Because of the irregular hours he kept, Harry became the lab's authority on finding a meal in Manchester at three in the morning. Rutherford, who was compulsive enough in his own right, favored the balanced life, with time for contemplation and relaxation. He talked of shutting the laboratory at night, but Moseley would have none of it. Working day and night by himself, young Harry got dramatic results with extraordinary speed.

In early November of 1913, Moseley wrote to his mother that he could now get a strong, sharp photograph of an X-ray spectrum in only five minutes. In four days' time he determined the atomic numbers of chromium, manganese, iron, nickel, cobalt, and copper, in addition to tantalum and silver. He told his mother that "the insides of the atoms are all very much alike, and from these results it will be possible to find out something of what the insides are made up of."

In papers from Manchester in 1913 and Oxford in 1914, Harry Moseley announced what he had learned by bombarding the elements from aluminum, with atomic weight 27, to gold, with atomic weight 197. Just as Bohr had predicted, the patterns of X-ray wavelengths showed a simple progression from element to element. Using Bohr's ideas for the X-ray spectra, Moseley found that he could determine the charge of each element's nucleus, an integer atomic number ranging from 13 for aluminum to 79 for gold. Cobalt is element number 27, and nickel 28; argon, number 18, does precede potassium, number 19. The chemistry hadn't lied, but atomic weight had been an imperfect guide to the order of the elements.

There was more. The distinguished French chemist Georges Urbain brought samples of the rare-earth elements to Moseley in Oxford. Moseley settled their order, and the deeper question of how many there actually were, in a matter of weeks. That task had occupied Urbain for years, and the chemist was extravagant in his praise. He pronounced that "Moseley's work substituted for Mendeleev's somewhat romantic classification a rigorously scientific precision." Rutherford ranked the discovery that the properties of an element are defined by its atomic number as "likely to stand out as one of the great landmarks in the growth of our knowledge of the constitution of atoms."

To account for the difference between atomic number and atomic weight, scientists incorrectly surmised that the nucleus contained both positively charged protons and negatively charged electrons. Until 1932, they built many ideas on the seemingly solid foundation of a nucleus populated by protons and electrons. They interpreted the charge of the nucleus—Moseley's atomic number—as the excess of positive protons over negative electrons. They took the atomic weight to be the total number of protons. Rutherford's partner, Frederick Soddy, was the first to realize that a single element might have several distinct atomic weights. Soddy called these breeds within the species "isotopes," and offered an explanation for their existence: adding a proton and electron to the nucleus would give a variant of the element with a different atomic weight, but practically identical chemical behavior.

Unlike the atomic weights, which increased by fits and starts from one element to the next, the atomic number increased regularly, reliably, one unit at a time. "Known elements," Moseley wrote, "correspond with all the numbers between 13 and 79 except three. There are here three possible elements still undiscovered." Like Mendeleev, Moseley was able to predict new elements: atomic numbers 43, 61, and 75. He could also say that those were the only missing elements between 13 and 79. There was no place for a missing family, like the noble gases that Mendeleev had failed to foresee. To fill the periodic table it was simply necessary to find the missing elements and prove that they had the expected chemical properties, à la Mendeleev, and, more important, the expected X-ray spectra, à la Moseley.

The Great War interrupted the search for Moseley's missing elements before it could begin. Like many of his young colleagues, Moseley volunteered for military service to stand up to the threat of Prussian militarism. Rutherford, a scientific generation older, was assigned to work on submarine detection. Basic research came nearly to a halt. On his way to combat in the Dardanelles, Moseley wrote a will leaving his worldly goods "to the Royal Society of London to be applied to the furtherance of experimental research in Pathology Physics Physiology Chemistry, or other branches of science but not pure mathematics astronomy or any branch of science which aims merely at describing cataloguing or systematising." It seems an ironic judgment from the man who had just ordered the elements of which the universe is made.

Following the failure on March 18, 1915, of the British naval attack against the Turks, Allied forces made landings on Gallipoli. Unable to advance, the troops remained in isolated beachheads along the peninsula. British commanders called for five more divisions, including the 13th, in whose 38th Brigade twenty-seven-year-old Harry Moseley served. The 38th Brigade landed at Anzac Cove, where the Australian and New Zealand Army Corps had established a foothold in April. The objective of the 40,000 men at Anzac was the ridge of Sari Bair. Its forbidding terrain was to be conquered at night against the well-entrenched foe. The misguided and futile attacks on August 8 and 9 failed utterly, and on the 10th the Ottoman forces counterattacked with devastating results. Harry Moseley, the man who numbered the elements, died of a bullet through the brain, one of 200,000 sacrifices to the obstinacy of the British command in the Dardanelles.

Efforts to complete the periodic table—Moseley's periodic table, now—began in earnest as research resumed after the war. In Berlin on September 5, 1925, Walter Noddack, Ida Tacke, and Otto Berg announced the discovery of elements 43 and 75, which lie below manganese, element 25, in the table. Element 75 they named rhenium, from the ancient name for the River Rhine. They called element 43 masurium, to commemorate the battles of the Masurian Lakes in East Prussia, where German armies had repeatedly defeated the Russians in the war. Rhenium was isolated by other chemists and within a few years could be bought for a few dollars a gram. The first edition of *Discovery of the Elements*, by Mary Elvira Weeks, printed in 1933, noted that masurium had still not been purified.

In the year after the discoveries of masurium and rhenium, B. Smith Hopkins of the University of Illinois announced that he had found element 61, which he named illinium. This first element discovered in the United States was an immediate sensation. Hopkins was the toast of Urbana, Illinois, and beyond. The March 21, 1926, *New York Times* devoted nearly a full page to illinium and its place in the history of chemistry. Miss Weeks was fairly transported by the achievement: "Just as in Tchaikovsky's *1812 Overture* the strains of the French hymn fade away in the triumphant swelling of the Russian anthem, so in Professor Hopkins's spectra, the neodymium lines faded out and the new lines of the absorption spectrum became stronger; moreover, the new lines of the X-ray spectrum were found to coincide with those predicted from Moseley's rule. Thus element 61, the last of the rare earths, took its place in the periodic table."

Alas, the triumph was largely illusion. Rhenium exists and is not so scarce, but there is no masurium and no illinium. Those empty spots in the periodic table were so inviting that eagerness outpaced critical judgment in the determined search for what was certain to be found. Noddack and his collaborators were wrong, and so was Hopkins. We know this today with great certainty. Forgotten, too, are virginium and alabamine, announced as elements 85 and 87. Everyone knew the elements had to exist, but few realized that nothing guaranteed that their atoms had to last long enough to be lying around waiting to be discovered. Noddack, Tacke, and Hopkins had not reckoned on radioactivity, which swept away their elements 43 and 61—and elements 85 and 87—in a geological blink of an eye.

Element number 43 and the others had to be made artificially, since they didn't exist in nature. Element 43 was not truly discovered until 1937, when Emilio Segrè and Carlo Perrier isolated a microscopic sample in a piece of junk from the Berkeley cyclotron. When we arrived as students in what was still called the Lawrence Radiation Laboratory thirty years later, Segrè was one of the resident gods of nuclear physics. Emilio was our link to the glorious pre–World War II days of European physics, full of stories about the giants who walked the Earth during his youth. He told the story of element 43 with particular relish.

Segrè first visited the Rad Lab in 1936, when he was starting his career at the impoverished University of Palermo. Touring the cyclotron, he noticed piles of radioactive scrap metal lying about, so he asked for samples to take back to Sicily. The scraps turned out to be rich in radioactive isotopes, including an unstable form of phosphorus that soon became a valuable tool for investigating biological activity by the method of radioactive tracers.

Early in 1937, Segrè received a letter from Ernest Lawrence containing more radioactive scraps, including a foil made of the metal molybdenum that had been part of the structure that steered the beam out of the Berkeley cyclotron. Thereafter, Segrè made it a habit to measure the radioactivity of any letter he received from Lawrence. Molybdenum is element 42, so Segrè guessed that bombardment with the cyclotron beam might have created the neighboring element 43. By dissolving the radioactive foil and separating all the known elements, Segrè and Perrier isolated a small amount of a new radioactive substance and showed that it had the chemical properties expected of element 43. After the war, when nuclear reactors generated element 43 in abundance, Segrè and Perrier proposed the name technetium for this first artificial element. Between 1939 and 1945, scientists in Paris, Berkeley, and Oak Ridge created and characterized the true elements 87 (francium), 85 (astatine), and 61 (promethium). Moseley's periodic table was complete. A new science, nuclear chemistry, was born.

It was no wonder that the earnest chemists who thought they had discovered masurium, illinium, and the rest, had gone astray. When Noddack and Hopkins and the others were prospecting for Moseley's new elements, beta-radioactivity was still mysterious, though thirty years had passed since its first observation in uranium. The basic phenomenon is this: a nucleus emits an electron, leaving behind a nucleus whose charge, that is, its atomic number, is one unit higher. Since the nucleus that decays and the one left behind are perfectly defined, the electron's energy should be a definite value, equal to the energy difference of the nuclei before and after. But the electrons emitted in beta decay have a continuous range of energies, never more than the expected value, almost always less. Where does this missing energy go? When repeated, careful measurements failed to find it, Niels Bohr ventured in desperation that energy might not be conserved within the nucleus. This time, Bohr was too quick to abandon the lessons of previous experience.

What proved to be the correct inference, no less daring, was put forward—tentatively at first—by Wolfgang Pauli. When he presented his interpretation of beta decay, Pauli was thirty years old, and already one of the gods of the new physics. As a student, Pauli wrote a classic treatise on the theory of relativity that won praise from Einstein himself. In his twenties, he made formidable contributions to atomic physics. Pauli was what physicists call a radical conservative, by which we do not mean a right-wing ideologue, but one who takes seriously great principles, follows their consequences to the logical conclusion, and refuses to sacrifice them lightly.

Pauli was all of that, and he also had a well-developed sense of priorities. When a conference of radioactivity experts was arranged in Tübingen in December 1930, two days before a student ball in Zürich, Pauli made the only choice he could. He stayed in Zürich to prepare for the ball, and dispatched a letter to his less fortunate colleagues.

"Dear Radioactive Ladies and Gentlemen," he began. "For the moment I dare not publish anything about this idea and address myself confidentially first to you." Pauli wrote that he had "hit upon a desperate remedy." He proposed that the missing energy in beta decay was whisked away by a light, neutral particle, later named the neutrino, the little neutral one. The neutrino interacted very feebly with matter, so it would escape undetected from any known experiment, taking with it some energy, which would seemingly be lost. With just the right touch of irony, Pauli signed himself "Your most humble and obedient servant." The neutrino was not detected until 1956, but Pauli's hypothesis quickly took hold.

When Pauli proposed the neutrino, the actual composition of the nucleus was still not known. To be sure, Moseley had long since numbered the elements and measured the charge on the nucleus of each element. But the prevailing portrait of a nucleus composed of protons and electrons was plagued by all kinds of problems. As Bohr had commanded, and the newer quantum mechanics had confirmed, electrons formed a cloud outside the tiny nucleus. They had no business inside the nucleus. Again the resolution came from Rutherford's laboratory, now the Cavendish Laboratory in Cambridge.

In 1932, James Chadwick announced his interpretation of a new, highly penetrating form of radiation emitted when beryllium is bombarded by alpha particles from a polonium source: a nuclear constituent quite like the proton, but without any electric charge, that he named the neutron. Over the next few years, the true nature of the nucleus emerged: it is made of protons and neutrons; there are no electrons in the nucleus. Chadwick's discovery made it clear that each element was distinguished by the number of protons in the nucleus, the atomic number. Adding the number of protons to the number of neutrons gives the atomic weight. Change the number of neutrons and you get a different isotope. Some isotopes will be stable; others will be more or less radioactive. Beta decay could be understood as the disintegration of a neutron into a proton, an electron, and Pauli's neutrino. The varying number of neutrons in the naturally occurring forms of each element could account for the fits and starts in Mendeleev's original periodic table, ordered by atomic weight.

Thus was the basic nature of tangible matter resolved. An atom consisted of a tiny nucleus and electrons orbiting at a distance, according to Bohr's dictum. The constituents of the nucleus were protons and neutrons. Quantum leaps from one permitted orbit to another accounted for the richness of atomic spectra. This new atom is not the impenetrable body the ancients conjured. Though it has moving parts, it is no grand piano. Physicists had come to a picture that was elegant, and simple, and potent.

And then a cosmic stork delivered to the physicists' doorsteps new forms of matter, foundlings that would upset this tidy arrangement.

4

HEAVENS

Ben Nevis rises to 4,408 feet, above Fort William at the western end of Scotland's Great Glen. Britain's highest mountain—M1 in Sir Hugh Munro's catalogue of 3,000-foot peaks—is a great weathered granite mass, with sheer cliffs that plunge as much as 2,000 feet. One derivation of its name translates from Gaelic as "Mountain with its head in the clouds." Guidebooks report that if the permanent cloud-shroud parts for a moment, a climber will be rewarded with a view of the Antrim hills in Ireland and the eleven Munros that make up the Black Cuillin Ridge on the Isle of Skye. Many climbers count themselves lucky just to find their way through the mists to the sturdy cairn that marks the summit. The upper reaches of Ben Nevis are often cold, windy, and wet; snow can fall any day of the year.

In September of 1894, a twenty-five-year-old physicist named Charles Thomson Rees Wilson spent a few weeks in the observatory that then existed on the summit. Since his teens, Wilson had been an ardent lover of nature and of the Highlands—an authentic naturalist. Whether C. T. R., as colleagues called him, ever saw the Cuillin or the Antrim Hills, we do not know. But in the fog around Ben Nevis, he found the inspiration for a new way of seeing the rays and particles of the subatomic world that literally changed the way we see nature.

"The wonderful optical phenomena," he wrote, "shown when the sun shone on the clouds surrounding the hill-top, and especially the coloured rings surrounding the sun or surrounding the shadow cast by the hilltop or observer on mist or cloud greatly excited my interest and made me wish to imitate them in the laboratory." When he returned to his part-time post as a demonstrator for Cambridge medical students, C. T. R. Wilson resolved to reproduce the dazzling coronas, glories, and Brocken spectres in J. J. Thomson's Cavendish laboratory, where Ernest Rutherford was a fellow student. His first step would be to understand how clouds form.

Clouds form not only in the sky, but all around us in daily life. A mist rises over a pond on a very cold day, or above a cup of hot coffee. Water vaporizes from the liquid into the warm air just above it. The warm, moisture-laden air cools as it mixes with surrounding air, so it can no longer hold all the water vapor it has taken up. Some of the water vapor condenses on the innumerable dust particles that we see as motes in a beam of sunlight. The small droplets quickly evaporate again as they are carried into the drier air nearby.

Wilson built a glass container fitted with a piston in which he could cool moist air by quickly expanding its volume without introducing contaminants from the outside. He learned to remove all the dust particles from a sample by a succession of small expansions that condensed water onto them so they fell out of the vapor as tiny drops of mist. With no dust particles on which droplets could condense, subsequent expansions produced no mists. Even in dust-free air, however, very large expansions produced a thin rain of drops that Wilson could never eliminate, no matter how many times he repeated the expansions. It was as if a small number of "condensation nuclei" were constantly being renewed.

When the air in the chamber was supersaturated with eight times the water vapor it could comfortably hold, large expansions produced a dense mist of very fine drops. When Wilson illuminated the mist with a beam of white light, it shone with brilliantly colored rings. Like the dancing colors in an oily puddle on a sunny day, the rings arise from interference effects that become prominent when the diameter of a droplet is nearly equal to the wavelength of the light. Wilson had succeeded in developing the apparatus that would enable him to recreate, then to understand, some of the natural wonders he encountered on Ben Nevis. He had also found a new question: what was the inexhaustible source of condensation nuclei in his cloud chamber?

Near the end of 1895 came news of Roentgen's discovery of X rays. Almost at once, J. J. Thomson explored the increased electrical conductivity of air exposed to the new rays, using a primitive X-ray tube made by his assistant, Mr. Ebenezer Everett. Thomson and Ernest Rutherford were both of the opinion that the increased conductivity was due to the production of ions—electrically charged atoms—in the gas. For his part, Wilson had learned that his condensation nuclei were not much larger than molecular dimensions, and speculated that they might be individual charged atoms. If Wilson and Thomson and Rutherford were right, then X rays should produce large numbers of condensation nuclei in the cloud chamber. In an atmosphere of great excitement, Wilson borrowed Mr. Everett's X-ray tube and irradiated supersaturated air in his cloud chamber. In dust-free air, with an amount of vapor that produced only a few droplets before irradiation, the X rays produced a dense cloud that took many minutes to fall. By establishing an electric field in the gas, Wilson could clear the air of the ions before he expanded the chamber, so that no cloud was formed. The cloud chamber, built to explore the coronas and glories around sun and shadow that entranced Wilson on Ben Nevis, now made manifest the ions that were generated in the air by X rays.

Wilson's triumph in understanding the nature of condensation nuclei added greatly to his reputation. By 1900 he was made a university lecturer and a fellow of the Royal Society. But another trip to the slopes of Ben Nevis had started him on a second track that was to intersect the cloud chamber with still more dramatic consequences.

During the summer of 1895, Wilson was caught in a severe thunderstorm on the flanks of Carn Mor Dearg, M7, the big red hill that joins with Ben Nevis in a vast horseshoe facing northwest. The summit was obscured by mist when he heard the first rumblings of distant thunder. Suddenly Wilson's hair stood on end, and he plunged headlong down the long scree slope in search of cover. Wilson wrote, "The storm broke overhead with a bright flash and loud thunder just after I had left the summit. This experience drew my attention very forcibly to the magnitude of the electric field of the thundercloud and its sudden changes." It also awakened his interest in the electrical conductivity of the atmosphere.

In a paper published in 1785, Charles-Augustin de Coulomb had observed that electric charge slowly leaks away from a metal sphere suspended by a long insulating silken thread. Coulomb was the gifted experimenter who formulated the inverse-square force law between two electrical charges, but he could not decide whether the charge leaked away over the surface of the silk or dissipated through the air. The matter was still unresolved when Wilson took it up more than a hundred years later.

Using a gold-leaf electroscope of his own design to measure the quantity of electric charge, Wilson established that the loss was due to the feeble residual conductivity of the air. He found that new ions were being generated continuously, even in closed vessels in a darkened room. Perhaps the continuously resupplied ions that made the air a feeble conductor were the condensation nuclei on which mists formed in the dust-free air of the cloud chamber.

Having linked X radiation with ionization in his cloud chamber, Wilson speculated that the residual conductivity of air was due to ionizing radiation, perhaps approaching the Earth from outer space. To test this idea, he measured the conductivity of air both out in the open and under hundreds of feet of rock in the Caledonian Railway tunnel near his home at Peebles in Scotland. Finding no appreciable difference, he concluded that an extraterrestrial source was insignificant. But he was misled. The confined air in caves and cellars conducts a little better than outdoor air because it is ionized by radiation from traces of the noble gas radon that seep constantly from the surrounding rock.

The problem of residual ionization attracted attention in many quarters. If the ionizing radiation came from the Earth, the atmosphere would be ionized most near the surface, because radiation from the soil would be absorbed before it reached high altitudes. During Easter week of 1910, the German Jesuit Theodor Wulf traveled from his teaching position at Saint Ignatius College in Valkenburg (Netherlands) to Paris. Like many tourists, he went to the Eiffel Tower. There he spent six days making meticulous measurements of the number of ions in the atmosphere, both at the base of the Eiffel Tower and on the spire nearly a thousand feet above. If the Earth were the source of all the ionizing radiation, the number of ions at the top should have been but a few percent of the number at the base. Father Theodor measured the ionization atop the tower to be much larger than predicted—more than half the value at ground level. The careful observations of his Easter vacation reopened the possibility that some of the radiation came from above. Wulf suggested that the matter might be settled by balloon-borne observations.

Victor Franz Hess, of the Institute for Radium Research at the University of Vienna, accepted Wulf's challenge. He undertook a series of remarkable balloon ascents to find how high into the atmosphere the ionization persisted. Following Wulf's model, Hess constructed electroscopes that would withstand the low temperatures and low atmospheric pressures encountered in flight. In 1911, he made two ascensions from the Royal Imperial Austrian Aeroclub, on a grassy plot near the giant Ferris wheel in Vienna's Prater amusement park, and found that the ionizing radiation was as strong at an altitude of 3,600 feet as it was near the ground. He concluded that there must be another source of the penetrating radiation, in addition to the radioactive substances in the Earth's crust.

A grant from the Imperial Academy of Sciences in Vienna enabled Hess to carry out seven more balloon ascents in 1912. The first of these took place during a partial eclipse of the sun on the seventeenth of April. Five more times Hess and his pilots lifted off from Vienna, at day and night, in fair weather and foul. The weak lifting power of the Vienna gas limited the altitude Hess could reach from there, so on the seventh of August he undertook a high-altitude flight from Ústí nad Labem (Aussig in German) in northern Bohemia. In a six-hour flight aboard the balloon *Bohemia*, lifted by nearly 60,000 cubic feet of hydrogen, Hess, his meteorological observer, and a pilot ascended to nearly 17,500 feet before returning to Earth east of Berlin.

For the first few thousand feet above the ground, Hess found that the radiation fell off a little. This decrease results from the absorption in the air of gamma rays emitted by radioactive substances on the Earth's surface. As the *Bohemia* rose higher in the atmosphere, the radiation levels increased—slowly at first, then more and more rapidly with altitude. At the highest point in the journey, the ionizing radiation was nearly three times as intense as on the ground.

Hess concluded, "The results of the present observations seem to be most readily explained by the assumption that a radiation of very high penetrating power enters our atmosphere from above, and still produces in the lowest layer a part of the ionization observed in closed vessels." Finding no reduction in radiation levels at night or during the solar eclipse, Hess eliminated the sun as the source of the hypothetical radiation. He noted that the absorption of the radiation coming from above follows an exponential curve, and predicted that the radiation would increase still more rapidly at greater heights.

Hess's report of the penetrating ionization at high altitudes provoked a spirited debate about the existence of extraterrestrial radiation. Many physicists regarded Hess's suggestion as pure fantasy, and it was not considered good form to speak seriously about rays from the Great Beyond. When a young PhD named Werner Kolhörster took up Hess's conjecture, it was an act of some courage, for he was risking not merely his life, but also his career.

Working at the University of Halle in a group that made balloon flights to perform meteorological and geophysical studies, Kolhörster further refined Wulf's electrometer to make it suitable for ascensions to very high altitudes. In a series of five daring ascensions during 1913 and 1914, he made precise measurements of the amount of ionizing radiation. On June 28, 1914, Kolhörster ascended 29,500 feet, approximately to the height of Mount Everest. He greatly extended Hess's observations and recorded a twelvefold increase in the radiations, confirming Hess's extrapolation. Within two months, before Kolhörster's results could be confirmed and digested, the world was at war.

Many skeptical scientists doubted both the observations and the interpretation of them as evidence of "ultra-rays" from outer space. C. T. R. Wilson, for one, speculated that the ionizing rays were electrons accelerated to great energies in the powerful electrical activity of thunderstorms.

In the mid-1920s, Robert Millikan's group at the California Institute of Technology devised an electrometer that recorded its measurements on film, without an operator.

Unmanned balloons could lift the instruments to much higher altitudes than before. Millikan had scoffed at Hess's claims that the radiation came from outer space. Now, his own measurements in unmanned balloon flights up to nearly 100,000 feet made it very difficult to maintain the view that the radiation originated in the atmosphere. By lowering instrumented probes into deep mountain lakes, Millikan confirmed that the radiation came from above. It had to have an external source. He changed his mind, promoted the term "cosmic ray" for the extraterrestrial radiation, and began to claim the discovery as his own. According to a famous Caltech story, a JESUS SAVES painted on a steam shovel working on campus in 1937 was amended to read, ". . . BUT MILLIKAN GETS CREDIT."

By 1930 it was generally accepted that energetic radiation rains down on the atmosphere from outer space. The nature and sources of the radiation were still unknown. We know now that most of the primary cosmic rays are protons, alpha particles, and heavier atomic nuclei, but mysteries remain about the sources and acceleration mechanism, especially for ultrahigh-energy cosmic rays.

∞

At the end of 1910, C. T. R. Wilson turned his attention back to the cloud chamber, now with a new goal. He had followed his demonstration that X rays could create the condensation nuclei that precipitate the formation of clouds with experiments that showed that ultraviolet light, alpha particles, and other agents produce identical condensation nuclei. Now Wilson hoped to reveal the track of a single alpha particle by condensing water droplets on the contrail of ionization in its wake. By March 1911, he was photographing wispy traces left by the passage of individual particles.

The impact on the scientific world was immediate and profound. It was as if the story of radioactivity had been written in invisible ink, and C. T. R. Wilson discovered how to render it visible. Wilson's student, Cecil Frank Powell, wrote without exaggeration that "The pictures of the tracks produced by X rays and alpha-particles which Wilson published in 1912 are among the most beautiful objects of experimental physics."

Rutherford himself declared Wilson's cloud chamber "the most original apparatus in the whole history of physics." All at once, physicists were not limited to recording where a particle hit a scintillating screen; now they could follow the path it traveled. Wilson, the great naturalist, made it possible for the nuclear scientists to gather specimens and distribute them for study.

When he was able to resume research after the Great War, Wilson published the results of his findings from 500 stereoscopic pairs of cloud chamber photographs. These pictures included the first images of subatomic reactions, revealed from start to finish, in three dimensions. Wilson photographed tracks of electrons knocked out of their atomic orbits by gamma rays, providing conclusive evidence for the phenomenon first studied by Arthur Holly Compton at the University of Chicago. Wilson's images confirmed the existence of the photon, or gamma ray, as a particle of light. In effect, the photon was the first particle discovered using the cloud chamber. Others would follow.

The cloud chamber lifted the veil on other nuclear reactions as well. Patrick Blackett became a research student in Rutherford's Cavendish Laboratory shortly after Rutherford had demonstrated the disintegration of the nitrogen nucleus under bombardment by alpha rays. Blackett's first project was to make a cloud-chamber picture of the events that occur during the transformation of a nucleus. In 1924, after examining photographs of 400,000 tracks of charged particles, Blackett found eight examples in which an alpha particle was swallowed by a nitrogen nucleus and a proton was seen leaving the collision site. What Frederick Soddy had dared call transmutation was there for all the world to see.

Cloud-chamber photographs displayed subatomic processes in all their richness and individuality. Cecil Powell said of his teacher's work, "Our whole conception of the world of atomic physics was strengthened and illuminated."

The cloud-chamber technique spread quickly to other laboratories, where it was extended and improved. In Leningrad, Dmitri Skobeltsyn placed a Wilson chamber in a magnetic field, so he could determine the momentum of each particle by measuring the curvature of its track. He was able to characterize the Compton effect in every detail. In 1927, Skobeltsyn noted extraneous tracks that were hardly deflected in the magnetic field, and that came not from his radioactive source, but from the skies. These were the first pictures of individual cosmic rays, as he rightly recognized, noting that their frequency in his photographs would account roughly for the ionization attributed to extra-terrestrial radiation, just as the balloon experiments had suggested. Collisions of primary cosmic rays with atoms in the atmosphere create great showers of secondary particles, some of which passed through Skobeltsyn's chamber. Not long after Skobeltsyn reported his discovery of cosmic-ray tracks in a cloud chamber, other groups began observations.

For the next three decades, the study of cosmic rays was the best way to look inside the nucleus by observing the debris of very-high-energy collisions. The cloud chamber yielded marvelous pictures of the tracks of individual particles. Under the right conditions, physicists learned how to determine from the sharpness of the tracks how much time had elapsed since the passage of the ionizing particle. From the density of the tracks they could say something about the ionizing characteristics—and therefore the identity—of the particle.

Beginning in 1930 Millikan and Carl D. Anderson designed a cloud chamber to operate inside a strong electromagnet that bent the paths of cosmic rays one way for positive particles like the proton, and the other way for negative particles like the electron. Anderson was looking at cloud-chamber photographs of cosmic rays on August 2, 1932, when he saw a track that was quite peculiar. The track's curvature showed that it had been left by a positive particle, but the positive particle did not seem to be a proton. Its faint track revealed it as a fast, light particle, not a slow, heavy one. In fact, it seemed to be about as light as an electron, but it had the opposite electric charge. On the basis of this one photograph, Anderson could assert the existence of a positive electron.

In a letter to the editor of *Science* published on September 9, 1932, Anderson announced his observation of "a positively charged particle having a mass comparable with that of an electron." A subsequent paper, "The Positive Electron," sent to *The Physical Review* at the end of February 1933, named the discovery the positron, in distinction to the normal negative electron.

Now the cloud chamber had revealed something entirely new, and apparently unexpected. Nevertheless, there was a hit-or-miss aspect to cloud-chamber research. A cloud chamber was expanded, illuminated, and photographed. With luck, the picture contained interesting tracks. Without luck, the picture would be blank. In Anderson's work, only about one photograph in fifty was worth looking at.

Back in Cambridge, Patrick Blackett continued his alpha-particle research until 1931, when Giuseppe (Beppo) Occhialini arrived at the Cavendish from Florence, schooled in new techniques of electronic circuitry. Occhialini and Blackett placed Geiger-Müller tubes (sensing elements of the famous counter devised in 1925 by Hans Geiger and Walther Müller) above and below the cloud chamber to register cosmic rays. When coincident signals from the counters above and below indicated that a particle had passed through their apparatus, Occhialini's electronic circuit—a primitive computer— expanded the cloud chamber and took a picture. The triggered chamber was an immediate success: nearly every picture contained particle tracks. Blackett and Occhialini placed their large chamber inside an electromagnet and photographed every expansion with two cameras to make stereo views that allowed a three-dimensional reconstruction of events. On the heels of Anderson's announcement, they immediately confirmed the existence of the positron, and made the first observation of electromagnetic "showers" of positrons and electrons emanating from a single point. It was the first observation of the creation of matter—and antimatter—from energy. Later in 1933, Blackett connected this observation to Paul Dirac's theory of pair production, or "twin birth."

The British theoretical physicist Paul Adrien Maurice Dirac had predicted in 1931 that there must be a positive electron, but Anderson was unaware of the prediction. Dirac invented the equation that describes the behavior of electrons of any energy. Solving his equation, Dirac encountered a puzzle akin to one of the leading causes of math anxiety. The quadratic equations that arise in "story problems" generally have two possible solutions, but sometimes only one of them makes any sense. Jack is two years older than Jill. The product of their ages is 120. How old is Jack? Well, Jack is twelve years old and Jill is ten years old works. So that's the answer. But suppose Jack is minus ten years old and Jill is minus twelve years old. That works, too, but it doesn't mean anything to be minus twelve years old.

When he took his equation at face value, Dirac found solutions for which the electron seemed to have negative energy. It seems like one of those cases where a mathematical solution makes no physical sense. But Paul Dirac struggled to find a sensible physical interpretation. A half-century later, Dirac asked rhetorically, "Now how does one make a prediction? The basic requirement is that one should have a theory in which one has a great deal of confidence, a very great deal of confidence. One must be prepared to follow up the consequences of this theory, and feel that one just has to accept the consequences, no matter where they lead one."

Dirac's equation told him something entirely new: the minus sign didn't mean that the energy was negative, but that there was a particle like the electron, but with opposite—positive—charge, and positive energy.

What will be the mass of this particle? The equation insisted that the positively charged particle was an antielectron, with the same mass as the electron. Dirac hesitated. He wanted to interpret the positive particle as the proton, whose mass is 1,836 times the electron's, and hoped that he had found in one equation the complete story of the hydrogen atom. At that time—before the neutron and the neutrino—people believed that the whole of matter had to be explained in terms of just two particles, electrons and protons. The Dirac equation would not bend, and Dirac came to accept the conclusion that he had predicted the existence of antimatter, something unknown in physics at the time.

A recollection from Chris: I met Dirac three or four times before I found the opportunity—and courage—to ask him about the positron. I wanted to understand what he could possibly have been thinking when he mistakenly identified the "negative-energy" solutions of his equation as the proton, instead of a new positive particle with the same mass as the electron. The occasion was a colloquium I gave at Florida State University, his retirement home. I found myself lecturing on quantum mechanics before one of the gods of modern physics, one who had shaped the heroic age of the 1920s and 1930s. I took it as a good—if ambiguous—sign that Dirac sat in the front row and smiled knowing smiles from time to time.

That evening, when the two of us were alone in a quiet corner at a cocktail party, I popped the question. Dirac was a master of the pregnant pause. One-on-one or before a group, he would stare down at his chest until he was sure he had everyone's full attention. This time he drew it out long enough for me to estimate just how long a line of young fools I had joined. Then, with precisely the right ironic wrinkle of an eyebrow, he said, "In those days, we were a little less ready to speculate about new particles." He had a point: theoretical physicists of the late 20th century were quick to take their equations very seriously indeed, and to propose new particles apparently without hesitation. But I think the truth was that even Dirac hadn't appreciated just how seriously to take the Dirac equation. He went on, "I just didn't dare to postulate a new particle at that stage, because the whole climate of opinion was against new particles."

The discovery of the positron and the observation of electromagnetic showers led to a fruitful period of using triggered cloud chambers to study the interaction of photons and electrons in the cosmic rays. At Caltech, Carl Anderson and Seth Neddermeyer adopted Blackett and Occhialini's invention to study the rate at which the cosmic rays lose energy when they pass through matter. They took about 6,000 photographs with a one-centimeter plate of platinum placed across the center of the cloud chamber. They found two groups of tracks: one consisting of particles that lose much of their energy on passing through the plate, and the other consisting of "penetrating" particles that lose a relatively small fraction of their energy. The tracks of the particles that lost much of their energy readily identified them as electrons and positrons. Anderson and Neddermeyer could explain the characteristics of the penetrating tracks by the hypothesis that they were charged particles like the electron or proton, with a mass in between the electron and proton masses. There is no evidence for such particles in ordinary matter, and so, Anderson and Neddermeyer remarked, "there must exist some very effective process for removing them."

Other physicists quickly confirmed the existence of these penetrating particles in cosmic radiation at sea level and narrowed the mass of the new particle to between one-tenth and one-seventh of the mass of the proton. In the scientific literature of the period, the new particle became known as the mesotron, or meson, to emphasize its in-betweenness.

We now call it the muon. Eventually, cosmic-ray measurements refined the mass to about one-ninth the mass of the proton, or 207 times the electron mass. At sea level, muons make up about 80 percent of the cosmic radiation. Dozens of cosmic-ray muons will pass through your body while you are reading this sentence.

In 1936, a year before the muon appeared in Anderson and Neddermeyer's cloud chamber, Hideki Yukawa, a Japanese theorist working in Kyoto, had proposed a theory of the powerful forces that bind protons and neutrons together in the nucleus. The small size of nuclei hinted that the range of the nuclear force was quite short, and that its influence was limited to about a millionth of a billionth of a meter (10^{-15} meter). Yukawa reasoned that the nuclear force was carried by a hitherto unknown particle, and he estimated the particle's mass from the size of the nucleus at about one-seventh of the proton mass. He speculated that the same force might also act as the agent of radioactive beta decay, changing neutrons to protons. To do so, the new particle itself would have to decay into an electron and Pauli's hypothetical neutrino. Yukawa suggested that his nuclear-force particle might be found outside the nucleus, in the cosmic radiation.

Yukawa issued his all-points-bulletin for the nuclear-force particle in a talk at a monthly meeting of the Physico-Mathematical Society of Japan at the

University of Tokyo. An article on his theory published in the society's *Proceedings* did not receive wide circulation outside of Japan, but he sent a copy to J. Robert Oppenheimer in Berkeley. When Anderson and Neddermeyer published their discovery, Oppenheimer and Robert Serber were quick to make the connection with Yukawa's work. Some publications from the period refer to Yukawa's force particle as the Yukon, and identify it with the Anderson-Neddermeyer particle. Evidence quickly appeared that the muon did disintegrate into an electron plus something neutral, fitting Yukawa's description and explaining why there are no muons in ordinary matter: they have all decayed.

The Second World War disrupted research as physicists were driven from their countries, or drafted into applied work, or merely forced to seek shelter. But small oases of basic research flourished amid the raging cataclysm.

To escape the wartime bombing in Rome, Oreste Piccioni removed his laboratory from the university, whose proximity to the San Lorenzo railway freight station made it vulnerable to "collateral damage," to a classroom, half below ground level, in safe precincts at the Liceo Virgilio near the Vatican. Using electronic equipment procured on the black market, Piccioni and Marcello Conversi studied cosmic-ray muons. In an ingenious, all-electronic-counter experiment, they measured the muon's lifetime to be about two millionths of a second.

Conversi and Piccioni continued their undercover work on the muon throughout the war. Once peace returned to Europe and their colleague Ettore Pancini returned from his role as resistance commander "Achille," they carried out the experiment that unmasked the Anderson-Neddermeyer particle. When negative muons encounter material, they are slowed by collisions with atomic electrons. If the muon were Yukawa's nuclear force particle, it would have a tremendous affinity for the nucleus, which would "eat it up before it had time to die in bed," as Enrico Fermi put it in a seminar. A muon would be captured before it could decay, as we now understand, into an electron and neutrinos. In December 1946, the three Romans reported that negative muons stopped in carbon were not absorbed by the nuclei, but decayed as usual. Anderson and

Neddermeyer's muon simply could not be Yukawa's force particle. Now there were two questions: What was the muon, and where was the Yukawa particle?

Isolated in Japan by the war, Yasutaka Tanikawa in 1942, and Shoichi Sakata and Takeshi Inoue in 1943, had proposed that the muon was actually a decay product of the true Yukawa particle. The same proposal was made by the American Robert Marshak in 1947 after the results of the Italian trio were known. The suggestion was that the muon was not just an imposter, it was actually the offspring of the true Yukawa particle. The decisive insight on the muon's nature came from another experimental technique, applied not in a subterranean shelter, but high in the mountains.

∞

The pioneers of photography learned early on that light was not the only agent that imprinted latent images on their temperamental photographic plates. Chalk and marble, cotton and feathers, all left their marks. Abel Niépce de St.-Victor, cousin of Joseph Nicéphore Niépce, the father of photography, was one of the first to make dry photographic plates that could be transported and stored for later use. Abel found in 1867 that uranium salts fogged his plates even when they were separated by layers of paper. He attributed the effect to the faint phosphorescent glow of the uranium compounds. Had he pursued the phenomenon, rather than taken care to avoid the problem, he might have advanced the discovery of radioactivity by thirty years.

Early in the 20th century, physicists found that photographic emulsions could record the passage of individual alpha particles. A photographic emulsion consists of myriad crystalline grains of silver bromide, with a small mixture of silver iodide, suspended in gelatin. A charged particle traversing the emulsion sensitizes grains of silver bromide lying in its path. When the plate is developed, the sensitized grains appear under the microscope as dark grains of metallic silver, like beads on an invisible string.

In the first applications, the photographic plate took the place of the scintillation screen. Each black spot on a developed plate corresponded to a scintillation whose position could be measured accurately. Only after the publication of Wilson's cloud-chamber pictures in 1911 did it occur to anyone to allow alpha particles to enter the emulsion end-on, so that physicists could find and follow individual tracks.

Following a flurry of work from 1910 to 1915, there was little progress for nearly ten years. Then Marietta Blau found tracks of protons that alpha particles had knocked out of hydrogen atoms in the gelatin of the emulsion itself. With her colleague Herta Wambacher at the Institute for Radium Research in Vienna, Marietta Blau devoted years of effort to improving the sensitivity of commercial emulsions by treating them chemically before exposure. During the '30s, Blau and Wambacher turned to cosmic rays. In 1937, they took photographic plates to the cosmic-ray observatory established in 1931 by Victor Hess on the Hafelekar near Innsbruck, at an elevation of 2,300 meters, to take advantage of the greater intensity of cosmic rays at high altitudes. In plates exposed for five months, they found a number of events they called stars, in which as many as eight or nine dense tracks emanated from a single point in the emulsion. Each star recorded the disintegration within the emulsion of an atom of silver or bromine under the impact of a cosmic ray, like a rack of billiard balls being broken apart by the opening shot of the cue ball. With Blau and Wambacher's photographic technique, an atom made a picture of itself being smashed.

When Marietta Blau began her work in Vienna, Cecil Frank Powell was a research student at the Cavendish Laboratory in Cambridge under C. T. R. Wilson and Ernest Rutherford. Powell shared with Wilson the instincts of the naturalist—the love for direct contact with the phenomenon and the gathering of specimens. At the Cavendish, he became a specialist in the use of Wilson's cloud chamber. In 1928, he was appointed research assistant to A. M. Tyndall, director of the newly established H. H. Wills Laboratory at the University of Bristol. After Cockcroft and Walton succeeded in artificially disintegrating the nucleus, Powell built a Cockcroft-Walton accelerator and plunged himself into the study of nuclear physics.

Powell's original intention was to use the cloud chamber as his detection device, but theoretician Walter Heitler alerted him to the success that Blau and Wambacher had achieved in recording nuclear reactions directly in emulsion. The photographic method looked so simple to Heitler that he suggested that even a theorist might be able to do it. He did his part, carrying a batch of emulsions to the 11,500-foot Jungfraujoch in Switzerland. In those plates, the Bristol team quickly found for themselves the stars showing many protons and alpha particles ejected in the disintegration of nuclei.

Powell's plan had been to take 30,000 cloud-chamber pictures over six months of accelerator running. Even a triggered cloud chamber is only sensitive for brief periods. A photographic emulsion never blinks; it records the track of every particle that passes through it. When Powell realized that he could gather better data in only a day or two using emulsions, he was converted.

The new photographic method offered other advantages. The photographic plate can record with precision the point of entry of a particle and its direction of motion. The character of the track a particle leaves behind reveals the particle's nature and energy. The faster the particle travels, the more widely spaced are the developed grains, since the faster particles have less capacity to ionize, and sensitize the emulsion. The distance between these grains is thus a measure of the particle's speed; consequently, the faster the particle, the more difficult it is to detect and measure in the emulsion. The commercially available emulsions of the time, designed for the quite different purpose of ordinary photography in visible light, were erratic in recording tracks of nuclear particles.

Beppo Occhialini, the pioneer of the triggered cloud chamber, had fled Fascist Italy to work in São Paulo, and later waited out the war as a guide in the Itatiaia mountains of Brazil. In 1945, he returned to Europe to join the Bristol group. Occhialini shared Powell's enthusiasm for the possibilities of the emulsion method and approached Ilford, Ltd., the maker of photographic plates, about improving the sensitivity of emulsion for nuclear research. Ilford found it possible to make plates that resulted in a very remarkable improvement in recording properties. Occhialini sent some pictures of proton and alpha-particle tracks to his former research student César Lattes in Brazil, and when Lattes expressed his enthusiasm for the new emulsion, Powell and Occhialini brought him to Bristol to join their team.

Occhialini not only had a great nose for physics, he also knew how to live.

Immediately he took a few small plates coated with the new emulsions to the cosmic-ray observatory at the Pic du Midi, about 9,400 feet high in the French Pyrenees, and left them to soak up cosmic-ray tracks while he went on a month's mountaineering vacation.

"When [the emulsions] were recovered and developed in Bristol," wrote Powell, "it was immediately apparent that a whole new world had been revealed. The track of a slow proton was so packed with developed grains that it appeared almost like a solid rod of silver, and the tiny volume of emulsion appeared under the microscope to be crowded with disintegrations produced by fast cosmic-ray particles with much greater energies than any which could be generated artificially at the time. It was as if, suddenly, we had broken into a walled orchard, where protected trees had flourished and all kinds of exotic fruits had ripened in great profusion." Powell set up a number of microscopes to search the plates and engaged the services of several young women as observers to scan the developed emulsion for interesting tracks. As soon as the observers found an event, such as a disintegration, they would call a physicist to examine it for noteworthy features. A few days into the scan, Marietta Kurz showed Powell a puzzling event: one meson that came to rest in the emulsion, its track becoming blacker and crooked as it slowed down, and, emerging from its end, a straight, light track, the signature of another meson, moving rapidly. One day later, a second double meson appeared.

Measurements of the tracks suggested that, if the second meson were identified as Anderson and Neddermeyer's muon, the penetrating particle of the cosmic radiation, then the primary meson must have a mass of about 265 electron masses. However, two specimens did not seal the interpretation. The Bristol group threw more observers into the hunt, but it was six or eight weeks before another double meson showed itself.

Lattes had a different idea for gathering more data: he would expose more plates on a higher mountain. He learned from geographers at Bristol that there was a meteorological station at about 17,700 feet above sea level, some fifteen miles by road from La Paz, the capital of Bolivia. He proposed to Powell and Occhialini that if they could finance the flight to South America, he would take care of exposing the plates on Mount Chacaltaya for a month. They agreed, and he left Bristol with several plates "plus a pile of pound notes sufficient to carry me to Rio de Janeiro and back." Tyndall, the director of the Bristol laboratory, urged Lattes to take a reliable BOAC flight, but Lattes followed his Brazilian heart and boarded a Varig Super Constellation. The reliable British aircraft crashed in Dakar, killing everyone aboard.

It is probably pure coincidence that Chacaltaya was the highest ski area in the world and the site of the first ski lift in South America, in operation since the early 1940s. With its 200-meter vertical descent, the slope was described as intermediate—but for the rarefied atmosphere, which made it distinctly advanced, if not hallucinatory. In any event, Lattes occupied himself for a month while the emulsions were on the mountaintop. He developed one plate in La Paz. Although the local water stained the emulsion, he quickly found a double meson of the same character as the Pic du Midi events. When developed and scanned in Bristol, the Chacaltaya plates produced about thirty double mesons to add to ten found in the plates Occhialini exposed in the Pyrenees. Chacaltaya remained an important cosmic-ray station for decades, and now is home to a climate observatory.

Further research confirmed that the initial tracks were made by Yukawa's nuclear-force particles; the secondary tracks were made by Anderson and Neddermeyer's muon. Powell and collaborators named the first track by the Greek letter pi (π), for primary, and the second track by the Greek letter mu (μ), for mesotron. Today we call them the pion and muon. The pion decays into a muon and a nearly massless neutral particle, Pauli's neutrino (about which much more later), which does not register in the emulsion.

The discovery of the pion in 1947 gave physicists a coherent picture of the atom and nucleus. Atomic electrons orbited the nucleus, in which Yukawa's pion bound protons and neutrons together. The neutrino, still unobserved, was the electron's partner in nuclear beta decay. And then there was the muon—first mistaken for Yukawa's particle—an apparent extravagance, the only one not present in ordinary matter.

5

FLORA

Hauling their cloud chambers and photographic emulsions up mountain peaks, the cosmic-ray pioneers were kindred spirits to the intrepid natural historians of the 19th century. In search of exotic plant and animal species, naturalists travel to out-of-the-way places like the Galapagos Islands, isolated from continental influences, or to extreme environments like the deep-sea vents where heat-loving bacteria nourished by chemosynthesis support communities of giant tubeworms and other novel creatures.

The cosmic rays that shower down everywhere on Earth are attenuated by the protective blanket of the atmosphere. That is why the cosmic-ray explorers mounted their own expeditions to lofty peaks around the globe: to the Jungfrau in the Bernese Oberland, to the Pic du Midi in the Pyrenees, to Mount Chacaltaya in the Andes. In place of botanical curiosities pressed between sheets of blotting paper or insects preserved in specimen jars, the particle prospectors brought home photographs of their own exotic species to examine and categorize. Harry Moseley, the man who numbered the elements, could deprecate "describing, cataloguing, or systematising," but if taxonomy is not science, it is a necessary prelude to science. First find regularities, then look for underlying principles.

An elegantly illustrated *Flora* depicts a plant species in meticulous detail—roots and leaves, flowers and seeds—perhaps with a composite time-lapse drawing that shows the key stages in the plant's development. The very best cosmic-ray pictures display the full life history of a particle—birth, passage through the cloud chamber or emulsion, and sudden demise. Transformations that are for a flower gradual—budding, opening, withering—are for a particle abrupt. A particle springs full-grown into being, lives out its life without aging, and suddenly is transmogrified. When a charged pion ceases to exist, a muon and neutrino nearly always appear in its place. The rebirth of most particles can take many forms. One charged pion in 10,000 is reincarnated not as a muon and neutrino, but as an electron and neutrino. Hundreds of alternative futures may lie before another particle species.

The shape of a pressed leaf and the network of its veins are there for all to see—and we have seen many of them in everyday life. The specimens collected by cosmic-ray physicists are not the stuff of common experience, but they give a reality—and immediacy—to the happenings in a world too small to see directly. Just as a budding natural historian must learn to read leaves for a plant's secrets, the cosmic-ray naturalist must learn to read the story in each particle trail. Most of those stories have become too familiar to warrant retelling, but occasionally a single track or a single photograph reveals a tale never heard before, and totally unanticipated.

George Rochester and Clifford Butler wrote the final chapter of the heroic age of cosmic-ray physics, when in 1945 they undertook a systematic study of the particles produced in cosmic-ray showers. Working in P. M. S. Blackett's new postwar laboratory in Manchester, they placed a large cloud chamber within the field of an eleven-ton magnet. The chamber was divided in half by a lead plate, an obstacle to test the penetrating power of the cosmic-ray particles. Using counters to trigger the expansion of the chamber, they made 5,409 stereoscopic pairs of photographs. In the early cloud chamber pictures and emulsions, hiding among the cosmic-ray dandelions and Queen Anne's lace—the pions and muons—were four-leaf clovers. Undoubtedly many cosmic-ray physicists had passed over them, either missing them entirely or failing to recognize their revolutionary importance. Sifting through their thousands of specimens, Rochester and Butler learned enough about the commonplace in all its variations to recognize the extraordinary.

On October 15, 1946, Rochester and Butler found the first example of a new species in a photograph of a penetrating shower. A bit below the lead plate, an inverted *V* appeared out of nowhere. Extensive analysis through the winter convinced them that the event represented the spontaneous decay of a neutral particle—invisible to the cloud chamber—into a positive track and a negative track with roughly equal energies. This one specimen showed that the "vee" particle was an entirely new unstable particle, perhaps a thousand times the electron's mass, much heavier than the mesons already known. Nevertheless, hoping that other vee particles would appear, Rochester and Butler did not publish their results at once.

On the twenty-fifth of May, 1947, a second unusual event appeared. This time, a track that entered the chamber from above suddenly veered off in a new direction. Rochester and Butler interpreted the kink in the track as the point where an unstable charged particle decayed into a charged particle and an unseen neutral particle. Again the disintegrating particle seemed to weigh about 1,000 times the electron's mass.

Rochester and Butler announced their discovery in December 1947 at the Dublin Institute for Advanced Studies, and detailed their findings in an article in *Nature*. The vee particles drew the interest of cosmic-ray physicists around the world, even as they fixed their primary attention on the newly understood pion.

For two years the Manchester group found no further examples of vee particles.

Urged on by Beppo Occhialini, Blackett sent his charges—and the eleven-ton magnet—off to the Pic du Midi, where they began running the cloud chamber in July 1950. By that time, the worldwide search for vee particles was in full swing. At the end of November 1949, Carl Anderson wrote Blackett from Caltech to report about thirty examples of "pothooks," as he called the forked tracks, and to support the interpretation of the events as decays of new unstable particles. Also in 1949, the Bristol emulsion group found another curiosity, a heavy, long-range particle that stopped in the emulsion and disintegrated into three pions. The mass of this new particle, called tau at the time, was 985 electron masses.

In their first six months on the Pic du Midi, Rochester and Butler found seven charged vee particles and thirty-six neutral vees. The neutrals were of two main types: one group disintegrated into a positive pion and a negative pion, like the first Manchester specimen. The other decayed into a proton and a negative pion. Cecil Powell's description of breaking into a walled orchard and finding a whole new world was no overstatement.

There was something highly odd about these new particles: they lived much, much too long. They were produced in collisions of cosmic-ray protons with nuclei, in the same strong nuclear interactions that produced a shower of pions. When the new particles decayed, they often produced strongly interacting particles like pions and protons, as well. And there's the rub.

Strong interaction processes typically take about the same time, a few trillionths of a trillionth of a second, as it takes light to cross a proton. The lifetimes of the vee particles are very much longer—more than ten trillion ticks of the strong-interaction clock. This is not a small matter: the discrepancy is 10,000 times greater than the ratio of the age of the universe to your age. Despite appearances, the decays of the new particles were not strong interactions, even though the particles involved were all strongly interacting. Something had to forbid the rapid strong decay.

Muons were occasionally among the particles observed to emerge in vee-particle decays. Unlike pions, muons do not experience the strong interactions. Here was further evidence that the decays of the vee particles were not strong processes, but weak interactions, which induce the more leisurely radioactive beta decays.

In 1953, theorist Murray Gell-Mann proposed a solution to this conundrum: simply declare the decays impossible! Endow the strange new particles with a new property, "strangeness." Vee particles carry strangeness, while pions carry none. If strangeness cannot be created or destroyed in strong interactions, then a vee particle cannot quickly turn into two pions. If weak interactions don't respect the rule of strangeness, then the vee particles can decay into two pions, but terribly slowly. How to account for the production of strange particles in strong collisions? Produce a particle carrying strangeness plus one in association with a particle carrying strangeness minus one. Altogether, no net strangeness has been produced.

This is a brilliant—and, in fact, correct—solution. Except that "strangeness" seems an entirely artificial, ad hoc concept. Where does this strangeness reside? Perhaps the pioneers never asked themselves this question. It was satisfying enough to have rules that successfully described the data, rules that could be reliably applied in new circumstances, and yet a small mystery—that strangeness—still kept a hint of intrigue in the field over the course of the next two decades.

The romantic cosmic-ray era of particle physics lasted until 1953, when the Cosmotron, a new accelerator at the Brookhaven National Laboratory on Long Island, turned on its beam of protons with an energy of 3 billion electron volts, 3 GeV. The Cosmotron's compact and well-controlled beam of protons offered overwhelming advantages over the diffuse and unpredictable sprinkle of cosmic rays. Foraging in the wild could not compete with reaping the harvest of the new Cosmotron hothouse, where physicists could control the experimental conditions and mount elaborate—and evolving—apparatus.

The last great cosmic-ray conference of the heroic era was held at the casino of Bagnères-de-Bigorre, in the French Pyrenees, in July 1953. Bagnères was a triumphant celebration of the discovery and understanding of strange particles. The street on which the casino sits is today called Boulevard de l'Hypéron, for it was there that the strange partners of the proton were given the name hyperons. But in his closing address, Cecil Powell, who with Lattes and Occhialini had first seen the pion-to-muon decay chain in emulsions exposed on the nearby Pic du Midi, declared, "Gentlemen, we have been invaded . . . the accelerators are here." Louis Leprince-Ringuet, a pioneer of the

French cosmic-ray program and Immortal of the Académie Française added, "*Mais nous devons aller vite, nous devons courir sans ralentir notre cadence : nous sommes poursuivis . . . nous sommes poursuivis par les machines !* (But we must move fast, we have to run without slowing our pace: we are pursued . . . we are pursued by the machines!)"

∞

When new PhD Donald Glaser departed Caltech in 1950 for a faculty appointment at the University of Michigan, he took with him a question written on the blackboard of his supervisor, Carl Anderson: "What have we done about the pothooks today?" Having measured the cosmic-ray spectrum at sea level using two cloud chambers and a large electromagnet for his thesis research, Glaser knew well the advantages and limitations of the cloud chamber. One of the frustrations was that even a large cloud chamber on a mountaintop could capture only about one vee particle specimen per day. Measurement errors complicated the interpretation of those specimens.

Glaser set out to develop a better detector that would record many more events and enable more precise measurements. He recalled years later, "My dream was that I could sit on a mountain top and check occasionally while the new gadget collected untold numbers of photographs of strange particles to be analyzed by armies of graduate students in every laboratory." Taking a cue from Geiger counters, nuclear emulsions, and cloud chambers, he looked for a medium in which the tiny energy deposited by the passage of a fast particle would trigger the growth of a recordable macroscopic response.

Glaser's solution, which he called the bubble chamber, is much like the negative image of a cloud chamber. A cloud chamber holds supersaturated vapor from which droplets condense along a charged particle's path. A bubble chamber holds a superheated liquid on the brink of boiling, in which bubbles form around the ions created by the passage of a charged particle. Rapidly lowering the pressure causes the latent droplets or bubblets to grow to visible size.

Once he proved that cosmic rays left well-defined tracks in a small chamber, Glaser built a six-inch model filled with propane, which he proposed to install at the Cosmotron. The request was rejected because "it was considered 'a misuse of public funds' for such a frivolous device" to be scheduled for beam time. He was permitted to set up his apparatus on a card table by a crack in the concrete shielding. When Glaser showed colleagues on the Cosmotron floor that a still-wet thirty-six-exposure strip of film contained twenty examples of the pion-to-muon-to-electron decay chain, which had been recorded rarely in emulsion stacks flown in balloons, the bubble-chamber revolution was launched.

The bubble chamber was much better suited to the rhythms of the new accelerators than emulsions or cloud chambers. Every few seconds, in synchrony with pulses of beam from the accelerator, a piston expands the chamber, reducing the pressure to let bubbles form in the wake of a charged particle. Lights flash and cameras record the image of where the particles have been.

To exploit the flood of pictures, bubble-chamber physicists recruited platoons of human "scanners" to find and measure the tracks, recording points along each curved path so that computers could reconstruct the paths in detail. Despite the new industrial scale—tens of thousands, even a million pictures might be analyzed for a single experiment—the instincts of the naturalist flourished. In bubble-chamber pictures, physicists found dozens of subnuclear particles—some revealed in a single specimen, others inferred from the painstaking identification of common characteristics in many events.

The Radiation Lab in Berkeley was a center of bubble-chamber culture when we arrived as theoretical graduate students in the late 1960s. Even as apprentice theorists, we were naturalists-in-training who had to learn the subatomic flora in preparation for our own fieldwork. The particles had to cease to be abstractions for us; they had to become as real as flowers and trees. We had to make their life cycles part of our own cultural heritage. The Rad Lab environment was wonderfully stimulating. Classic bubble-chamber pictures decorated the corridors, and we could always slip into the crepuscular scanning rooms to watch scanners winnow new batches of specimens. When an unusual picture appeared, the lucky finder would show it off on the scanning table and discuss interpretations with anyone who would listen.

Total immersion in a new culture encourages rapid learning, but even an eager explorer can benefit from the right guidebook. To provide a convenient and authoritative *Flora*, an authoritative reference for those who hadn't the memory for botany, Walter Barkas and Arthur Rosenfeld began in 1958 to issue wallet-card compilations of particle masses and lifetimes. By the time we reached the Rad Lab, the Particle Data Group was a Berkeley institution and a fixture of the international particle physics community. The 1967 edition had grown to three full sheets that folded into wallet cards, accompanied by a fifty-one-page article. The 2022 *Review of Particle Physics* is a 274-page booklet excerpted from a 2,270-page two-volume work published in the Japanese journal *Progress of Theoretical and Experimental Physics*!

Consulting the Rosenfeld Tables for 1967, we could see at least ten different ways a positive kaon (the modern name for a charged vee particle) could decay at the end of its ten-nanosecond life. The likely ending was as a muon and a neutrino, the outcome 63.4 percent of the time. In contrast, the probability that the kaon would turn into an electron and a neutrino was only about a fiftieth of a percent. All sorts of decays into pions, or pions plus a neutrino plus an electron or muon, accounted for the remaining 36 percent plus.

The tables of particle properties offered challenge upon challenge for the student. Could you explain why one decay was fast and another slow? Why one was common and another rare? Why would the kaon prefer to decay into a muon and a neutrino, while avoiding nearly completely the electron-plus-neutrino alternative?

It was not enough to learn the theoretical explanations for the decay patterns, where there were explanations. We needed to develop intuition, a feeling for the particles. Intuition is empathy, a gift granted if you work hard enough to put yourself in nature's place—inside her head—to know what she would do. It means knowing how things should turn out before they do. Developing intuition means learning to hear the poetry of nature, finding the right metaphor to express the essence of a particle, the right image to turn over in your mind.

As a Harvard undergraduate, Bob took a course on atomic physics taught by E. Bright Wilson, Jr., a distinguished physical chemist. Nearly thirty years before young Mr. Cahn turned up in his classroom, Professor Wilson coauthored a famous textbook on quantum mechanics with the legendary Linus Pauling. The pages of *Introduction to Quantum Mechanics with Applications to Chemistry* are filled with memorable diagrams showing baroque, lobed structures, the orbitals of atomic electrons that give rise to the structure of molecules. Who better to ask how to picture the electron than a man on intimate terms with these intricacies for three decades? Still, it took some courage to pose the question to the great man: "Professor Wilson, how do you picture the electron?" "I think of electrons as little yellow balls," Wilson replied.

No one knew better than Professor Wilson that electrons can't be yellow because they are smaller, by far, than the wavelength of yellow light. It isn't proper to think of them as balls since we have no evidence that the electron has any size or shape at all. Still, we all need images to manipulate, tokens more immediate than the solutions to mathematical equations. Even a cartoon image can be helpful, as long as you remember that it is a cartoon.

Now that we have been on intimate terms with the intricacies of the electron for many years, how do we think about it? In many different ways, as we would think of a good friend. An electron can seem silvery, like the surface of a clean metal whose electrons reflect light perfectly. It can seem fuzzy, like the electron cloud that determines the chemical properties of an atom. When a lightweight electron races through matter, it is buffeted by the electric field of every atom it encounters. At every turn, the electron sends out a beacon of light or X rays that may beget a shower of additional electrons and antielectrons. When we see a spray of flowers, or the train of a bride's gown, we think of an energetic electron passing through matter. And, yes, we've learned enough that we, too, sometimes think of electrons as little yellow balls.

As we learned the behavior patterns of the particles, we began to develop instincts about them. A particle called Delta-zero (Δ^0) decayed to a proton and a negative pion or a neutron and a neutral pion. So did the strange particle, Lambda (Λ), but it did so a million-billion times more slowly. Even if we couldn't put our fingers on strangeness, it was something real, something we had to internalize. The strange particles, the kaon, Lambda, Sigma (Σ), and Xi (Ξ), also called "Cascade," wore an exotic aura that their nonstrange partners π, N, and Δ did not. The strange particles carried a nearly ineradicable property that could only be shed in the weak interactions. But strangeness itself remained an abstraction. Electric charge is something you can feel in your fingertips. The crackling that ensues when you take off a sweater in the winter dryness of a heated home announces the reuniting of positive and negative electric charges previously separated. Strangeness is as real as the vee-shaped tracks in the cloud chambers, but it never emits a sound, never causes an everyday attraction or repulsion.

Strangeness was nevertheless essential to particle taxonomy. First came the two great divisions: the strongly interacting particles—hadrons for short—and the others—the leptons. Among the hadrons were two classes: baryons, including the proton and neutron, and mesons, including the pions. Each class was further split into families by strangeness.

The proton and neutron have no strangeness. The Λ and Σ have strangeness equal to minus one. Among the mesons, the positive K meson has strangeness plus one, while its antiparticle, the negative K meson, has strangeness minus one. Among the baryons there were even particles known with strangeness minus two, Ξ^0 and Ξ^-. At the final level of classification are specific charge states. The pion has positive, neutral, and negative versions. The proton and neutron, whose masses are nearly identical, form a pair, positive and neutral. There is only one neutral Lambda, but three species of Sigma, just like the pions.

Organizing the particles was a bit like doing the diagramless crossword puzzle in the Sunday *New York Times*, in which you are given a rectangular grid without any of the black squares that, in ordinary puzzles, indicate where the words begin and end. You must determine not only the answers to the clues, but also which squares to blacken, and thus the very shape of the puzzle itself.

The particle puzzle was more difficult still. First, you couldn't be sure what the clues actually were. It seemed that the across clues were descriptions like "two nonstrange baryons with similar mass." That is pretty easy: neutron and proton. Much less clear was the nature of the down clues, which might relate strange particles to the nonstrange. Moreover, since the clues were simply the known facts about the particles, some clues were probably missing or even wrong. It was not likely that all the known particles would fit into one single array, so actually there were several diagramless puzzles to solve at once. Particle physicists of the early 1960s had to figure out which clues went with which puzzle and, indeed, how many puzzles there were. There was no guarantee of a solution. And they couldn't look up the answer in next Sunday's *Times*.

The puzzles were solved independently in 1961 by Murray Gell-Mann at Caltech and Yuval Ne'eman at Imperial College in London. They found that baryons fit in a hexagonal array. In the top row, the neutron and proton are the answers to 1-across. Beneath them, in 2-across are the three Sigmas: Σ^-, Σ^0, and Σ^+. They are not aligned in columns as in a crossword puzzle, but with Σ^0 centered beneath the neutron and proton. In the next row down, the answer to 3-across is the pair Ξ^- and Ξ^0, lined up below the neutron and proton. What determines the organization in the down direction? The particles in the top row have zero strangeness; those in the second row, strangeness minus one; and those in the third row, strangeness minus two. The overall theme of the puzzle was that all the entries were baryons with spin 1/2. There is one more such baryon, the Λ^0, with strangeness minus one. Since Σ^0 had the same electric charge and strangeness, Λ^0 shared its space at the center of the puzzle. As an extra bonus, reading down a diagonal from upper left to lower right, all particles have the same electric charge.

Spin is a property of particles that has no precise counterpart outside the realm of quantum physics. The nearest analogue is the angular momentum possessed by matter in motion about an axis—a turning wheel or the moon in motion around Earth, for example. From their studies of atomic spectra, the early quantum scientists discovered that electrons bound to a nucleus also may carry angular momentum, in multiples 0, 1, 2, . . . of a quantum called $\hbar$ (pronounced h-bar). This fundamental unit of angular momentum is so exceedingly tiny that we never notice discontinuous jumps of angular momentum in everyday life. A long-playing record turning at thirty-three and one-third revolutions per minute carries something like one followed by thirty zeroes times $\hbar$.

At the end of 1924, Wolfgang Pauli brought new order to the structure of atoms and the patterns of atomic spectra. Seemingly out of thin air, he enunciated his exclusion principle, the notion that no more than one electron can occupy a particular quantum state. Audacious enough—but to reproduce all the observations, he had to assume, without explanation, that electrons could exist in two distinct states. The following summer, two graduate students in Leiden, Sam Goudsmit and George Uhlenbeck, mused that an electron might carry half a unit of $\hbar$, in the form of a new internal rotation—a kind of spin. Spin pointing up or down could supply Pauli's two distinct states. It's a crazy idea, in terms of our everyday experience. What is spinning? Why doesn't it lose energy and spin down like a top? And yet it is true, as experiments would soon be understood to show. All particles with half-integer spin (or odd multiples thereof), including the electron and the proton, must obey Pauli's exclusion principle, about which we will have much more to say later. We call such particles fermions.

The hexagon with six spin-½ particles around the periphery and two at the center is a pleasing design, but was it anything more than a picture? The pattern of blackened squares in a diagramless puzzle must conform to the rules of a good crossword puzzle: The pattern must be symmetric, with the top and bottom, and left and right, reflections of each other. The rules of the particle puzzle also required that the solution have symmetry—but the symmetry wasn't known in advance, it had to be learned from the clues. Part of the puzzle was to determine that symmetry. The hexagon with an extra particle at the center is an allowed pattern for a symmetry known as SU(3), the special unitary group of degree 3.

Could the same rules—the same symmetry pattern—organize the mesons? At the time, early 1961, just seven mesons with zero spin were known, and they too fit in a hexagonal array. The three pions take the positions of Σ^-, Σ^0, and Σ^+. The two mesons with strangeness plus one, K^0 and K^+, could sit above the pions (by the rule that strangeness increases going up), and the antiparticles of these, K^- and K^0-bar, with negative strangeness, could sit below. An eighth spin-zero meson, named eta (η), appeared in late 1961 to share the neutral pion's spot, confirming the pattern.

The two hexagons brought order to all the known hadrons that are stable, decay by emitting photons, or decay slowly through weak interactions. But physicists had discovered other particles, or resonances, that are momentary liaisons of the "stable" particles. Enrico Fermi established the first of these, back in 1954. Using primitive beams of pions, he and his coworkers found that pions and protons had a strong affinity for each other when their combined energy was just right. The affinity indicated that the pion and proton actually joined to form a new particle. With no strangeness to inhibit decay, the particle disintegrated in the natural time for nuclear processes. A similar affinity was present at the same combined energy for a pion of any charge combined with either a proton or a neutron. A positive pion plus a proton made Delta-plus-plus (Δ^{++}), a positive pion and a neutron made Δ^+, a negative pion and a proton made Δ^0, and a negative pion and a neutron made Δ^-. Physicists studying the decays of Δ realized that unlike the proton and neutron, which have spin 1/2, the Δ has spin 3/2.

Other spin-3/2 particles, cousins of the Sigmas and Cascades, were discovered in 1960 and 1961. The next year Gell-Mann organized the spin-3/2 baryons into an SU(3) pattern that resembles an array of ten bowling pins. In the top row lie the four Deltas, with strangeness zero. In the next, three Sigma resonances, with strangeness minus one. In the third row, two Cascade resonances, with strangeness minus two. The headpin was missing. Gell-Mann announced that there had to be a tenth particle, negatively charged, with strangeness minus three, which he named Omega-minus (Ω^-). The rules of SU(3) allowed for a pattern of ten particles, but not nine. Unlike its nine partners, the hypothetical particle would not decay rapidly, since its extreme strangeness would prevent strong decays.

At Brookhaven National Laboratory, Nick Samios and thirty-two collaborators mounted an experiment to search for Gell-Mann's missing particle. They directed a five-billion-electron-volt beam of negative kaons at their new eighty-inch bubble chamber, which was filled with 900 liters of liquid hydrogen. The team took pictures night and day beginning on December 14, 1963, accumulating 50,000 by the end of January. On January 31, Samios saw the entire life cycle of the Omega-minus on the scanning table. An Omega-minus, produced with a K^+ and an unseen K^0, left a short track before it decayed into a Xi-zero particle and a pi-minus meson. The sequence of decays Ω^- into Ξ^0 and pi-minus, Ξ^0 into Λ^0 and pi-zero, Lambda into proton and pi-minus, pi-zero into two photons that materialized into electron-positron pairs, makes a beautiful picture. It is an icon, like the best cloud-chamber and emulsion photographs: a single specimen that tells a rich story.

Almost at the moment that Samios found the first Omega-minus, Gell-Mann at Caltech and, independently, George Zweig at CERN (the European Laboratory for Particle Physics in Geneva) made a proposal of audacious simplicity. They observed that the curious SU(3) patterns of octets and decimets could be explained if the hadrons weren't elementary particles after all, but were instead built from three fundamental objects. Zweig wanted to call his new constituents aces, but the word Gell-Mann coined—quarks—was catchier.

Quarks also acquired a mythic literary heritage in the exclamation, "Three quarks for Muster Mark!" from Humphrey Chimpden Earwicker's dream in *Finnegans Wake*. Even Murray, who by his own admission knew everything, conceded that he hadn't had Joyce's line in mind when he chose the sound.

Three quarks, which we now call up, down, and strange, could explain all the hadrons known in the 1960s. A proton is two ups and a down, a neutron two downs and an up, an Omega-minus is three strange quarks. Combining a quark and an antiquark gave mesons like the pion and kaon.

One feature of the proposal was especially odd. The quarks carry fractional electric charges. The up quark has a charge 2/3, the down and strange minus 1/3 each. No one had ever seen particles that didn't carry a whole number charge: one for the proton, zero for the neutron, minus one for electron, and so on. Perhaps this is why Gell-Mann repeatedly emphasized that the quarks might be convenient mathematical fictions. Zweig's view was more literal: he wanted the quarks to be real. When young Zweig described his conviction to Professor Gell-Mann, his senior colleague responded, "Oh, the concrete quark model . . . that's for blockheads."

Whether quarks were real or not, the diagramless puzzles of the mesons and baryons could be built from a simpler, more fundamental quark puzzle. In that puzzle, the down and up quarks are the answer to 1-across. 2-across is simply the strange quark, centered beneath the down and up. The triangular quark pattern is an individual tile in the ten-particle triangle that the Ω^- completes. Antiquarks form the same triangle flipped upside down. The anti-strange quark is 1-across, while anti-up and anti-down fill 2-across. By pasting together these triangles you could reproduce all results of SU(3) symmetry painlessly. The ten-pin triangle and the hexagons were now child's play.

Quarks not only explained the patterns of the particle families, they also demystified strangeness. The strange particles contain strange quarks or antiquarks. In a collision of two strongly interacting particles, like a pion and a proton, quarks can pass from one particle to the other, in a sort of exchange of genetic material. In high-energy collisions, a quark and its antiquark can be produced together. If a strange quark is created, so is its antiquark, so that on balance no new strangeness appears. A strange quark doesn't disappear on its own in a strong interaction. That is why a kaon, which contains a strange quark or antiquark, doesn't turn into a pion, which contains no strange quarks, through the strong interaction.

The weak interaction is an agent of change that turns one type of quark into another, or one kind of lepton into another. When a neutron decays into a proton, an electron, and an antineutrino, the weak interaction changes one down quark into an up quark, and creates an electron and antineutrino with a combined charge of minus one. Altogether the electric charge is unchanged. A positive pion is made of an up quark and an anti-down quark. When the pion decays, the quark and antiquark annihilate, and a positive muon and a neutrino appear in their place. The weak interaction can change a strange quark into an up, or an anti-strange quark into an anti-up.

All this falls into place perfectly. The only problem was that no one had seen anything that looked like a quark. It was as if someone very clever had studied plants and animals very carefully and had concluded that they could be understood only if their cells contained chromosomes made of DNA, and then, when people looked carefully at cells, they found no chromosomes, no DNA. Such a perfect model couldn't be ignored, but neither could it be fully believed.

One more reason to take the absent quarks seriously appeared in 1968. The Stanford Linear Accelerator Center was built to repeat, on a truly grand scale, the experiment Geiger and Marsden performed for Rutherford at Manchester. Its two-mile accelerator, aimed right at the Stanford campus, is enclosed in what may be the world's longest building, depending on how you define a building. The machine shoots straight through the foothills that lie between the Santa Cruz Mountains and the flatlands near San Francisco Bay. SLAC, as the lab is called among high-energy physicists, disturbs this bucolic scene only slightly with its cluster of low buildings crowned with red-tile roofs, the architectural emblem of the university.

In the experiments done by Jerry Friedman, Henry Kendall, Dick Taylor, and other physicists from MIT and SLAC, the two-mile-long electron accelerator stood in place of Rutherford's alpha source. Instead of Rutherford's gold foil there was a hydrogen target. The phosphor-coated scintillation screen used in Manchester to view the scattered alpha particles was replaced by electronic detectors inside electromagnets as big as trucks, mounted on scaffolding that could rotate around the target, to view the scattered electrons from various angles. Though most everyone expected that there would be few events in which the electrons—with energies as large as twenty-one-billion electron volts—scattered through large angles, once again the "fifteen-inch shell" bounced back from the tissue paper. Inside the proton, just as inside the atom, was something small and hard.

To the prepared mind, the high rate of large deflections showed that there were tiny charged bodies within the proton. No mind was more prepared to take the leap than Richard Feynman's. Feynman presented his interpretation at a SLAC colloquium that occasioned Chris's first pilgrimage across the Bay from Berkeley.

The colloquium was then held in the evening after what has been described to me as a vintner's dinner. Whatever the reason—maybe it was the physics!—I remember both speaker and audience as extremely exuberant. If an electron scattered from one of the hypothetical tiny charged bodies, not the whole proton, it was easy to understand why the rate of large-angle scatters was so large. Instead of measuring the delicacy of the proton, the MIT and SLAC experimenters were measuring the hardness of the little bits. Feynman wasn't prepared to say what the tiny charged parts of the proton were, so he called them "partons." Everyone in the room must have thought, "Quarks?"

When Bob joined SLAC as a postdoctoral fellow in 1972, his first task was to learn the new language that Feynman and James "Bj" Bjorken at SLAC had created to describe electron-proton collisions. Bj was an awesome figure, not so much because he stood six-and-a-half feet tall but because of his technical prowess and amazing intuition for physics. Once when he was descending from the summit of the Grand Teton with two companions, a storm came up and they sought shelter in a small cave. Lightning struck above the mouth of the cave, burning the three climbers and knocking one unconscious. Generations of SLAC postdocs speculated that the stroke established a direct line from Bj to nature herself and that this explained his uncanny insight. The beauty of Feynman and Bjorken's picture was that it made it simple to understand the experiments at SLAC. Inspected closely, a proton was seen to be a collection of quarks and antiquarks. The MIT-SLAC experiments determined precisely how the various quarks and antiquarks shared the proton's energy.

Despite the data from SLAC and the elegant simplicity of the parton model, quarks still seemed to most physicists to be a bookkeeping device, as atoms had seemed to their scientific ancestors. They satisfied the urge to find order, to organize the proliferation of species—the mesons and baryons, the strange particles and nonstrange particles—but their steadfast refusal to make themselves seen discouraged many physicists from taking them literally. Quarks, it seemed, would provide a secure underlying principle for particle physics only if they would deign to make an appearance. Then came the November Revolution!

6

REVOLUTION!

Monday, November 11, 1974
The University of Washington — Bob

"Call me or call SLAC. Dave."

The message was waiting for me when I arrived in the turret that was my office at the U, right in the vacant spot where the ceaseless Seattle rain had a week earlier seeped through the roof and dissolved a sheaf of my calculations. Dave Jackson's imperative tone couldn't be ignored. Whatever had happened, it must have been at the Stanford Linear Accelerator Center—SLAC. My fingers dialed the number automatically: 9-1-415-926-3300. At extension 2266 Sharon Jensen answered, "Would you like to speak to Haim?"

Fermi National Accelerator Laboratory — Chris.

"Looks like charm is found." Wearing a dazed grin, Ben Lee stepped uncertainly into my office, making a determined effort to appear nonchalant.

Then, looking across the winter-pallid prairie into the sun of a November morning and blinking his eyes in wonder, he confessed his amazement. "Am I dreaming?"

My tweeds and Ben's navy blazer gave us the appearance of a pair of East Coast professors, as we had been only a few months before. Just after that summer's International Conference on High Energy Physics in London, we had both moved to the new Fermi National Accelerator Laboratory, where Ben was leading the Theoretical Physics Department. I was a month from my thirtieth birthday, four years past my Berkeley PhD. Ben, who had come to the United States from Korea as a college student, was ten years older, an established figure in the particle-physics pantheon, and not easily flummoxed.

Through the windows, across an unpaved parking lot, we could see a hint of what had brought us to Batavia, Illinois: a raised ring of earth precisely 2π kilometers around that modeled the prairie like a giant gopher's burrow. The ring marked the location, thirty feet underground, of the most powerful particle accelerator on Earth.

With the ritual "May I?" that physicists pronounce before wiping a colleague's work from the blackboard, Ben shifted an unlit Meerschaum pipe to his left hand, which already clutched a page of notes on a sheet of lined yellow tablet paper, and started to clear the board.

"I just got a call from SLAC . . ."

Bob

Sharon was right. Haim Harari was certainly the one I wanted. I had met him during my year and a half as a postdoctoral fellow at SLAC, where he seemed to spend as much time as he did at his official home, the Weizmann Institute of Science. Haim's lectures at the annual SLAC summer schools, delivered in his unvarnished Israeli accent, were classics—models of logical thought spiced with wit. They drew devoted legions of graduate students and new postdocs like myself. Most lecturers at the summer schools used felt-tipped pens to write equations and diagrams on clear sheets of film that could be projected onto a screen. These overhead transparencies were an aid for the students, who preferred them to scribbles on a blackboard, and a comfort for the lecturer, who could use them as a text. Haim didn't simply follow a linear script; he had made transparencies an art form. His talks unfolded like mystery stories, with empty spaces on early transparencies filled in by overlays, as he first constructed, then resolved, each summer's puzzles in particle physics.

Haim's intensity was always set on maximum in the lecture room, in informal hallway conversations, or at the lunch table, where wrongheaded physicists and politicians were skewered daily. It's not enough to say that Haim was sure of himself. He eradicated all contrary thinking with devastating criticisms. But above all, Haim was a raconteur, and when I reached him that Monday morning he had a tale worthy of his talents.

The news came from SPEAR (for Stanford Positron Electron Accelerating Ring), a new instrument at the end of Stanford's two-mile-long linear accelerator, in which bunches of electrons could be made to collide nearly head-on with their antimatter counterparts, bunches of positrons. When two beams cross, most particles pass without disrupting each other. Sometimes particles scatter, deflecting each other like billiard balls. Occasionally, something spectacular happens: the bits of matter and antimatter annihilate, briefly concentrating all their energy in a tiny volume until it materializes back into an electron and positron, or into other known forms of matter, or into something new. In SPEAR, the electron and positron beams looped around and around in opposite directions, giving the particles a new chance to collide every time the beams crossed—almost two million times a second.

While we almost never meet antimatter in everyday life, the conversion of energy into matter and antimatter is more routine in accelerator laboratories than the exchange of currencies in a Swiss bank. And why not? No commission is charged for the transaction, as Albert Einstein understood. Since matter and energy are convertible currencies, it is clumsy to account for one in US dollars and the other in British pounds. We measure both energy and mass—the quantity of matter—in electron volts. Like high financiers, high-energy physicists deal in millions, billions, and even trillions of the basic denomination.

At one of SPEAR's crossing points, about three dozen physicists from SLAC and the Lawrence Berkeley Laboratory had mounted a complex detector to record and analyze the products of electron-positron collisions. The detector elements formed a cylindrical structure about three-and-a-half meters in diameter and three-and-a-half meters long. For more than a year, the group had been conducting a methodical survey, setting the energies of the electrons and positrons first to one value and then to another, and measuring the rate of annihilations.

The construction of each new accelerator or detector is a major event in particle physics, with each step marked by formal reviews, progress reports at conferences, and seminars given by the participants at universities and other labs. Some puzzling preliminary results from SPEAR had already been presented at conferences, but it was hard to know whether they were the hint of discovery or mere glitches that would go away after the experimenters got to know their detector better.

The problem began innocently enough. When the combined energies of the electrons and positrons were set to 3.2 billion electron volts (3.2 GeV), annihilation events came 30 percent more often than expected. At neighboring energies of 3.0 and 3.3 GeV, there was nothing out of the ordinary. Repeated measurements at 3.1 GeV only confused the picture: Six runs out of eight were perfectly normal, but in the other two, annihilations occurred at three and five times the expected rate. What could make the measurements so frustratingly erratic?

In the early days, setting the energies of SPEAR's colliding electron and positron beams was like setting an oven temperature. When a chef dials 375°, he expects that his oven will not give 325° or 450°, but the success of a recipe doesn't require the temperature to be exactly 375° instead of 374°, or even 380°. However, if one day the 375° setting produces 355° and another day 395°, the texture of a soufflé Grand Marnier might change from custard to cotton candy.

Could it be that SPEAR was a temperamental oven and electron-positron annihilations made a hypersensitive soufflé?

To test this idea, the experimenters built new controls that could adjust SPEAR's energy more precisely, in steps like 3.100 GeV, 3.102 GeV, and so on. Like a chef choosing to bake one soufflé at 374.8° and the next at 375°, the SLAC team tuned SPEAR's energy up and down in small steps on Saturday and Sunday.

Normally, the detector was triggered every few seconds, creating a cascade of tiny sparks in the apparatus. Usually the trigger was activated by the passage of a cosmic ray, or by an electron and positron knocked off their tracks by a close encounter, or by electronic noise—"garbage" to the experimenters. The real quarry of the experiment, an electron and positron annihilating to produce other kinds of matter, appeared every minute or so. In the SPEAR control room, the electronic storm set off by the pulsing of high voltage on the spark chambers caused an audible click over the PA system. The experimenters could hear their detector at work: *tick . . . tick . . . tick . . . tick . . .*

The soufflé electron-positron was exquisitely sensitive. There was nothing out of the ordinary at 3.100 GeV, but at 3.105 GeV the PA system sounded like a Geiger counter ticking as fast as it could, more than once a second. The collision rate rose to seventy times its normal value! Almost every event the online computer displayed was an annihilation. It was as if a soufflé baked at just the right temperature would rise twenty-three feet instead of four inches.

As the "temperature" of the beams increased beyond 3.105 GeV, the barrage subsided. By 3.140 GeV, the annihilations fell to merely twice the normal rate.

For physicists, the sudden rise and fall of the annihilation rate could mean only one thing, a resonance. Resonances are ubiquitous in nature. Blow across the top of a Coke bottle at the right speed and the air within reverberates at a characteristic pitch, giving a muted imitation of a foghorn. A more highly evolved musical instrument, like a flute, can be played at several resonant frequencies. The resonances of particle physics are unstable compound particles with characteristic masses and lifetimes. The mass corresponds to the resonant frequency or pitch, and the lifetime is related to the purity of the tone.

When you stop blowing across the top of the Coke bottle, the sound dies out quickly, but a tuning fork rings long after it is struck. The purity and length of its tone go hand in hand. A particle resonance decays abruptly, disintegrating into two or more particles that suddenly fly out from the spot where the resonance had been. The average lifetime of a resonance is related to its width, the indefiniteness of its mass.

It was to make sense of the subnuclear particles and resonances that George Zweig and Murray Gell-Mann had independently invented what we now know as the quark model. Short-lived resonances were momentary liaisons of three quarks, or of one quark and one antiquark.

Though quarks—idealized as structureless and indivisible—could explain much of what was known about the subnuclear world, physicists regarded the quark model with varying degrees of skepticism, for no one had ever seen an isolated quark. Physicists had looked in vain for free quarks at high-energy accelerators. In hopes that quarks might be concentrated in biological systems, others had ground up oyster shells and looked there. The quarks didn't turn up, but intrepid quark hunters could enjoy the oysters!

When the electron and positron annihilated at 3.105 GeV, they fused to make an ephemeral compound particle. But a compound of what? It had to be a fourth quark and its antiquark. It had to be charm.

Chris

As Ben finished describing the SLAC results, it was my turn to look dazed.

When we came to the lab that morning, we already knew hundreds of particles—the neutrons and protons in the nucleus, the pions and kaons first seen in cosmic rays, the Lambdas, Sigmas, rhos, and the rest produced in accelerators, so many that they nearly exhausted the Greek and Latin alphabets. So why lose our composure over one more? The SPEAR particle stood out because it was heavier than any we knew, more than three times the proton's mass.

The increase in the annihilation rate was stunning. But what truly astonished us was that purity of tone, the sharp resonance peak that was a thousand times narrower than an ordinary particle of the same mass would be. SPEAR had not found a needle in a haystack, but a flagpole standing in a meadow.

These extraordinary properties were impressive to me, but they were even more astounding for Ben. If the emerging theory of the weak force—the interaction that governs radioactivity and related phenomena—made any sense, Gell-Mann and Zweig's system had to be incomplete. There had to be a fourth quark, which was given the name "Charm" because it warded off diseases, or experimental embarrassments, in the theory of weak interactions. Ben had just completed a paper called "Search for Charm" with Mary K. Gaillard and Jonathan Rosner in which they described a resonance with hidden charm, made of charm plus anti-charm. Hidden charm would have a mass around 3 billion electron volts (3 GeV) and a width of only two million electron volts (2 MeV). Most of the quark-antiquark resonances we know are very brief encounters of the constituents. This hypothetical particle had an unusually long lifetime—a narrow width—because it could only decay after the charmed quark and antiquark ran into each other, annihilated, and rematerialized into pairs of electrons, or pairs of muons (heavier cousins of the electron), or into other known particles. The news from SLAC was all that Gaillard, Lee, and Rosner had asked for—and more.

"One more thing . . ." Ben shook his head as if to clear it, and blinked again. "This is incredible . . . Sam Ting has it, too."

A continent away from Stanford, at the Brookhaven National Laboratory on Long Island, MIT professor Sam Ting and a dozen colleagues had picked the same particle out of the splash of stuff created when a beam of protons hit a target made of several slices of the metal beryllium. If the key to the SLAC experiment had been precisely controlling the combined energy of the electron and positron that went into the collision, the key to the Brookhaven experiment was precisely measuring the combined energy of the electron and positron that came out of the collision. Many of the electron-positron pairs generated in the Brookhaven collisions had a combined energy very close to 3.1 GeV. Though the effect was dramatic, Ting's team had kept it under wraps while they changed experimental conditions to make sure it was real.

Fate put Sam Ting on a plane to San Francisco just as the data that sealed the West Coast discovery were being recorded. He was on his way to a meeting of the Program Advisory Committee (P.A.C.) that recommended experiments and scheduling priorities to Wolfgang Panofsky, the director of SLAC. For high-energy physics laboratories, the ritual of the P.A.C. combines peer review, consultation with experts outside the lab, and a network for disseminating information to the wider community. On Monday morning, before the meeting opened, Ting stopped to tell Panofsky and Burt Richter, the builder of SPEAR, about his new resonance, only to be greeted with the news of SLAC's new resonance—the same one! Only the names chosen by the discoverers were different: the Californians had chosen the Greek letter *psi*, which has the form of a trident, ψ, while the Brookhaven group had chosen capital *J*. It was amusing to learn later of the purely coincidental orthographic similarity of the letter *J* and the Chinese character pronounced "ting."

Ben interrupted his fantastic tale. "You're on the P.A.C.," he accused. "Aren't you supposed to be in California?"

I should have been. The November meeting was to be my first as a member of the SLAC P.A.C., but I had asked to be excused when the agenda looked more ceremonial than scientific. Still settling into my new life at Fermilab, I was eager to finish an article for *The Physical Review* before driving to Pasadena for a month's visit to Caltech. I confessed that when Ben veered into my office, I had been congratulating myself for displaying wisdom beyond my years by avoiding a pointless meeting.

He savored the irony: "Looks like the joke is on you!" News of the double discovery was coming out in hastily arranged seminars at the very P.A.C. meeting that I had so cunningly avoided. Physics had been turned on its head in the space of a few days and, instead of witnessing history, I was listening to a second-hand account of a telephone call! I vowed never to miss another committee meeting. I am still waiting for my belated virtue to be rewarded.

Bob.

Chris and I were both students of Dave Jackson. The name John David Jackson is seared into the consciousness of physics graduate students, for every student's training includes a course in electricity and magnetism taught from Dave's book. Each chapter of *Jackson* concludes with a very long selection of excruciating problems. Solving a *Jackson* problem requires not just complete mastery of the concepts, but also some minor mathematical virtuosity. The psychic pain these exercises inflict on students around the globe has inspired a variety of testimonials. When students at the University of Colorado ran a contest to carve a Halloween pumpkin resembling their tormentor, the winner was a merciless brute. Students at Cornell composed a song to the tune of ("How do you solve a problem like . . .") "Maria," from *The Sound of Music*:

> How do you solve a problem out of *Jackson*
> In any less than geologic time?

Some students try to salvage their self-esteem by fantasizing that even Jackson couldn't do all the problems in *Jackson*. They are deluded.

Dave learned of the SLAC discovery in a phone call to his Berkeley home on Sunday afternoon. He asked about a couple of details, and then sat down to calculate. In a few lines, he showed that in an idealized experiment with perfectly controlled energies the peak would have been thirty times as narrow and thirty times as high as it appeared at SPEAR. It was an analysis worthy of a *Jackson* problem. The perfectly baked soufflé would rise 700 feet! This conclusion so startled the SPEAR physicists—and their theoretical advisors at SLAC—that they omitted it from the paper announcing their discovery.

As I spoke with Dave, I tried hard to be a skeptic. After all, the charmed-quark interpretation was simply too pat. The new particle occurred just about where Gaillard, Lee, and Rosner had predicted. It was amazingly narrow. It was all too perfect. It shouldn't be charm, that was too obvious. There were other possibilities to investigate. And yet . . .

Of course I had believed in quarks. I had calculated using them. I had written papers about them. But a quark had been an abstraction, an artifice. As atoms had been to scientists in the mid–19th century, quarks were an economical way to understand experimental phenomena, but it wasn't necessary to think of them as physical. Rather than admit to relying on constituents that no one had ever seen, you could conceal the quarks behind a facade of equations. You did a calculation using quarks and at the end you erased the quarks and said, "The quarks might only be mathematical fictions, but I believe the answer."

Now all that had changed. On November 11, 1974, quarks became real.

A binder on my desk was devoted to calculations I had been doing the past few weeks on the hypothetical hidden-charm particle. A graph of data taken at Fermilab by a Seattle colleague showed an obscure and uncertain bump, a pebble on a hillside. I had been trying to decide whether it could be a subtle sign of charm. I looked at the infinitesimal bump in the Seattle data and thought about the 700-foot soufflé at Stanford. It was monumental; it rose suddenly from a plain to a pinnacle, then plunged down again, making a logarithmic ski jump on the back side. I closed the binder and put it back on the shelf, never to be reopened.

Chris

Before long, my office overflowed with colleagues who had already heard the news or were drawn by the growing din of our shouts and exclamations. We moved out into the corridor and gathered before a floor-to-ceiling blackboard that would soon be covered with bulletins about the new physics.

The new particle that Ben described was unlike anything we had seen before in experiment, yet it looked familiar. It realized, in extreme form, possibilities we had seen before in our collective fantasies, without knowing how seriously to take them. Already in the year before the discovery, one quarter of the theory papers written at Fermilab dealt with themes the J/ψ brought to the fore. For most of us, the news crystallized the research agenda: quarks were real and the ideas about the fundamental interactions that had demanded there be a charmed quark were no longer just theorists' dreams; they now had to be worked out and tested in every detail.

When we dispersed, some of us rushed back to our phones to share the excitement and spread the word or ask for more details. Others started a line-by-line study of "Search for Charm." Still others began to consider how Fermilab experiments could be bent to the new task. The truth is, most of us tried to do all three at once. Before we went to lunch, where we did far more talking than eating, we knew that our world had changed.

Even on a Monday during the school year, more than half the experimentalists at Fermilab are visitors from universities: graduate students and postdoctoral fellows who live at the lab, or professors who have arranged their teaching schedules so they can spend long weekends there. The cafeteria in Fermilab's soaring atrium was a souk, a marketplace of news, comment, and gossip from everywhere. Everyone was talking about the new particle. The first question on everyone's lips was, "How could this be wrong?" Though experimenters and theorists all asked the same question, the different ways we thought about it are revealing. A theorist presented with an oddball result will search for signs of human error or try to come up with a mundane interpretation. An experimentalist, experienced in the ways of the real world, will ask how the apparatus could misbehave and lie to its builders.

The skeptical reception that greets astonishing results doesn't reflect stodginess or a slavish commitment to the established order. It grows out of a discipline, born of the knowledge that science is not a collection of isolated phenomena with unrelated explanations, but a web of interconnected observations given form by unifying themes. It is also nourished by experience. In particle physics as in life, most things that *seem* too good to be true *are* too good to be true.

But the results from SLAC and Brookhaven were too good to be false. Two different groups using entirely different techniques had seen the same resonance. It was real. So the question every experimenter wanted to answer was, "How can I get my hands on it?"

Bob

Seattle and Fermilab were two ganglia in a far-flung neural network that joined Berkeley, SLAC, Cambridge, and Princeton, to Geneva, Hamburg, Paris, and Rome, and on to Moscow and Tsukuba. The few thousand high-energy physicists formed a community in which everyone knew everyone else and no data could long remain secret. Scholarly journals propagated reports of research along sluggish pathways, while the real news sped by the fastest means possible. In 1974, before the days of electronic mail and the World Wide Web, this meant by telephone. Monday morning, November 11, the global village of high-energy physics was alive with the ringing of phones.

The miraculous new particle was an instant international hit. It was undeniably of fundamental significance, even if its exact nature was uncertain. By Friday it had been recreated in Frascati, Italy, where it required pushing a collider called ADONE beyond its design limits.

Ten days later, when we had time to catch our breath, a few colleagues and I prepared a seminar for the entire Physics Department, to describe the J/ψ and explain how it would fit with charm, but also to mention alternative interpretations. The day of our talk, just as we were about to give a thoroughly evenhanded presentation of all the possibilities, the SPEAR group found a second narrow resonance near 3.695 GeV. This amazing development was like hearing a musical overtone of the J/ψ. It was perfectly natural if we were seeing a figurative atom composed of a charmed quark and charmed antiquark, but an incredible coincidence if we were not. Quarks hadn't been liberated, but the signs that they were real, mechanical objects inside the new resonances were unmistakable.

I struggled to suspend judgment until more data were in, but I could see where it was leading and I said to myself, "If this is really charm, I'll change the way I do physics."

∞

Within hours of learning of the double discovery, we became agents of a vast collective intelligence. Around the physics world, alternative hypotheses were formulated, tested, and debated. Implications of charm were worked out. New experiments were conceived and critiqued. Results propagated instantaneously, without being communicated through journals, or even preprints. It seemed as if everyone knew what everyone else was doing. For a time, the solidarity was so complete that it became a point of honor not to spend time writing a formal paper.

The next weeks were a jam session that went on and on, with no one tiring or losing interest. Who needed a score? One group after another would embellish what was there before or play a variation on a theme. A soloist might improvise a new line that provoked a response, a dialogue, and—if it seemed interesting or provocative enough—development by the entire band. A chorus of good-humored brays, as from trumpets played through Wow-Wow mutes, greeted sour notes—dead ends already ruled out by experiment. New experimental results drove the evolution of the music, interrupting one line, encouraging another, and introducing new material with irresistible urgency. The ideas that survived were so simple and powerful that anyone who didn't know the music could pick it up after someone hummed a few bars.

Complete acceptance of the new idiom was the only rational response to the data that appeared in the next days, weeks, and months. In the end just about everyone played in harmony. The obdurate few who refused to convert were consigned to irrelevance.

Before the November Revolution, particle physicists spoke many languages. Afterward, there was only the language of quarks—not the mathematical-fiction quarks, but the literal quarks you could reach out and nearly touch. The triumphant language was universal. It was spoken everywhere and it addressed all topics. The division of particle physicists into separate tribes that studied electron interactions, proton interactions, and neutrino interactions was abolished.

Around the world, people who hadn't known they had anything to say to each other before now had reason to listen to one another—first about the new particles, each bringing specific talents and experience—then more broadly. Everything we knew began to fit together. The joint discovery of the J/ψ by two teams, one using electrons, the other protons, was emblematic.

Though we knew from the first news of the J/ψ discovery that our world had changed, we didn't know how profound and lasting the change would be. We didn't know that it would take a year and a half for the charm interpretation to be vindicated. We couldn't know that the ideas that took center stage on November 11 would pass every test and emerge as the conceptual foundation of particle physics that guides our thinking and aspirations today. We couldn't know the extraordinary lengths to which particle physicists would go to challenge and extend those ideas. We had no idea that, less than ten years later, we would be planning accelerators that would dwarf the one outside Chris's window to complete the unfinished business of the November Revolution. Now, it is easy to see that the direction of our research changed at once, our worldview over a couple of years. It is no exaggeration to say that the direction of the whole field was set on that November day in 1974.

Just because you are closing in on the secrets of nature is no reason to take yourself too seriously. The ersatz atom made of an electron and a positron instead of an electron and a nucleus is called positronium. In analogy, the new charm-anti-charm resonances were dubbed "charmonium." In early January, Chris and Martin Einhorn sent off a short manuscript, "On the New Narrow Resonances," to the highly selective—and unapologetically stiff—journal *Physical Review Letters*, announcing "our nearly inescapable conclusion that the new particles are quark-antiquark bound states of a new, fourth quark. We call the quantum number carried by the fourth quark *Panda*, and suggest the name Pandaemonium for the new particles." A footnote explained that they had chosen this name "because of the panda's well-known shyness, and tendency to stay among his own kind. The great mass of the giant panda has also influenced our thinking." This bit of whimsy made perfect sense. The new quark was rare and had only been seen mated with its own antiquark, whereas the familiar up, down, and strange quarks had been observed in all quark-antiquark combinations. It was also a heavyweight, three to five times the quark-model masses of up, down, and strange. Six weeks later, the editors sent back a standard letter of rejection and a review so carefully measured in its levity that the secret of their sense of humor was preserved. Naturally, pandaemonium got more attention in the press than any of Einhorn and Quigg's serious work. The punishment fit the crime.

November 1974 marked the beginning of a new era in which the boundary of science would be expanded, making accessible questions that before had been beyond the compass of physics. Ever since the 1920s, quantum mechanics had made it possible to describe ordinary matter. You inserted the masses of the electron and proton and their electric charges into a simple equation and out came the properties of hydrogen. But knowing that the proton's mass is 938.272 088 16 MeV and the electron's is 0.510 998 950 00 MeV doesn't explain them. Where did those masses come from? What equation gave them? The events of November 1974 were the turning point in making such questions part of science, opening the way to a new level of understanding of the physical world.

Within two or three decades, the most esoteric scientific advances, if they are truly important, are assimilated into the standard school curriculum. Fifteen years after the drop-everything moment of Monday, November 11, 1974, multicolored charts displayed the charmed quark and its partners in science classrooms around the world. Perhaps students regard the chart as petrified knowledge, no more remarkable than an adjacent topographic map of the continents. But for those who participated in the great events of November 1974, the chart speaks of revolution.

And the revolution is not yet ended.

7

STORYTELLERS

Some years ago, we visited a harpsichord maker in his New Mexico atelier. The artisan—a retired Caltech physicist named Bob Walker—showed us his materials, explained the various stages in the construction of a harpsichord, then treated us to a recording of Igor Kipnis playing Antonio Soler's *Fandango* on one of his instruments. But the object he took the greatest pleasure in showing was not a finished harpsichord; it was a block plane—clean, precise, and utterly apt—that he had built in order to sculpt soundboards of surpassing beauty and eloquence.

This maker of exquisite instruments was also a maker of peerless tools.

We thrill to the playing of the virtuosi of our age. Violinists and their audiences revere the master luthiers of Cremona—the Amati, the Guarneri, and Stradivari—who centuries ago invented the modern instrument and built the most magnificent examples. We wonder which tools those great craftsmen might show us with special pride if we could visit them at work. Musicians and composers are just as much intertwined with the artisans and craftsmen who make the instruments as are scientists—both theoretical and experimental—with their toolmakers. Our own esteem for scientific toolmakers shows in the attention we have paid to the instruments they made and the discoveries they enabled. In this chapter, we will celebrate makers of theoretical tools while assembling the pieces of the best description of nature that we have, the standard model of particle physics.

Along with attentive scrutiny of nature and the instruments that extend our senses, surely language is an indispensable tool for making sense of the physical world. Among human languages, mathematics has such a special claim on our attention that many thinkers have remarked on its role.

In notes for a treatise on painting, the iconic Renaissance man Leonardo da Vinci (1452–1519) advocated the mathematization of experience—the exact sciences of arithmetic and geometry leading to perspective and then to astronomy. "No human investigation may claim to be a true science if it has not passed through mathematical demonstrations, and if you say that the sciences that begin and end in the mind exhibit truth, this . . . must be denied for many reasons, above all because such mental discourses do not involve experience, without which nothing can be achieved with certainty."

Galileo Galilei is lauded as a harbinger of modern science, not only for his contributions to three revolutionary developments but also for his way with words. He helped complete the Copernican revolution, establishing that humans do not occupy a privileged location in the universe; he rejected authority, learning to interrogate nature by conducting his *cimenti*—experiments or trials—and he emphasized that answers to majestic questions would be woven out of small truths gleaned from carefully prepared investigations. "Philosophy," he wrote, "is written in this grand book, the universe, which stands continually open to our gaze. But the book cannot be understood unless one first learns to comprehend the language and read the letters in which it is composed. It is written in the language of mathematics, and its characters are triangles, circles, and other geometric figures without which it is humanly impossible to understand a single word of it; without these, one wanders about in a dark labyrinth."

When observations of the 1919 solar eclipse confirmed that the sun's gravitational influence deflected light rays from distant stars just as Albert Einstein's

General Theory of Relativity had foreseen, the *New York Times* of November 9 exulted, LIGHTS ALL ASKEW IN THE HEAVENS: MEN OF SCIENCE MORE OR LESS AGOG OVER RESULTS OF ECLIPSE OBSERVATIONS. EINSTEIN THEORY TRIUMPHS. The forty-year-old physicist instantly became a household name and the most famous scientist on Earth, a celebrity status that endured throughout his life, and indeed beyond. Einstein spoke aphoristically and with nuance in a 1921 address to the Prussian Academy of Sciences, declaring, "As far as the laws of mathematics refer to reality, they are not certain; and as far as they are certain, they do not refer to reality."

We quote our celebrated scientific ancestors not to settle anything by decree (*Nullius in verba!*), but to exhibit the diversity of opinions. How do *we* characterize the role of mathematics in a few words? Bob remembers declaring to his Harvard classmate, Roger Howe—now a distinguished mathematician and highly decorated mathematics educator—that Mathematics tells us everything that might be, while Physics tells us what actually is.

Chris sees mathematics as one of the languages in which we have learned best to listen to nature. It is a refiner's fire, distilling and purifying our thoughts so we may apply them reliably beyond the settings in which we first formed them and, not infrequently, find new wisdom.

It is far from intuitively obvious that the electric field created by a battery or the magnetic field of a compass needle has any relation to light. But that is what James Clerk Maxwell found in the middle of the 19th century. His equations show that the electric and magnetic fields can work together to make waves—light waves, radio waves, microwaves, and X rays. Heinrich Hertz, the first to observe radio waves, said of the equations that predicted them, "One cannot study Maxwell's marvelous electromagnetic theory of light without sometimes having the feeling that these mathematical formulae have an independent existence and an intelligence of their own, that they are wiser than we are, wiser even than their inventor, that they give back to us more than was originally put into them."

Contemplating the evolution of a language, we are tempted to identify an inflection point that marks a before and after, an era when the old is supplanted by a richer and more expressive new. Most English speakers would place the emergence of our modern idiom in the age of Shakespeare. Shakespeare invented many of our words and locutions, and he spoke to universal themes of the human condition. Pivotal figures in the development of other European tongues include Dante for Italian, Cervantes for Spanish, Molière for French, and Goethe for German.

So it is with the evolving languages of science. We consider Isaac Newton to be the Shakespeare of classical physics. Born a quarter century after the Bard's death, Newton invented the language of physics, both the arguments and the mathematics, that in large measure we and our colleagues use today. To be sure, he wrote his 1687 masterpiece, *Philosophiæ Naturalis Principia Mathematica (Mathematical Principles of Natural Philosophy)*, in the ancient scholarly language of Latin, couching many of his demonstrations in the geometrical lexicon that Galileo favored. And yet his work changed the future of science.

The law of gravity that Newton announced in the *Principia* is one of the great triumphs of human intelligence: no deviations (apart from small refinements due to General Relativity) have been observed from astronomical distances down to small, but still macroscopic, distances. Not only did he explain the elliptical orbits of the planets, he also showed that the same force of gravity was responsible for the motions of the satellites of Jupiter and Saturn, the movement of the moon about the Earth, the tides in the sea, and the fall of an apple from a tree. In other words, he emphasized the *universality* of the law of gravity, an indication that the same laws of nature hold at all times and in all places. Along with consistency, which is the basis for our conception of nature as orderly, rather than capricious, universality makes the universe a place we may hope to comprehend.

Sub rosa at first, Newton introduced a new mathematics based on change, rather than the cherished traditional ideal of stability. Newton called his creation fluxions; with it, we describe what is in flux—which includes most interesting phenomena—as an accumulation of infinitesimal—barely perceptible—changes. What we now call *calculus* remains an indispensable mathematical language. Moreover, its invention created the possibility for physicists to analyze with exactitude a great sweep of phenomena.

On the continent, the German polymath Gottfried Wilhelm Leibniz independently conceived the calculus, formalizing its methods and devising many of its enduring symbols. The Marquise Émilie du Châtelet propagated his conventions—more transparent than Newton's—in the extensive commentary she joined to her translation of the *Principia* into French. We accord Newton primacy because he was first to explore so many physical problems using calculus, whereas Leibniz had a genius for notation and formalism. We can't exclude the possibility that if our mother tongue were other than English, we might come to a different assessment.

For Newton, a precisely defined starting point always led to an unambiguously determined future: you just had to follow his laws of motion and solve the equations. In his "General Scholium," an essay appended to the *Principia*, he argued that "the Planets and Comets will constantly pursue their revolutions in orbits given in kind and position, according to the laws above explain'd." He needed to invoke a higher power to account for the starting point. "But though these bodies may indeed persevere in their orbits by the mere laws of gravity, yet they could by no means have at first deriv'd the regular position of the orbits themselves from those laws." The solar system, he contended, could only have proceeded "from the counsel and dominion of an intelligent and powerful being." Newton's methods, if not his cosmogony, held sway until the 19th century.

A Newtonian equation is a script; the physicist who writes the equation a playwright. The cast might be a cannonball or a heavenly body or, more interestingly, an ensemble of heavenly bodies. It is a versatile script, without a prescribed cast. The physicist chooses the actors and the script tells them what to do at any moment. We think of the equation—as script—whispering in the actors' ears, "do this, then do that." Newton's obedient actors are obliged to follow the equation's direction to the letter; there is no room for improvisation. The notion that the evolution of a physical system was precisely predetermined persisted, even after Einstein swept away absolute space and time, refined

Newton's equations to account for rapid motions, and gave a new conception of gravity. Newton's framework remains the preferred way to analyze most of our everyday experience outside the laboratory.

The 20th century's new world of atoms and quanta demanded new ways of making predictions. When dealing with nature on a submicroscopic scale, we simply cannot define a precise starting point. How can this be plausible? Consider the case of one single electron. The more closely you try to monitor its position—say by bouncing light of ever shorter wavelength from it—the more you change both its location and its velocity, because the light gives it a kick. Making a measurement disturbs the system, so that determining position first and then velocity is not the same as determining velocity first and then position. Werner Heisenberg's "uncertainty principle," which limits how well we can pin down both location and motion, blurs the Newtonian construct of precise starting conditions. But there is more.

Physicists learned in the 1920s that to understand why a table is solid or why a metal gleams, we need to explore a world a billion times smaller than our own—the quantum realm of atoms and molecules—to discover the rules that prevail there. That realm is ruled not by the customs of everyday life, but by the laws of quantum mechanics. It is hard to imagine that any amount of contemplating the properties of matter at human dimensions would have led to the insights that today govern not only our understanding of nature, but much of commerce. Going to the nanoworld is akin to visiting a foreign land, in which the language and customs cannot necessarily be gleaned from our familiar experience but must be observed and judged by their internal consistency. It seems weird and wonderful that what happens in the quantum world determines so much of the character of the world we experience every day.

This insight is an early instance of what we consider to be the great lesson of 20th-century science: the human scale—by which we mean our everyday notions of time and distance—is not privileged for understanding nature and may even be disadvantaged. It's a lesson not just for physics, for which we are able to give it a precise and powerful mathematical form, but for all of science. It is all the more potent for the parallels in biology, cosmology, and ecology that range over distances and time spans both exponentially larger and exponentially smaller than the human scale. Take an example from the animal kingdom: We may wonder at the appearance of a giraffe, so different from our own. But we do not understand what makes a giraffe a giraffe, or how it relates to other creatures, unless we study the giraffe's genetic heritage—its DNA. Nor do we really understand what a giraffe is until we observe the behavior of herds over many seasons.

We contend that our new understanding of scale—our new sense of proportion—is as important as the enlightened sense of place we owe to Copernicus for his discovery that we are not at the center of the universe, to Einstein for his observation that there is no privileged gridwork of space and time, and to Darwin, Mendel, and the pathfinders of the genetic code for their insights into the nature of life.

The quantum domain is subtler by far than the clockwork universe born out of 17th-century physics. Even if we posited precisely defined starting conditions, we couldn't predict the final outcome with certainty. The rules of the game were changed completely. The most probable outcome might seem to follow from Newtonian reasoning, but the result of a quantum calculation would always be fuzzy. All we could do was give probabilities for alternative outcomes. To calculate those probabilities required entirely new methods and new ways of thinking that grew up in the 1920s through the work of a gifted generation led by Werner Heisenberg, Erwin Schrödinger, Wolfgang Pauli, Paul Dirac, and Max Born.

The form in which we most often apply those methods is a brief equation for the propagation of matter waves that Schrödinger developed—according to his testimony—during a winter-holiday tryst in the Swiss alpine village of Arosa as 1925 turned to 1926. Whatever sparked his creativity, the Zürich professor's equation was, and remains, a triumph. Its solution is a "wave function" that prescribes the probability that a measurement will reveal a particular outcome at various times and places. The Schrödinger equation gives a highly satisfactory account of many simple systems, including the discrete energy levels of the electron bound to a proton—a hydrogen atom. It has never gone out of fashion; a half century after its conception, it reentered particle physics in a big way as the tool of choice for describing quarkonium, an "atom" composed of a heavy quark and heavy antiquark, such as the J/ψ particle made of charm and anti-charm that kicked off the November Revolution. We have honed our understanding of the force between quarks by comparing predictions and observations of quarkonium spectra, and we have learned more about Schrödinger's equation by applying it to this new problem.

Like Newton's formulation, Schrödinger's has no predetermined cast of characters. But Schrödinger's quantum actors have much greater freedom to improvise than Newton's classically trained troupe. From opening scene to final curtain, they may do whatever they please, so long as the outcome of the drama is the requisite set of probabilities. As a memorable story may draw together many interlinked narratives, the conclusion of a Schrödinger production is woven from an infinite number of histories.

To account for the allowed energy levels of atoms beyond hydrogen, Schrödinger's universal script needed to specify that the electron cast members carry half a quantum of spin, as imagined by Sam Goudsmit and George Uhlenbeck, and to require that only a single electron with particular properties could occupy a quantum state, as mandated by Wolfgang Pauli's exclusion principle. Without these stipulations, Schrödinger's theory of matter waves had no hope of explaining how atoms with more than one or two electrons would behave. We will see how the exclusion principle shapes the everyday world in our concluding chapter, and how very different our world would be without it.

The English theorist Paul Dirac wondered why nature should have conferred this peculiar attribute, which he called "duplexity," on the electron. Perhaps because he was a man of few words, Dirac expected the laws of nature to reflect a less-is-more aesthetic. At the beginning of 1928, he published the sparest quantum script that he could imagine—consistent with Einstein's special theory of relativity—for a one-character show starring an isolated electron. Spare it was, shorter than a Shakespearean line of iambic pentameter. Dirac thought that the equation he had written down had the right look to it, but the author himself had to puzzle over its meaning.

Once deciphered, Dirac's equation turned out to be one of those precious formulae that give back more—much more—than was originally put in. First, it contains everything there is to know about a single electron. The two requirements of relativity and quantum theory require the electron to carry half a quantum of spin as an outcome of the theory, not a feature to be added on by hand. In the abstract to a paper in which he applied the new theory to atoms with many electrons, Dirac announced, "The underlying physical laws necessary for the mathematical theory of a large part of physics and the whole of chemistry are thus completely known, and the difficulty is only that the exact application of these laws leads to equations much too complicated to be soluble."

To this day, historians, philosophers, and chemists debate whether Dirac's immodest claim was justified, and whether the rise of computational chemistry has fulfilled his desire for practical approximate methods of applying quantum mechanics. Having heard Dirac lecture, we would not exclude the possibility that—in the spirit of a summation over histories—he was at once making a sober assessment and indulging in just a little bit of drollery.

On further reflection, Dirac discovered that he had not written a one-character show after all. There was no escaping that his minimalist equation gave the electron a costar with one unit of positive electric charge—just opposite to the electron's. Timorous at first, Dirac was forced to accept that in his relativistic quantum theory, the electron must have an antimatter partner, as we recounted when we described Carl D. Anderson's discovery in cosmic rays of the "positive electron." We know now that all of the basic structures that make up matter have their own antimatter counterparts (discussed further in chapter 18, "Trading Places").

In the first reading of Paul Dirac's script, electromagnetism seems to enter only as an external influence—the attraction between an electron and proton in the hydrogen atom, for example. Even in that guise, it reveals a new aspect of the electron. Because the electron has both an electric charge and a spin, it acts like a miniature magnet, a subatomic compass needle. The magnet's strength is a stupendously small number in human-sized units, so atomic physicists define a quantity they call the *g*-factor, for which Dirac's equation prescribes the value 2, twice what physicists might otherwise have guessed. And 2 is the value, subject to some small refinements, that nature has chosen for its electron.

Those refinements begin to show themselves once we identify electromagnetism—light—as an actor in its own right. When he introduced light, Dirac gave it a minimal stage direction: Look for electric charge. Light's assignment specified how the electron was to interact with any electric field—including its own! That terse instruction inspires a set of interrelated vignettes that are linked with the saga of the hydrogen atom: two electrons colliding and then going their individual ways, for instance. The simplest story of two electrons colliding is straightforward: at some point in its history, one electron sends out a photon that is absorbed by the other electron. It's natural to imagine a richer story in two acts in which the second electron emits a photon that is absorbed by the first one.

But this is the quantum realm, in which improvisation is not just tolerated but compulsory. A different tale in two acts would have the photon emitted by one electron provide a little entertainment before reaching the second electron, turning momentarily into an electron and positron that would recombine into the photon that finds the second electron. Even the most imaginative stage director can conceive only a few two-act dramas between electrons according to Dirac's rules. Quantum theory requires you to put them all together to make a coherent story and a reliable prediction. Just how to put them together consistently is a challenge of considerable subtlety that physicists did not sort out until the late 1940s.

One aspect of the procedure involves systematically removing infinite contributions to measurable quantities such as the electron's mass. Most of us accept this program of "renormalization" as sound, even if inelegant, but Dirac could not abide the warts on the theory of photons and electrons. Into his final years, he exhorted his colleagues to do better, insisting that there must be a technique that did not require the slaying of dragons—infinities— at intermediate stages. It would be lovely if his intuition were again correct.

At three, four, five acts and more, the number of possible scenes in a Dirac drama grows rapidly. To keep order, a quantum stage director requires a choreographic map of all the possible sequences of encounters between the players. Symbolic manipulations offer one path to such a map. This was the approach of the Harvard theorist Julian Schwinger, an exquisitely elegant thinker and expositor, and one of those who built a robust calculational framework out of Dirac's insights. In 1947, he found that what we have called the second act increases the electron's g-factor ever so slightly, from 2 to 2.002324, in agreement with newly precise postwar measurements. Theorists have now calculated through act six, predicting the value 2.002 319 304 360 50, while the most precise measurements yield 2.002 319 304 361 46, each with uncertainties of only a few parts per trillion. For theory and experiment to line up through the eleventh place of the decimals can only be called stunning agreement.

Schwinger's approach gave a central place to an abstraction called the quantum field. Michael Faraday, the English experimenter whose meticulous researches laid the foundation for electromagnetic theory, introduced the field concept in 1845 as an attribute of space that provokes physical consequences. To illustrate the idea, he showed how to map the lines of force that reach from one end of a bar magnet to the other. Place a sheet of paper on top of a magnet and sprinkle iron filings in a thin layer onto the surface. The filings align along nested arcs connecting the magnet's north and south poles. Seeing with one's own eyes the effect of the magnetic field, could anyone doubt its reality? In general, a field has a value and perhaps a sense—a direction or other orientation—everywhere in space. Quantum field theory weds this construct from 19th-century physics with quantum mechanics and special relativity. In this picture, which is the most fruitful description that we have yet found of the forces of nature, a particle such as a photon or an electron is an excitation—a quantum—of the corresponding field.

For all but the simplest stories involving electrons and photons, most mortals—theoretical physicists included—find Schwinger's approach cumbersome. We rely instead on a brilliant innovation introduced by Richard Feynman, another hero of quantum electrodynamics, as the polished theory is called. Like many innovators before him, Feynman gave us a new language—this time a graphical language of great clarity and precision.

Feynman's diagrams are minimalist renderings that describe how subatomic action unfolds. They are built of simple graphical elements that represent individual particles. An electron is a thin line bearing an arrow; a photon is a wavy line. A dot marks the spot where an electron absorbs or emits a photon. One way to represent the collision of a photon and electron is this: An electron and photon enter stage left and approach each other. Where they meet, the photon—absorbed by the electron—vanishes, the transition marked by a dot. The electron continues across the stage and at some point, marked by another dot, sends out a new photon. The straight line and the wavy line exit stage right.

The Feynman diagram is much more than a cartoon; it carries by implication the full apparatus of quantum field theory. Explicit rules tell us how to read a diagram and build from it a probability amplitude by multiplying together specific factors for each line that enters or exits, for each dot, for each line between dots. Evaluating a Feynman diagram is not quite as simple as multiplying a few numbers together, so mastering the art is a significant rite of passage for students, one that confers great power. Regarded from different perspectives, the diagram we have just described can also represent two photons colliding to create an electron and positron, or vice versa. Before we write a single formula, we see relationships among reactions.

So much did Richard Feynman delight in his diagrams that when he customized a new 1975 Dodge Tradesman Maxivan for family excursions, he adorned it with more than a dozen of them. Julian Schwinger continued to esteem the approach that gave primacy to the fields, rather than the particles. Speaking at a time when pocket calculators first made instantaneous arithmetical calculations widely available, Schwinger remarked that "Like the silicon chips of more recent years, the Feynman diagram was bringing computation to the masses." We confess without apology that we are among the grateful masses! The diagrams define not only the way we calculate, but the way we think and the way we discuss.

Feynman diagrams make it straightforward—a famous weasel word of physics textbooks—to construct reaction mechanisms at three acts and beyond. Simply use the basic elements to draw all the distinct diagrams you can with the desired incoming and outgoing particles and a specified number of dots, or interaction vertices. The possibilities mount quickly; act six of the electron's magnetic moment is composed of 12,672 diagrams! Much of the work has been automated, but such a calculation still demands steady nerves, uncommon stamina, and superhuman powers of concentration.

Although theoretical physics is a cerebral undertaking, it is not lacking in tactile pleasures. Long before graduate school, we prepared for our craft by learning to write comely Greek letters, draw graceful mathematical symbols, and draft elegant equations on paper or a blackboard. Later we practiced making pleasingly proportioned Feynman diagrams—evocative and easy to parse. We owe beautiful thoughts the honor of beautiful expression.

8

FUNCTION FOLLOWS FORM!

In an 1896 essay, "The Tall Office Building Artistically Considered," the American modernist architect Louis Sullivan argued that a structure should not merely recapitulate classical forms, but should express the building's intended uses.

"All things in nature have a shape, that is to say, a form, an outward semblance, that tells us what they are, that distinguishes them from ourselves and from each other . . .

"Whether it be the sweeping eagle in his flight or the open apple-blossom, the toiling work-horse, the blithe swan, the branching oak, the winding stream at its base, the drifting clouds, over all the coursing sun, form ever follows function, and this is the law."

The "law" that Louis Sullivan espoused, usually expressed in the alliterative design principle "form follows function," seems self-evidently apt in the context of a Greek temple, a Roman bath, an office tower—or, for that matter, a citrus reamer or a graphical user interface. But is it the only productive way to think?

Over the past century, physicists and mathematicians investigating the fundamental forces of nature have come to understand the value of inverting Sullivan's modernist motto.

We have learned that certain symmetries of nature's laws dictate interactions, which is to say that *function follows form*. This principle underlies our understanding of the strong, weak, and electromagnetic interactions and the gravitational interaction prescribed by Einstein's General Theory of Relativity. It is truly the key that unlocks our understanding of these fundamental forces.

To Isaac Newton, force meant a push or pull on a body that changes its state of motion—its speed or direction. We understand force in a more general sense, as an agent of change: an interaction among two or more colliding particles that can alter their energy, momentum, or type. An interaction can also cause one isolated entity—a single particle or a complex object such as an atom—to transform spontaneously.

The journey that led the architects of modern physics to the new conception of forces dictated by symmetries advanced by fits and starts over several scientific generations, beginning with two visionary contributions in the year 1918.

German-born Hermann Weyl held the chair of mathematics at the Swiss Federal Institute of Technology in Zürich during the years when Albert Einstein was developing his general theory, which explained gravity in geometrical terms. Weyl was dazzled by his colleague's achievement, writing in the preface to his 1918 lecture notes, *Space—Time—Matter*, that "[i]t is as if a wall which separated us from Truth has collapsed. Wider expanses and greater depths are now exposed to the searching eye of knowledge, regions of which we had not even a presentiment. It has brought us much nearer to grasping the plan that underlies all physical happening."

Weyl was fascinated by symmetry in art and ornamentation, in nature organic and mineral, and in the mathematical expression of physical law. He described it as "arising from the somewhat vague notion of harmony in proportions." Symmetry means sameness. It is on display in objects we can see and hold that are unchanged in appearance under reflections, rotations, or displacements: a mirror image identical to its unmirrored self; a sphere that looks the same no matter how we rotate it; a crystal in which the neighborhood of one atom is identical to the neighborhood of any other. The "form, the outward semblance" of a theory is reflected in the symmetries its equations exhibit.

In 1918, Weyl made the bold conjecture that he could extend Einstein's theory to incorporate electromagnetism, so that the two forces that describe classical physics would arise from the geometry of spacetime. Introducing a measuring scale, or gauge, for distances and time intervals, he required that the laws of nature not change even if the scale were chosen independently at each point in spacetime. That attempt came up short, but over the following decade, with advice and constructive criticism from colleagues, he proposed a symmetry that did lead to the theory of electromagnetism. It is the most arcane of the symmetries we shall encounter and could only be imagined after the invention of quantum mechanics in the 1920s, so bear with us for a moment.

To understand Weyl's strategy, consider an exercise from the art of navigation. It is conventional to refer headings to North, defined by some absolute standard. We could just as well refer headings to some other direction—east-southeast, for example. That might seem perverse, but if everyone agreed to the new global convention, we would all specify our heading in degrees from the new standard. If we go further and allow navigators at different locations to choose their reference directions independently, confusion follows unless some messenger carries news of the local conventions to others.

We remind our students that there is a similar invariance, an independence of convention, in quantum mechanics. "Imagine," we say, "an electron moving undisturbed through space and time." The students know that in quantum theory the electron's position is described by a wave function that is represented at every location in spacetime by a complex number, which is to say, by a real value and an imaginary value. We remind them that they learned in their first course in quantum mechanics that the overall phase of the wave function—an angle that measures the relative size of the real and imaginary parts—does not have any measurable consequence.. That means that if we decide to add some amount to what we originally called the phase throughout spacetime, physics doesn't change.

We have noted that if navigators set the reference directions independently at different places, orderly navigation requires some means to inform all vessels of the local conventions. Again, a similar need arises in quantum theory. How can physics outcomes stay the same if we give everyone permission to pick a phase standard independently everywhere in space and time? Only if we introduce an interaction where none was present before; it must be carried by a new force particle—a messenger—with one unit of spin. The new interaction is electromagnetism, and the force carrier is the massless photon, the seeker of electric charge.

Deriving a force from a symmetry is the remarkable outcome discovered by Hermann Weyl. Our students might think us eccentric, but one cannot resist being impressed by this first example of a symmetry dictating an interaction. The essential ingredient is local phase symmetry. To acknowledge the origins of the argument in Weyl's geometrical program of 1918, we call it a *gauge symmetry*.

A quarter century would pass before Weyl's construction became more than a curiosity, a parlor trick to recover a law of nature that was already established. Too little noticed at the time, a 1918 paper by the German mathematician Emmy Noether set out in great generality how symmetries can imply conservation laws and interactions. Her two great theorems undergird the way we think about nature's laws.

Amalie Emmy Noether was born in Erlangen, a university town north of Nuremberg in Germany, in 1882. Her father Max, professor of mathematics at the University of Erlangen, was an accomplished scholar elected to numerous learned academies.

Notwithstanding Max's position, Emmy's path to a university education and on to a career in mathematics was impeded by the prevailing discrimination against women in academic life. In 1898, for example, the Erlangen Academic Senate held that the admission of women "would overthrow all academic order."

Nevertheless, she persisted. Like many young women of her milieu—aspiring middle class, with intellectual inclinations—she attended the Municipal

Secondary School for Girls, which prepared her for a respectable career as a teacher of English and French. Emmy Noether had other ideas. While following the traditional course of ladylike study, she was taking private lessons in Stuttgart and Erlangen on the mathematical curriculum reserved to young men in the *Gymnasium*. This preparation enabled her to gain permission to audit university lectures and in 1903 to pass the university qualification. Erlangen remained closed to women, but the University of Göttingen was a bit more open-minded. During a semester there, she heard lectures by the mathematical luminaries Karl Schwarzschild, Hermann Minkowski, Felix Klein, and David Hilbert.

After one semester, Erlangen relented and accepted Noether and one other woman in a class of about a thousand, and so she was able to enroll there as a student of mathematics. She received her doctoral degree summa cum laude in 1907—only the second woman in Europe to earn a doctorate in mathematics—for a dissertation that was heavy on tedious computation. She would later describe her thesis topic as *Mist* (dung), hardly an expression of pride; she aspired not merely to compute, but to invent.

Now *Doctor* Emmy Noether continued to work in the Erlangen Mathematical Institute from 1908 to 1915, effectively playing the role of a faculty member, though with neither compensation nor status. She became a member of the German Mathematical Society in 1909, and in the same year became the first woman to lecture at the society's annual meeting. Her research turned from computation to abstract mathematics, for which she began to display a prodigious talent.

In 1915, Felix Klein and David Hilbert invited Emmy Noether to join their institute in Göttingen, looking forward to her advice on some aspects of general relativity. The opportunity was irresistible for her; to a mathematician, Göttingen was Mount Olympus. Her hosts were two of the Göttingen gods. Felix Klein is known to popular scientific culture for devising a generalization of the Möbius strip, the Klein bottle, whose outside surface is also its inside surface. His 1872 "Erlangen program," named after the city of Emmy Noether's birth, introduced symmetry considerations into the study of geometry. David Hilbert's own work touched many aspects of modern mathematics and raised the mathematical literacy of physicists. He promulgated a manifesto of two dozen problems that kept many mathematicians busy over the next century.

As a welcoming gesture, the Mathematics and Science Department of the philosophical faculty put her forward for the *Habilitation* lecture to become a *Privatdozent*, licensed to give courses in exchange for fees paid directly by students. However, the Historical-Philological Department blocked the move, citing "concern that seeing a female organism might be distracting to the students." Hilbert subverted the blackball by announcing courses under his name that were, in the event, given by his "assistant," Fräulein Noether.

In July of 1918, Emmy Noether dedicated a report on "Invariant Variational Problems" to Felix Klein on the occasion of the fiftieth anniversary of his doctorate. The two theorems she presented in that one article ensured her immortality among physicists. She showed how notions of symmetry brought new constraints—and new insights—to applications of nature's laws. Theorem One applies to global symmetry operations such as our navigation example—the same change of convention, say, made everywhere in spacetime. It teaches us that a conservation law arises from every symmetry operation—built up from infinitesimal changes—that does not alter the form of the equations that express nature's laws. Weyl's phase symmetry, for instance, implies that electric charge cannot be created or destroyed.

Other far-reaching conservation laws follow from symmetry principles that are considerably easier to visualize. The science of mechanics, which describes motion, developed through inspired trial and error as well as mathematical virtuosity. Even a result as fundamental as the Law of Conservation of Energy—the rule that energy cannot be created or destroyed, but only changed in form—appeared to be little more than an empirical regularity, a useful construct lacking a clear basis. Emmy Noether's Theorem One reveals that energy is conserved if the laws of nature do not change over time. Similarly, conservation of momentum follows if the same laws hold at all locations. If no direction in space is preferred over others, then angular momentum, a measure of rotational motion, must be conserved. The correspondence between symmetries and conservation laws shows that broad assertions about how nature behaves have specific, useful, and unexpected consequences. It tells us that a conservation law violated means a symmetry not respected. And Theorem One is only the beginning.

Emmy Noether's second theorem concerns symmetry transformations carried out independently at different locations in spacetime. It contains the seeds of all gauge theories ("Symmetries dictate interactions") and exhibits their kinship with general relativity. It turns Hermann Weyl's inchoate intuition into a powerful general strategy. For any conjectured gauge symmetry, Theorem Two yields a theory of interactions carried by massless spin-one force particles, their number and characteristics dictated by the symmetry. As the photon seeks electric charge, the force carriers in general seek the conserved charge that corresponds to the hypothesized gauge symmetry. The electrically neutral photons do not interact directly with other photons. Consequently, unless electromagnetic fields are very strong, they pass right through each other. All manner of electromagnetic waves coexist benignly—think of the AM, FM, TV, Wi-Fi, 5G, Bluetooth, and light waves that suffuse our surroundings.

Just because you can guess a symmetry and be assured of constructing a theory that describes an interaction doesn't mean that your choice has anything to do with nature. But other factors inhibited a great wave of theory-building that might have been enabled by Noether's work. First, her paper gathered far less notice than we now know it merited.

Emmy Noether herself moved on to abstract mathematics, becoming an inspirational figure—"The Mother of Modern Algebra." Second, how many fundamental forces could there be? Neither radioactivity nor the structure of the atomic nucleus was understood well enough to point to what we now call the weak and strong interactions. Third, physicists had plenty of other fish to fry—explaining the properties of atoms beyond hydrogen and helium in the periodic table, applying lessons of the atom to bulk materials in the beginnings of solid-state, or condensed-matter, physics, learning about the nucleus, and more. And the immense explanatory power of symmetries was not yet widely understood among physicists.

Although Emmy Noether flourished mathematically in Göttingen, she never enjoyed an academic position commensurate with her achievements. After Germany's defeat in the Great War, the Weimar Republic permitted women to teach in universities. In 1919, Noether was granted *Habilitation* on the basis of "Invariant Variational Problems," and could serve as a sort of adjunct faculty member beginning in 1922. She gathered a devoted following of students, mostly men, called *die Noetherknaben* (the Noether Boys), who trailed her through the streets of Göttingen in a boisterous band, debating mathematics all the way. Recognition began to come her way in the form of prizes and prestigious lectureships. Then Germany fell under the pall of National Socialism, and her life in Göttingen fell apart.

In 1933, the culture minister decreed that anyone of Jewish background had to be put on leave from the university. Not only was Emmy Noether Jewish, her pacifist inclinations and contacts with mathematicians of the Russian school marked her as an undesirable to the Nazi authorities. She was among the first six faculty members rusticated from the University, and the implications of that ominous order soon became inescapable.

As the crisis deepened, Hilbert and other sympathetic colleagues scrambled to find safe harbors for the eighteen mathematicians who, by the end of 1933, left or were driven from the faculty at the Mathematical Institute in Göttingen. Thanks to the Emergency Committee in Aid of Displaced German Scholars established by American academic leaders, Noether landed a two-year visiting professorship at Bryn Mawr College in Pennsylvania.

Founded in 1885, Bryn Mawr was among the earliest institutions established in the United States to open higher education to women. It offered students rigorous intellectual training, including postgraduate study, and the opportunity to engage in original research in the tradition of European universities. Noether's academic program included weekly trips to the Institute for Advanced Study, fifty miles distant in Princeton, where she gave seminars and lecture courses. That affiliation gave her contact with other illustrious immigrants, among them colleagues she had known in Germany, including Hermann Weyl.

Whether or not the Bryn Mawr women who followed "Miss Noether" on brisk walks were as disorderly as the *Noetherknaben* in Göttingen, they were just as engaged and effervescent. Emmy Noether herself was looking forward, full of curiosity about American ways, stimulated by her students and her interactions at Princeton, and generally full of life. During the spring vacation of 1935, she had abdominal surgery that was expected to be routine. She seemed to be recuperating well but suffered complications and died within a few days. Albert Einstein wrote an "Appreciation of a Fellow-Mathematician" to the *New York Times*, celebrating Emmy Noether as "the most significant creative mathematical genius thus far produced since the higher education of women was begun." Speaking at her memorial service, Hermann Weyl offered a very long hymn of praise to the colleague "I knew to be my superior as a mathematician in many respects."

Two decades would pass after Emmy Noether's death before physicists began to exploit the full power of Theorem Two. James Chadwick's 1932 discovery of the neutron led eventually to the prominence of gauge theories and to deserved recognition for Noether's insights.

Together with the proton, the neutron is a constituent of the atomic nucleus. The proton carries one unit of positive charge, just opposite to the electron's charge, and the neutron carries none. Each carries half a unit of spin, as the electron does. They are nearly identical in mass; in the units that atomic physicists use, the proton's mass is 938,272,088.16 electron volts, while the neutron's is 939,565,420.52, each with an uncertainty of less than one electron volt. (In round numbers, each is about one billion electron volts, 1 GeV.)

With the two masses differing by a mere seven-tenths of a percent, it is not much of a leap to speculate that the proton and neutron might be two faces of one particle, called the nucleon. Werner Heisenberg, the wunderkind of quantum mechanics whom we encountered in chapter 3, did that and more. The relation between proton and neutron, he surmised, might be similar to that between electrons with spin up and spin down. The new quantum number distinguishing proton from neutron is—extending the analogy with spin—called isospin; we say that the proton corresponds to isospin up, the neutron to isospin down. Through the 1930s and 1940s, as physicists learned more about the force that binds nuclei together and about nuclear reactions, the case for a symmetry linking proton and neutron grew richer, and isospin symmetry became a valuable tool.

Werner Heisenberg had a deep respect for the power of symmetry. In conversation with his disciples late in life, he proclaimed, "'In the beginning was the symmetry,' that is certainly more correct than the Democritean thesis, 'in the beginning was the particle.' The elementary particles embody the symmetries, they are their simplest representations, but they are above all a consequence of the symmetries." And yet, it never occurred to him to combine his new symmetry with Noether's theorem, perhaps because—having been preoccupied with the development of quantum mechanics—he never read Emmy Noether's paper.

Applying Theorem Two to isospin symmetry would wait until 1954, when Chen Ning Yang and Robert L. Mills, working at Brookhaven National Laboratory, framed the issue in very physical terms. Assuming that electromagnetic interactions can be put aside, what one chooses to call a proton or a neutron is purely arbitrary—conventional. It is not a matter of either/or; we could instead assign the name proton to some mixture of the familiar proton and neutron. That arbitrariness reflects the isospin symmetry that points—following Emmy Noether—to isospin as a conserved quantity. That ground had been covered before. But Yang and Mills protested that it seems needlessly restrictive to say that once one chooses what to call a proton or a neutron at one spacetime point, one is then not free to make any choices at other spacetime points. They explored the possibility of adopting independent conventions for proton and neutron everywhere in spacetime. This is the isospin generalization of the phase-symmetry challenge we gave to our students.

Our students who accept the Yang-Mills challenge will find that the algebra needed to carry out the derivation fills a few pages instead of a few lines. That is because the outcome of two distinct isospin symmetry operations (there are three) depends on the order in which they are performed. If you have ever worked to solve a Rubik's Cube, you have already experienced such "noncommuting" operations. Here is a simpler illustration—a lab exercise, if you like. Place a book on a table before you with the front cover showing and the spine on your left. Rotate the book a quarter turn counterclockwise, so the spine has moved to the bottom. Next, lift the edge on your left (now the top edge of the book) and rotate a quarter turn clockwise until the book is standing on its bottom edge. You'll be looking down on the top edge, with the spine toward you, oriented vertically and right side up. Now, let's do the same operations in the opposite order. Place the book on the table in the starting position, but this time begin by lifting the spine and rotating a quarter turn clockwise until the book is poised on the fore-edge, where the pages open freely, and you are looking down at the spine, oriented vertically. Rotate the book a quarter turn counterclockwise, keeping the fore-edge on the table, and the spine—still on top—will be oriented horizontally. If it is a typical American book, the letters will be right side up. Identical manipulations carried out in the opposite order lead to different outcomes.

Theorem Two assures us that the kind of theory that Yang and Mills imagined can be constructed; with a little toil, our students can reproduce it. It is a theory of interactions among protons and neutrons carried by three massless force particles with one unit of spin and charges +1, 0, −1. Particles with spin measured in integer units are generically called bosons. The Yang-Mills force particles—gauge bosons—seek isospin. Because these force particles carry isospin, they interact among themselves. This is a big difference with respect to electromagnetism; it will turn out to be very consequential.

By deriving an interaction not previously known, Yang and Mills changed the way we think about the forces of nature. Their example of isospin symmetry made explicit the power of Emmy Noether's Theorem Two: symmetries dictate interactions; function follows form, inverting Louis Sullivan's famous dictum. Exciting stuff!

But not so fast! We discussed in chapter 4 that the most important part of the nuclear force is mediated by Hideki Yukawa's pion, which carries spin zero—not spin one—is massive, not massless, and has been observed in experiments. The elegantly conceived Yang-Mills theory cannot serve the purpose for which it was intended. Why, then, do we pay attention?

In contrast to biological evolution, unsuccessful lines in theoretical physics do not become extinguished, never to rise again. Scientists have memories and our institutions have libraries, society's collective memories. We are free to borrow potent ideas from the past and to apply them in new settings, to decisive effect. Yang and Mills applied the right idea to the wrong problem, because the right problems hadn't yet presented themselves. We'll next present those right problems and explain how the reasoning of Noether and Weyl, Yang and Mills showed physicists how to describe the strong interactions and more.

9

A NEW LAW OF NATURE

If Yang and Mills had not succeeded in making a successful theory that would explain the nuclear forces among protons and neutrons, many of their colleagues found the idea that symmetries dictate interactions—that function follows form—was too good not to be true. Through the following decade, theorists explored ways to give masses to the gauge bosons that carry the isospin force while maintaining the central place of symmetry. Those investigations kept the Yang-Mills strategy in the common consciousness. Whether it was to be just a playground for theoretical recreations or to emerge as the key to understanding the fundamental interactions would not become clear for another decade.

The prevailing wisdom did not encourage optimism. The great Russian theorist Lev Landau wrote in 1959, "It is well known that theoretical physics is at present almost helpless in dealing with the problem of the strong interactions . . . This is evident . . . particularly from Dyson's assertion that the correct theory will not be found in the next hundred years." Freeman Dyson's work integrating diverse approaches to quantum electrodynamics gave his pronouncement the ring of authority.

In quantum electrodynamics, a simple one-act play typically portrays a good first approximation to a physical phenomenon that can be refined by considering subsequent acts, or multiple interactions. This convergence to a precise result occurs because electromagnetism is not a particularly strong interaction. The strength of the nuclear force means that a one-act "first approximation" can be overwhelmed by subsequent acts that make a calculation spiral out of control.

During our time as graduate students, Berkeley was the Mother Church of what was envisioned as a pragmatic alternative to quantum field theory. The S-matrix bootstrap theory championed by Geoffrey Chew denied that any of the dozens of strongly interacting particles—nucleons, pions, and the rest—should be seen as elementary. It conceived a reality in which general principles would constrain all observables to fit together in a unique way, so that predictions about the strong interaction would emerge from demands of self-consistency. This approach stimulated much imaginative work, including the development of theoretical tools we rely on today, but it did not solve the problem.

The Yang-Mills invention did ultimately provide a comprehensive theory of the strong interactions. That triumph came only after physicists identified the right building blocks—quarks, rather than protons and neutrons—and the right gauge symmetry.

Physicists did not become aware that protons and neutrons are made of something until the mid-1950s. The first hint came from Robert Hofstadter's team at Stanford University, in experiments that scattered electrons from hydrogen at higher energies than before. Hofstadter concluded that a proton's positive charge is dispersed throughout a fuzzy sphere about a millionth of a billionth of a meter across.

We saw in chapter 5, "Flora," that, by positing the existence of three new elementary constituents of subnuclear matter, George Zweig and Murray Gell-Mann were able to anticipate the patterns of all the mesons and baryons known in the mid-1960s and to predict the patterns of related particles still to be discovered. Three quarks—up, down, strange—and simple rules for combining them could account for the entire spectrum of strongly interacting particles. A meson, they asserted, is a quark plus an antiquark; a baryon is made of three quarks.

Theorists—especially those fascinated by the Yang-Mills schema—were not quick to embrace the idea of a new level of structure. Some dismissed it as a cheap trick; others, a bit more open-minded, conceded that it might be a convenient mnemonic. Only a few early adopters took the idea at all seriously; even they were careful to present their calculations "as if" the mesons and baryons were composed of quarks, just as chemists had once viewed atoms as convenient fictions. No doubt the quarks seemed a little alien because they carried electric charges that were fractions of an electron's or proton's charge. A greater obstacle was that no one had ever seen an individual isolated quark, despite what came to be prodigious efforts.

By the late 1960s, the SLAC-MIT experiments that we discussed in chapter 5 took pictures of the proton's interior with finer spatial resolution and shorter exposure times than ever before, finding that protons and neutrons contain tiny charged particles that Richard Feynman and Bj Bjorken called partons. Inside the nucleon, these objects—soon to be identified as quarks—seemed to behave as free and independent, and yet they never were seen to escape. Until this paradoxical state of affairs could be explained, quarks would remain a story, not a theory.

Even as a story, the quark model didn't make perfect sense. The moment that you are tempted to treat quarks as real, the ghost of Wolfgang Pauli appears, demanding that the quarks, as particles with a half unit of spin, obey his exclusion principle. The famous Omega-minus particle, Ω^-, illustrates the challenge. In the original versions of the quark model, Omega-minus is composed of three identical strange quarks, all with spins aligned and the same zero isospin, which is to say three strange quarks, all in the same quantum state. The resolution emerged from proposals by Oscar W. Greenberg, Yoichiro Nambu, and Moo-Young Han. It seems almost too simple to be serious: evade the exclusion principle by decree. Each flavor of quark must come in three distinct varieties.

It is suggestive to label the different varieties in analogy to the primary colors of light—red, green, and blue. Then an Omega-minus made of a red strange quark, a green strange quark, and a blue strange quark has no net color, since red plus green plus blue adds up to white. Similarly, mesons must be colorless. This useful metaphor leads to the conjecture that only colorless states can exist in isolation. When physicists began to make their "as if" calculations with colored quarks, they discovered that three colors gave better agreement with measurements than colorless quarks.

If color is more than an excuse to evade the exclusion principle, it deserves to be taken seriously. In the same way that naming a proton or neutron is arbitrary (putting aside electromagnetism), the color labels must be a matter of convention; that suggests a color symmetry akin to isospin symmetry. A natural guess is SU(3)—already familiar to particle physicists from Gell-Mann and Ne'eman's classification of mesons and baryons—now applied to color instead of flavor. Whereas flavor symmetry applies to external attributes such as charge and strangeness, color symmetry applies to an internal attribute that is not directly observable. The meson and baryon families that Gell-Mann and Ne'eman recognized contain one or eight or ten members; the families are called SU(3)-flavor singlets or octets or decimets. Because all the known quark-antiquark mesons and three-quark baryons are neutral in color—color singlets—it was natural to speculate that any more complicated combinations of quarks and antiquarks that might exist as isolated particles would have to be color singlets, too.

Having come this far, why not take the next step and emulate what Yang and Mills did for isospin symmetry, now choosing color-SU(3) as the gauge symmetry? In this case, the derivation leads to eight massless force particles, each carrying both a color and an anticolor. We call these force particles gluons, because their role is to glue the quarks and antiquarks together to make hadrons. As photons seek electric charge, gluons seek color. Force particles are agents of change. A photon can change the energy and momentum of a charged particle. A gluon can do more: it can change a quark's color. As colored objects, gluons interact among themselves, in contrast to the electrically neutral photons. Like quarks, gluons should be confined within color-singlet objects.

When theorists began to tinker with this color gauge theory, they did so gingerly. Murray Gell-Mann, who with his collaborators was among the first to explore a theory of colored quarks and gluons, expressed a marked restraint in his talk summarizing the 1972 International Conference on High Energy Physics. "[I]t may well be possible," he opined, "to construct an explicit theory of hadrons, based on quarks and some kind of glue, treated as fictitious, but with enough physical properties abstracted and applied to real hadrons to constitute a complete theory."

Why would theorists be so timorous? No free quarks had ever been observed, and the concept of confinement was merely an assertion. Even if the theory gave the correct equations, the very strength of the strong interaction meant that calculations in multiple acts, the technique we call perturbation theory, could not be trusted. That concern actually goes even deeper. In quantum theory, the strength of an interaction depends on the resolution with which it is measured. In quantum electrodynamics, the more closely a charge is examined, the greater is its strength. This is because an electron, for example, is surrounded by an ever-changing quantum cloud of photons, electrons, and positrons—"virtual" particles that wink in and out of existence. The electron attracts the virtual positrons and repels the virtual electrons, so the charge seen from afar is reduced a little bit by the excess of positive charges gathered near the electron. In fact, at shorter and shorter distances, the strength of electromagnetism grows larger and larger, tending toward infinity, so the theory ceases to make sense. This is the behavior that so troubled Lev Landau.

In quantum electrodynamics, that calamity occurs at such infinitesimal distances—equivalently, such high energies or short wavelengths—that for many physicists it was easy to imagine that some new phenomenon might intervene to rescue the theory. Only those as sensitive as Hans Christian Andersen's princess, who detected a pea under a stack of twenty mattresses and twenty feather beds, were paralyzed by this knowledge. It would be a far different story for a theory of strong interactions, where the infinity would come into play at nearby distance scales or energies. It is no surprise that Gell-Mann might be alert to this problem, as he was one of the first to analyze the scale-dependence of coupling strengths in the 1950s.

In any event, this behavior is contrary to what would be needed for a theory to incorporate both the parton model and confinement. For a time, it became a matter of great theoretical sport to show that quantum field theory was inconsistent with the parton model, however intuitively appealing that picture might be. Whether such demonstrations would be taken as evidence against the parton model or against a field theory of strong interactions was a matter of taste, or perhaps of preconceptions. Example theory after example theory behaved like quantum electrodynamics: the shorter the distance, the stronger the interaction.

Then, in the spring of 1973, David Gross and his student Frank Wilczek at Princeton and David Politzer, a student at Harvard, separately announced a stunning result: a gauge theory of colored quarks and gluons displays behavior opposite to quantum electrodynamics (QED). The interaction strength becomes smaller at high energies or short distances. The new element compared to QED is that virtual gluons tend to disperse a quark's color, so that to examine a quark more closely means to measure a smaller color charge.

The phenomenon that the strength of interaction between quarks and gluons diminishes steadily as we probe shorter and shorter distances is called asymptotic freedom. It immediately raised the prospect of making reliable calculations for the strong interactions, well before Dyson's prophecy of a hundred years in the wilderness had been accomplished. Because of the central role of color, the newly respectable theory became known as quantum chromodynamics, or QCD.

Right away, what was dubbed the QCD-improved parton model advanced from a cartoon picture to a tool that made precise predictions for experiments that scattered electrons from protons. In 1974, Politzer and Tom Appelquist speculated that a bound state of a hypothetical heavy quark and antiquark might resemble an atom. This came to pass with the discovery of charmonium that we recounted in chapter 6, "Revolution." As experiments moved to higher energies at Fermilab's Tevatron and CERN's Large Hadron Collider, theorists learned to evaluate Feynman diagrams of increasing number and complexity containing quarks, antiquarks, and gluons. By considering not just one-act descriptions, but two-, three-, and even four-act productions, they could make predictions with uncertainties of only a few percent that had the essential virtue of agreeing with experiment.

Although quarks and gluons are not observed as free particles in the laboratory, a SLAC experiment found in the early 1970s that electron-positron annihilations typically yielded back-to-back sprays of pions and other strongly interacting particles, or hadrons. These jets, as they are known, behave precisely as if they are tracers for a produced quark and antiquark that materialize into the observed particles. In 1979, experiments at the electron-positron collider PETRA in Hamburg, working at higher energies, observed three-jet events that have all the characteristics of a quark, and antiquark, and a gluon. All but the most recalcitrant were persuaded that gluons are real!

QCD realized and enriched Bjorken's and Feynman's intuition and meant that, at least in the realm of high energies, the strong interactions were no longer intractable. And there was more to come.

∞

In the opening lines of his *Principia Mathematica* (1687), Isaac Newton defines Mass as "the quantity of matter . . . arising from its density and bulk conjointly." That notion of mass as an intrinsic attribute of matter, extended by Newton's law of universal gravitation and his laws of motion, is a foundation of classical physics. Mass, for Newton, is both the source of gravitational attraction and the measure of inertia—the tendency of a body to remain in uniform motion (or at rest) unless an external agent should act upon it. Under Newton's definition, the mass of an object is the sum of the masses of its parts, in agreement with everyday experience. It is then inescapable that mass is conserved. But in the classical worldview, mass does not arise, it simply *is*.

In the 18th century, Mikhail Lomonosov in Saint Petersburg and Antoine Lavoisier in Paris extended the law of conservation of mass to the realm of chemical reactions. Their careful analytic work was central to the development of chemistry as a quantitative science, leading—through the work of John Dalton and others—to the empirical underpinnings of the modern atomic theory.

Mass remained an essence—part of the nature of things—for more than two centuries, until scientists began to speculate that the mass of the newly discovered electron might arise from its own electric field. We trace our modern conception of mass to Albert Einstein's pregnant question of 1905, "Does the inertia of a body depend upon its energy content?" and his revolutionary response that the mass of a body is its energy at rest divided by the speed of light squared.

Confirmation of Einstein's mass-energy relation came in 1932 from the atom-smashing experiment carried out by John Cockcroft and Ernest Walton. Bombarding lithium-7 with protons, they transmuted the proton plus lithium into two alpha particles accompanied by a release of energy. The following year, Irène and Frédéric Joliot-Curie found that an energetic photon passing through matter could materialize into an electron and an antielectron, demonstrating the conversion of energy into matter.

Not only is Einstein's a precise definition of mass, it invites us to consider the origins of mass by coming to terms with a body's rest energy. For the first steps down the quantum ladder, the difference between Einstein's conception of mass as rest energy and Newton's assertion that the mass of an object is the sum of the masses of its parts is subtle but telling. We understand the mass of an atom or molecule by adding the masses of the atomic nuclei and electrons, minus small corrections that account for the attraction of negatively charged electrons to the positively charged nucleus. "Small corrections" is a bit of an understatement. The energy liberated when methane burns with oxygen to form carbon dioxide and water corresponds to a fractional mass difference between the reactants and combustion products of about one part in 40,000. The law of conservation of mass holds to an impressive degree, and yet, our fossil-fuel economy—not to mention our personal metabolism—feeds on the tiny deviations.

In precise and practical terms—if not quite from first principles—the masses of all the nuclei follow from the proton mass, the neutron mass, and our semi-empirical knowledge of nuclear forces. The mass of the deeply bound alpha particle, or helium nucleus, is only three-quarters of a percent less than the individual masses of two protons and two neutrons. Even in the nuclear realm, the notion that the mass of an object is the sum of the masses of its parts is a reliable first approximation. On a macroscopic scale, that small mass defect is a proxy for the prodigious store of energy that powers the sun or lies in wait at the heart of thermonuclear weapons.

And what of those formerly elementary particles, the nucleons? A proton's properties are set by its quark constituents, two ups and a down, while the neutron's are determined by two downs and an up. Quantum chromodynamics tells us the origin of the proton and neutron masses.

We have seen that the property of asymptotic freedom means that the strong interaction among quarks and gluons becomes weaker at distances smaller than the size of a proton. That property empowers us to make calculations in multiple acts and arrive by successive approximations at reliable predictions for high-energy collisions. The traditional strategy of multiple acts gives no hint that the colored objects—quarks and gluons—must never be seen in isolation, but are permanently confined within colorless objects, as experiments consistently indicate. But the other face of asymptotic freedom is that the strong interaction increases in strength at large distances comparable to the size of a proton.

By defining quantum chromodynamics on a spacetime grid—a lattice—rather than a continuous spacetime, physicists have created mathematically rigorous methods to extract reliable predictions when the interaction is strong. These "Lattice QCD" calculations, which were pioneered by the Cornell theorist Kenneth Wilson, attack the calculation all at once, not act by act. They reach far beyond what anyone might achieve with pencil and paper; they require elaborate simulations carried out on supercomputers. The results shed light on the confinement of quarks and gluons, predict the sizes and masses of the nucleons and other strongly interacting particles, and tell us how quarks and gluons might behave in neutron stars, in high-energy collisions of heavy nuclei, and even in cataclysmic neutron-star mergers.

While the lattice QCD simulations and many pieces of circumstantial theoretical evidence make it a very credible working hypothesis to suppose that QCD confines quarks and gluons within color-singlet states, that conclusion has not yet been proved with mathematical rigor. So significant is that gap in our understanding that the Clay Mathematics Institute in Cambridge, Massachusetts, has set the challenge of completing a proof as one of its seven Millennium Prize Problems. An unimpeachable demonstration would earn the eternal thanks of a grateful company of physicists and a worldly compensation of one million dollars. Whether in the future color confinement should be proved, remain a working hypothesis, or be disproved, the experimental search for free quarks and gluons must continue, to test our understanding and look for surprises.

Seen through the lens of lattice calculations, quantum chromodynamics has shown us that the proton and neutron are a novel kind of stuff. In contrast to macroscopic objects, and to a degree far beyond what we observe in atoms, molecules, and nuclei, most of the mass of a nucleon is not given by the sum of its parts. The masses of the three quarks that make up a proton or neutron account for no more than a few percent of a proton's or neutron's mass. The lion's share is created by the strong mutual attraction of gluons, which generates the mass and size of nucleons. To excellent approximation, the nucleon's mass arises from the relativistic energy of the quarks in motion within the nucleon and the energy stored up to confine the quarks into a tiny volume. This makes nucleon mass the very embodiment of Einstein's conception of mass as the energy of the nucleon at rest.

Moreover, in explaining the nucleon mass, quantum chromodynamics accounts for nearly all the visible mass of the universe, since the luminous matter is essentially made of protons and neutrons in stars and clouds. This is a most remarkable achievement! It is no exaggeration to say that quantum chromodynamics is a new law of nature.

Yet a haunting mystery remains: What sets the masses of particles such as the electron and the quarks? If these particles have no perceptible size and are not, so far as we can tell, made of any smaller bits, why do they weigh? We have reason to think that the answer lies in a theory of the weak and electromagnetic interactions, another theory built on the now-familiar theme that symmetries dictate interactions (function follows form)—but with a twist. We will develop that theory as we continue on our journey and see how far we have come toward understanding the origin of elementary-particle masses. As a first step, let us consider the path that led physicists from beta decay to a provisional theory of weak interactions.

10

BETA VERSION

The pioneers of radioactivity—Henri Becquerel, Marie and Pierre Curie, Ernest Rutherford, and Frederick Soddy—encountered the spontaneous metamorphosis of one element into another in two separate ways. In the first one, an alpha particle flies out of a radioactive nucleus, leaving behind a nucleus with two fewer protons and two fewer neutrons than the parent. The escaping alpha particle, or helium-4 nucleus, was already present in the parent; it escapes by the quantum tunneling phenomenon described by George Gamow that we saw in chapter 1. No protons or neutrons are created or destroyed in the process. In the second kind of transmutation, a nucleus emits a beta ray, raising or lowering the electric charge of the nucleus by one unit without changing the sum of protons plus neutrons.

An example of alpha decay familiar to the Curies is the decay of radium-226 (88 protons and 138 neutrons) into radon-222 (86 protons and 136 neutrons) plus an alpha particle (2 protons and 2 neutrons). The simplest example of beta decay is the transformation of a neutron (charge 0) into a proton (charge +1) and electron (charge −1), plus—we now know—an unobserved (anti) neutrino, a very light particle with spin-1/2 and charge 0. One nucleon is present before the transition, one after. Alpha decay redistributes protons and neutrons without altering their species; beta decay changes one kind of nucleon into the other.

In the summer of 1930, Enrico Fermi traveled from Rome to deliver a series of lectures on the quantum theory of radiation at the University of Michigan's Symposium for Theoretical Physics. Summer schools have a rich history as informal but intense gatherings at which leading scholars initiate students into a field and bring experienced researchers up to speed with the latest developments. The most influential lecture notes serve as texts for university seminars, or even full courses, in the years that follow and stimulate new waves of original research. Chris's colleague, Andreas Kronfeld, has commented that a summer school should provide a lifetime of homework—if not for the students, certainly for the lecturer.

Fermi's Michigan lectures, published in 1932 in the American journal, *Reviews of Modern Physics*, became an instant classic—one that we were urged to comb through decades after its publication. Fermi recast Dirac's theory of photons and electrons in a form more transparent than the original and presented illuminating applications to important problems.

When Wolfgang Pauli made known his neutrino hypothesis, Fermi was quick to see that the quantum theory of radiation might help him think about the emission of beta rays. As an excited atom leaps to a lower quantum state, a new photon is created. That is to say, the total number of light quanta is not fixed. Fermi hypothesized by analogy that when a nucleus undergoes beta decay, an electron-plus-neutrino pair is created as the neutron becomes a proton. Unlike the alpha particle, the electron and neutrino did not reside in the nucleus before the transition. The alternative was to imagine that the neutron was built up of a proton, an electron, and a neutrino. Werner Heisenberg's idea that proton and neutron are two states of the nucleon supported Fermi's analogy to quantum jumps in atoms.

Pauli had named his lightweight, neutral, penetrating particle *Neutron* (in German), but when the neutral counterpart of the proton showed itself, that took the neutron name. Inspired by the Italian suffixes *-one* (big) and *-ino* (little), Fermi's group called the proton's partner *neutrone* and Pauli's particle *neutrino*, for the little neutral one. We use neutrino as a generic term, but call the particle emitted with an electron the antineutrino, to signal the creation of a lepton-antilepton pair analogous to an electron and positron.

During our student years in Berkeley, Emilio Segrè delighted in recounting the 1933 winter holiday that Fermi's Rome group took to the Dolomites, between Christmas and New Year's. One evening, after a full day of skiing, Fermi invited his colleagues to his room in the Hotel Oswald in Selva, Val Gardena, to explain the essence of his ideas about beta decay. Emilio and two other acolytes crowded along the edge of a narrow bed, craning their necks to follow along as the Pope, as Fermi was known, sketched his ideas on a sheet of paper. He imagined that the interaction among proton, neutron, electron, and neutrino took place at very tiny distances, which he idealized as a geo-metrical point. Although his theory of beta-ray emission—we would now say, the weak interaction—had a structure similar to Dirac's electrodynamics, it did not include a long-range force particle akin to the photon. He showed how the formalism of photon emission and absorption could be carried over to beta decay. Some mathematical details flew by too quickly for the listeners to master at once, but then Segrè's retelling did suggest the presence of Chianti. Fermi showed how to estimate the beta-decay lifetimes of several radioactive nuclei and how to calculate the energy spectra of the beta rays, the continuous distributions that had motivated Pauli to propose the neutrino.

Fermi had submitted an announcement of his new theory to *Nature*, where (as reported in his *Collected Works*) "the manuscript was rejected by the Editor of that journal as containing abstract speculations too remote from physical reality to be of interest to the readers." [All that Fermi had done, we can now see, was to define a new force of nature, alongside electromagnetism, gravitation, and the vaguely defined strong nuclear interaction. We can also look back on his picture of beta decay as the seed of what is now called flavor physics, the transformation of one quark species into another through the action of the charge-changing weak interaction.] He was able to publish a four-and-a-half-page "Attempt at a theory of the emission of beta rays" in *La*

Ricerca Scientifica, the journal of the Italian National Research Council, at the end of 1933. Early in the New Year, Fermi published extensive accounts in the Italian *Nuovo Cimento*—note the echo of Galileo's *cimenti*—and the German *Zeitschrift für Physik*.

Our own most legendary ski-trip adventure was when Bob opened an après-ski game of Scrabble by laying down the word $A_1 X_8 I_1 A_1 L_1 L_1 Y_4$, scoring a semi-infinite number of points and rendering the contest instantly noncompetitive. An unforgettable triumph, but not a match for witnessing the birth of a new force of nature!

Giving mathematical form to a physical picture not only refines the ideas, it also encourages applications to new settings. In the language of Feynman diagrams, we can look at a reaction forward, backward, and sideways. Fermi's theory invites the possibility that—in the right nuclear setting—a proton could be seen to turn into a neutron, emitting a positron and an unobserved neutrino. In 1934, Irène Curie and Frédéric Joliot reported that they had bombarded aluminum-27 (thirteen protons and fourteen neutrons) with alpha particles, yielding a neutron plus phosphorus-30 (fifteen protons and fifteen neutrons), which they observed to emit a positron. Their experiment is celebrated as the discovery of artificial radioactivity, because phosphorus-30 does not exist in nature; its half-life is only two-and-a-half minutes.

On the heels of the Curie-Joliot discovery, Hans Bethe and Rudolf Peierls sent their own letter to the editor of *Nature* from the University of Manchester on February 10, 1934. Adopting the Roman nomenclature for Pauli's particle, they titled their note "The 'Neutrino.'" They observed that the discovery of an artificial positive-beta decay strongly supported the idea that the beta particle and neutrino were created at the time of emission, "as one can scarcely assume the existence of positive electrons in the nucleus." Then, they argued that if neutrinos can be created, they can be annihilated. The most interesting case would be a neutrino absorbed by a nucleus that emits a positive or negative electron while the charge of the nucleus changes by one unit to compensate. But the probability of such an interaction would be very small. A neutrino emitted in beta decay would have "a penetrating power of [ten million billion] kilometers in solid matter," which is to say, one thousand light-years. The Manchester theorists concluded that it is "absolutely impossible to observe processes of this kind with the neutrinos created in nuclear transformations."

Fermi's theory has the curious feature that the rate of interactions increases in proportion to the neutrino energy; even so, it seemed highly improbable to Bethe and Peierls that, even for cosmic-ray energies, the interaction rate would become large enough to allow the process to be observed. If, therefore, the neutrino has no interaction but what Fermi attributed to it, "one can conclude that there is no practically possible way of observing the neutrino." Their verdict is consistent with a confession traditionally ascribed to Pauli, that he had done a terrible thing by postulating a particle that cannot be detected.

Bethe and Peierls had good reason to be pessimistic about the prospects for observing the neutrino, but the interaction rates they estimated guided the design of experiments that would, in time, exploit new technologies to make possible the study of neutrino interactions. More than two decades would pass before the reaction they imagined—an antineutrino absorbed and positron emitted—could be observed using the intense flux of antineutrinos produced in nuclear transformations in a reactor core. Today, detectors as large as a cubic kilometer routinely record interactions of neutrinos produced in cosmic rays, even of neutrinos arriving from extraterrestrial sources.

But for a single—very consequential—modification, the theory of beta decay that Fermi exposed to his colleagues in the Val Gardena is what we learned in school three decades later, and what our colleagues in nuclear and particle physics still rely on today in many circumstances. The need for that modification grew out of the most memorable "Asking for a friend" anecdote in the history of modern physics, one that emphasizes the recurring importance of the challenge, "How do we know that?"

First, a little background. The theories of classical physics do not privilege particular locations in space, instants in time, or orientations. They are symmetric under shifts, or translations, in space and time and under rotations in three dimensions. Those symmetries correspond, through Emmy Noether's theorems, to conservation laws for momentum, energy, and angular momentum. The classical theories also respect discrete symmetries, along with the continuous symmetries of translation and rotation. An allowed motion, for example, remains an allowed motion if we exchange our notions of past and future, running a movie of the motion in reverse. The same is true for spatial inversion: this means not just the familiar mirror reflection that exchanges left and right, but a three-dimensional reflection that exchanges left and right, up and down, forward and backward.

As the experimental study of atomic spectra matured in the 1920s, physicists noticed striking regularities in the catalogue of transitions that were observed, so seemed to be "allowed," and those that were absent, as if "forbidden." Quantum mechanics, expressed by the Schrödinger equation, helped make sense of these patterns. Schrödinger wave functions for atomic levels are of two types. *Even* wave functions have the same value when the electron's location is flipped to the other side; *odd* wave functions change sign under inversion.

Atomic spectroscopists found that the strongest transitions involving the emission of a single photon always coupled a state corresponding to an even wave function to one with an odd wave function. This made sense if the inversion property of the wave functions could be attributed to the atomic states themselves and if this property, now called *parity*, did not change overall in the transitions, taking into account both the atomic states and the photon. The photon prefers to carry odd parity, so a transition from an even-parity level must end in an odd level, the odd destination parity times the odd photon parity giving even parity overall, which matches the parity of the departure level. These regularities, and many like them, validated inversion symmetry—parity conservation—in the atomic realm. To this day, the parity rule remains the unbroken law for electromagnetism, quantum as well as classical. Nothing seemed to stand in the way of regarding parity conservation as a law of nature, without qualification.

When physicists began to identify numerous unstable particles through their decay patterns in the 1940s, it was natural to assign to each particle a parity—along with mass, charge, and spin—and to invoke parity selection rules to classify the decays into lighter particles. The case of two strange particles, then called by the names tau and theta, stood out as puzzling, even vexing. Tau was the name given to a meson that decayed into three pions; close analysis indicated that it carried negative parity. Theta referred to a meson with positive parity that decayed into two pions. Could it be mere coincidence that both particles had the same mass, about one thousand times the electron mass, and equal, or closely equal, lifetimes? To compound the problem, the same mass and lifetime were observed for other decay products containing a muon or electron. Was this a case of five nearly identical but distinct particles . . . or four . . . or three . . . or two—or a single particle behaving weirdly? Whatever it might be was a hot topic at the 1956 Rochester Conference on High Energy Physics, held in early April.

A young Duke University experimenter, Martin Block, was sharing a hotel room with Richard Feynman. One evening, Block asked him why everyone was so hung up on the parity rule, and why tau and theta couldn't be the same particle. "What would be the consequences if the parity rule were wrong?" As recounted in *Surely You're Joking, Mr. Feynman!*, the theorist responded, "It would mean that nature's laws are different for the right hand and the left hand, that there's a way to define the right hand by physical phenomena. I don't know that that's so terrible . . . Why don't you ask the experts tomorrow?"

We both knew Marty Block for many years. As a student at Columbia in the late 1940s, he worked with Marietta Blau on cosmic-ray stars recorded in nuclear emulsions. He rode the tide of new experimental techniques all the way to the Tevatron Collider at Fermilab. Turning to theoretical analysis, he continued to publish in *Physical Review* beyond his ninetieth birthday. We never detected the slightest hint of reticence. Indeed, he was an ebullient character who sometimes veered toward bull-in-a-china-shop mode. Nevertheless, on that occasion in 1956, despite Feynman's reassurance that his question was not utterly misguided, Block demurred. "No," he said, "they won't listen to me. You ask."

The next day, following C. N. Yang's introductory lecture on "Theoretical Interpretation of New Particles," Feynman did ask, on Block's behalf. The scientific secretary reported, "Could it be that the theta and tau are different parity states of the same particle which has no definite parity, i.e., that parity is not conserved. That is, does nature have a way of defining right- or left-handedness uniquely?" Yang stated that he and Tsung-Dao Lee had looked into the matter without arriving at any definite conclusions. One participant opined that "the most attractive way out is the nonsensical idea that perhaps a particle is emitted which has no mass, charge, and energy-momentum and only carries away some strange space-time transformation properties"—code for parity. The transcript does not reveal whether this remedy, even more desperate than Pauli's neutrino, was offered in earnest or in jest. The session ended inconclusively; none of the experts cited a terrible fate that would accompany parity violation. Perhaps others had already dared to challenge parity conservation in their private thoughts; Martin Block's little heresy brought the question into the open.

T. D. Lee of Columbia University and C. N. Yang of the Institute for Advanced Study took on the issue of parity conservation in the fundamental interactions with new urgency. Both were born in China in the 1920s and had immigrated to Chicago at the end of World War II to study in Enrico Fermi's circle at the University of Chicago. Yang earned his PhD in 1948, advised by Edward Teller; Lee received his degree in 1950, guided by Fermi himself. A few weeks after the Rochester Conference, Lee and Yang began to realize that, while parity conservation was securely established for the strong and electromagnetic interactions, no experiment had ever put parity violation on trial in weak decays. As they set about devising decisive tests, Lee sought the counsel of his Columbia colleague Chien-Shiung Wu, whose meticulous postwar measurements of the energy spectrum of electrons emitted in beta decay helped establish Fermi's description of weak interactions. This body of work cemented her reputation as the leading experimental authority in beta decay and weak interaction physics.

Wu had arrived in the United States from China in 1936 and enrolled in graduate school at Berkeley. Her advisor-of-record was Ernest Lawrence, but her 1940 dissertation on fission products of uranium was supervised by Emilio Segrè. In 1952, she became the first woman to hold a tenured faculty position in Columbia's physics department. A correlation between the direction of the electron emitted in beta decay and the spin of the radioactive nucleus would establish parity violation, because the momentum is inverted under parity, but spin—a form of angular momentum—is not. Wu advised that the beta decay of cobalt-60 (twenty-seven protons and thirty-three neutrons) into nickel-60 (twenty-eight protons and thirty-two neutrons) would be a perfect candidate for this study. The spin of the cobalt nucleus could be aligned at low temperatures in a strong magnetic field, providing a reference direction for the emitted electrons.

On June 22, Lee and Yang submitted "Is Parity Conserved in Weak Interactions?" to the *Physical Review*. Many physicists evince a strong aversion to questions as titles; they assert that the answer is always, "No," which for this paper would have been a revolutionary verdict on parity conservation. Our colleagues' negative bias exists despite the evidence of Einstein's mass-energy connection that we cited in the preceding chapter, for which the answer was a resounding "Yes!" In this case, it was Sam Goudsmit, editor-in-chief of the American Physical Society, who objected to the question mark. Goudsmit, one of the discoverers of electron spin, was a legendary defender of clarity in thought and expression, as well as a zealous skeptic of neologisms and cuteness. The seminal paper, with the acceptable title, "Question of Parity Conservation in Weak Interactions," appeared in the October 1, 1956, issue.

Lee and Yang argued that the tau-theta puzzle could not by itself be taken as evidence that parity conservation is violated in the weak interactions because the strange particles were still too poorly understood to serve as reference points. Echoing Feynman's formulation at Rochester, they declared that to "decide unequivocally whether parity is conserved in weak interactions, one must perform an experiment to determine whether weak interactions differentiate the right from the left." The first experiment they listed—the beta decay of the cobalt-60 nucleus—came from the Lee-Wu discussions. Another promising approach concerned the decay of a charged pion into a muon and neutrino, followed by the decay of the muon into an electron and two neutrinos. If parity were violated in the first step, the muon would emerge with its spin aligned along its direction of motion. A correlation of the muon spin direction with the electron direction would again be a signature for parity violation.

Over the summer of 1956, Chien-Shiung Wu designed her experiment. She knew that cobalt-60 nuclei could be polarized—their spins aligned in a particular direction—if they were embedded in certain crystals and cooled nearly to absolute zero under the influence of a magnetic field. She enlisted the collaboration of specialists in the low-temperature alignment of radioactive nuclei at the National Bureau of Standards led by Eric Ambler.

During the autumn and winter, Wu shuttled between her laboratory in Columbia's Pupin Hall and the NBS campus in the upper northwest section of Washington, DC. By late December, the team had overcome many challenges concerning the cobalt-laden crystals and the measurement apparatus. When they started to see a strong indication of parity violation—the beta rays were preferentially emitted opposite to the direction of the nuclear spin—rumblings of their progress began to circulate at Columbia.

A storied Columbia physics tradition was the Friday lunch at the Shanghai Café, animated by T. D. Lee. On January 4, 1957, having curated the menu of the day with his usual showmanship and care, T. D. announced that Wu had telephoned to say that her preliminary data showed a huge effect. This got everyone's attention, especially Leon Lederman's. Leon had considered the pion-muon-electron sequence during the fall and concluded that the effect should be too small to detect. But if Wu was seeing a big effect in cobalt-60 decay . . .

Driving home for dinner before joining his graduate student, Marcel Weinrich, for a shift at the Nevis Cyclotron in Irvington-on-Hudson, Lederman had a small epiphany: the apparatus that Weinrich was using to study muons could be modified—or cannibalized, depending on one's viewpoint—to search for a large parity-violating effect. He recruited Richard Garwin, at the age of twenty-eight already a known wizard at the IBM Watson Scientific Research Laboratories, to complete the team. Working through the weekend and overcoming their own litany of experimental challenges, they had a definitive result by Tuesday morning. Garwin called Lederman, who had gone home for a little sleep, to report a twenty-two-standard-deviation signal for parity violation, and Lederman passed the word on to Lee.

A standard deviation is a measure of statistical improbability—a chance fluctuation that misleads. For what is called a normal statistical distribution, there is a 32 percent chance that a result lies more than one standard deviation from the true value, which is no big deal. Only one fluctuation in a thousand will be off the mark by 3.3 standard deviations, only one in a million by 4.9 standard deviations. If you find a result that misses the true answer by 22 standard deviations, either you did something wrong in executing the experiment or (are you listening, financial analysts?) you blundered by assuming a normal distribution of errors—a bell curve.

At 2:00 A.M. on Wednesday, the cobalt-60 collaborators had confirmed and refined all measurements and calibrations to their satisfaction; they celebrated with a 1949 bottle of Château Lafite Rothschild quaffed from paper cups. Wu reported that in the morning, other NBS scientists, seeing the Bordeaux-stained detritus in the wastebasket, knew without being told that parity conservation was no more.

Columbia organized a press conference for January 15, highlighting the contributions of Lee & Yang, Wu and her Bureau of Standards collaborators Ambler, Raymond Hayward, Dale Hoppes, and Ralph Hudson, and the Nevis group of Garwin, Lederman, and Weinrich. A two-column story atop the next morning's *New York Times* front page proclaimed, BASIC CONCEPT IN PHYSICS REPORTED UPSET IN TESTS. The story continued inside the paper, accompanied by a sidebar on "The Meaning of Parity" and the full text of the Columbia press release. Humans now had a means, readers were informed, of communicating to intelligent beings on another world what we understand by left-handed and right-handed. Both groups submitted letters to the editor of *Physical Review* at the time of the press conference; they appeared back-to-back, cobalt-60 followed by pion-muon-electron, in the February 15 issue. A manuscript from Jerome Friedman and Valentine Telegdi of the University of Chicago reporting their nuclear-emulsion evidence for parity violation in the pion-muon-electron decay chain was received by *Physical Review* editors on January 17, but "for technical reasons" could not be published alongside the other two and appeared in a later issue on March 1.

So far as we know, extraterrestrials haven't weighed in on the discovery, but the physics community was electrified. Wolfgang Pauli had scoffed at the notion of parity violation, but he accepted nature's verdict with a mourner's dignity. He addressed a black-edged card to Victor Weisskopf, bearing the inscription (we translate from the German):

> It is our sad duty to announce that our loyal friend of many years PARITY went peacefully to her eternal rest on the 19th of January 1957, after a short period of suffering in the face of further experimental interventions.
>
> For those who survive her, e μ ν [electron, muon, neutrino]

In common with most living physicists, we learned about parity violation in school, so it requires some conscious effort to recapture the surprise and wonder that attended the discovery. As Chris wrote in his textbook on gauge theories, "Perhaps our familiarity with parity violation in the weak interactions has dulled our senses a bit. It seems to me that nature's broken mirror . . . qualifies as one of the great mysteries. Even if we will not get to the bottom of this mystery next week or next year, it should be prominent in our consciousness—and among the goals we present to others as the aspirations of our science."

There was more to be said: a careful analysis by Lee, Yang, and Reinhard Oehme, carried out before experiments demonstrated that parity is not a symmetry of the weak interaction, showed that the pion-muon-electron effect would only be observable if both parity and charge conjugation—a kind of particle-antiparticle transformation—invariances were violated. The results of Garwin and company meant that *two* discrete symmetries did not hold for the weak interactions. The combined operation of charge conjugation (C) and parity (P), known as CP invariance, did survive and, with it, time-reversal invariance (T), since the only consistent descriptions of nature—then and now—are invariant under the combined operations of C and P and T. This analysis has important consequences for the decays of neutral K mesons and their antiparticles, which we will encounter in chapter 18. An anti-cobalt-60 experiment may never be practical, but experiments that compared the decays of negatively charged muons and their positively charged antimatter counterparts confirmed maximal violations of both parity and charge conjugation.

Early in 1957, physicists at the Kamerlingh Onnes low-temperature laboratory in Leiden (Netherlands) reported that in the decay of cobalt-58 (twenty-seven protons and thirty-one neutrons) into iron-58 (twenty-six protons and thirty-two neutrons) plus a positron and neutrino, the emitted positron tended to follow the spin direction of the parent nucleus—just opposite to the behavior of the electron emitted in cobalt-60 decay. The particle and antiparticle behaved differently!

All of these implications follow if the beta-decay interaction is left-handed: an electron emerges in beta decay spinning counterclockwise. In general, spin-one-half particles such as the electron and proton are combinations of left-handed and right-handed pieces. The weak interaction responsible for beta decay acts only on the left-handed components of particles, and on the right-handed components of antiparticles. An ingenious experiment carried out at Brookhaven National Laboratory by Maurice Goldhaber, Lee Grodzins, and Andrew Sunyar determined the handedness of the feebly interacting neutrino without detecting it. They monitored the decay of europium-152, one of the rarest of the rare-earth elements (sixty-three protons and eighty-nine neutrons), into samarium-152 (sixty-two protons and ninety neutrons). This is a variant of beta decay in which an electron of the europium atom is captured by the nucleus. The nucleus responds by emitting a neutrino and turning into an excited state of a samarium nucleus that radiates a gamma ray as it relaxes to the ground state. By measuring the sense of the photon's circular polarization, the team could infer the spin direction of the neutrino. The emerging neutrinos are purely left-handed!

Handedness is a familiar concept for macroscopic objects. The pairs of hands on a human body are mirror images of one another; they are not identical, and one cannot be substituted for another. No manipulation can cause the two to coincide.

The 1848 discovery of crystal handedness was the first great achievement of twenty-five-year-old Louis Pasteur, shortly after his graduation from École Normale Supérieure.

Scientists at the time were fascinated by the interplay between light and matter, and specifically by phenomena involving polarized light. Light emitted by the sun, a candle, and many artificial sources is unpolarized, its electric field pointing equally in all directions perpendicular to the direction of motion. Most of the light reflected from a horizontal surface such as water or pavement only vibrates in one plane. Polaroid filters used in sunglasses only allow light vibrating in a single plane to pass, so a carefully chosen orientation suppresses reflections. Many computer displays are coated with polarizing filters to cut down glare. Rotating Polaroid sunglasses while viewing such a display through them, you will find an orientation in which the screen is nearly black.

In the first decades of the 19th century, the French physicist Jean-Baptiste Biot studied the interactions between polarized light and various biological substances, including turpentine vapor and solutions of tartar, crystalline deposits that accumulate at the bottom of wine vats. He established that the plane of polarization—the plane in which light vibrates—was slightly rotated as light passed through the columns of liquid or vapor. In some way, the substances were "optically active"; they distinguished left and right. Biot attributed the activity to properties of the molecules that composed his samples—long before atoms (and therefore molecules) were accepted as real. Biot's observation grew into a puzzle when the German chemist and crystallographer Eilhard Mitscherlich reported that solutions derived from tartars extracted from wine vats were indeed optically active, but chemically identical solutions derived from industrial production did not rotate the polarization of light.

Once he read this report, the young Louis Pasteur could not let the question go: Why should two chemical substances, apparently identical, affect polarized light so differently?

When he evaporated Mitscherlich's optically inactive liquid Pasteur noticed that it left behind crystals that were not perfectly symmetrical but "dissymmetrical," as were crystals derived from Biot's optically active liquid. On close examination, he found that crystals from the active liquid were all of one kind, which he called left-handed. The optically inactive liquid yielded two distinct shapes of crystals, left-handed and right-handed mirror images of one another, which he separated with tweezers. When dissolved, the left-handed crystals behaved as they had for Biot, but the right-handed crystals rotated the plane of polarization in the opposite sense. When he mixed the two kinds together, to emulate the industrial preparation studied by Mitscherlich, the two influences canceled out! Not only the chemical composition of substances but also their shape as displayed in crystals determines their effects. Pasteur's insight led to the new science of stereo-chemistry, or spatial chemistry, and to the ball-and-stick models that depict bond angles and interatomic distances for molecules from water and carbon dioxide to Buckyballs and DNA.

Pasteur abstracted from his explorations what is sometimes called the Law of Life Dissymmetry: "Only products that originate under the influence of life are dissymmetrical, because their development is governed by cosmic forces that are themselves dissymmetrical." Indeed, it is true that many organic products—sugars, for example—display a pronounced handedness that reflects the dissymmetry of the biological factories that make them. For Pasteur, dissymmetry marked the frontier between the organic and mineral worlds. A fascinating open question is whether the right-handed predisposition of organisms resulted purely by chance during the course of evolution or whether the odds were tipped by the character of the fundamental forces of nature.

What Pasteur called dissymmetry we today call chirality, a coinage due to Lord Kelvin, president of the Royal Society, in an 1893 lecture on molecules and crystals to the Oxford University Junior Scientific Club. "I call any geometrical figure, or group of points, *chiral*, and say that it has chirality, if its image in a plane mirror, ideally realized, cannot be brought to coincide with itself." [The term is drawn from the Greek for hand, χειρ (*kheir*).] The chirality—or handedness—of an electron or neutrino has nothing to do with shape. It is an intrinsic property of these infinitesimal particles; like spin, it resides in the spacetime symmetries that appear to govern our universe.

The insight that the beta-decay interaction is left-handed was given mathematical expression by Richard Feynman and Murray Gell-Mann, and by Robert Marshak and George Sudarshan. They proposed a specific form for the weak force called "vector minus axial vector" (colloquially V minus A), which was quickly validated by a burst of experimental activity. Once the muon was shown to be partnered with its own muon neutrino, distinct from the electron neutrino, a universal structure fell into place: the weak interaction could turn an electron neutrino into an electron (and vice versa), a muon neutrino into a muon (and vice versa), or a proton into a neutron (and vice versa), all with essentially the same strength.

Could the preference of weak interactions for left-handedness be the reason that all the amino acids in living organisms come with just a single handedness as well? A tantalizing idea, but quite improbable since the weak interactions are indeed feeble compared to molecular forces.

The essential structure of the Universal Fermi Interaction, as it came to be known, has survived until today as an "effective theory," reliable over a limited range of energies. Over time, quarks supplanted the proton and neutron, and a new lepton family joined the electron and muon. One sign of the limitation is that quantum corrections to the Fermi interaction—Act II and beyond in the language we used for the other interactions—are out of control. Instead of giving refinements to the one-act calculation, the quantum corrections simply explode in magnitude. The looming problem is illustrated by the growth with energy of neutrino scattering noted by Bethe and Peierls. The probability for a muon neutrino to strike an electron, producing an electron neutrino and a muon is quite modest for the neutrino energies produced in our accelerator laboratories, but as the neutrino energy approaches a billion-billion electron volts, we calculate a probability greater than 100 percent, which is crazy talk.

Something must be done to keep the interaction probability within the bounds of reason and, with it, to tame the uncontrolled behavior of quantum corrections. To see how the symmetries-dictate-interactions strategy can save the day, it will profit us to explore the realm of the ultracold. There we will encounter some unexpected properties of matter and find inspiration for a gauge theory with massive force particles.

11

DOCTOR DEEPFREEZE

Ninety degrees north latitude! Ninety degrees south! The unconquered poles drew them like magnets, the daring men propelled by their own muscles, with skis, and sleds, and dogs. The trackless kingdoms of ice and snow lured adventurers racing to stand at the mathematical points that define the very ends of the Earth. But when Robert Peary reached the North Pole in April 1909 and Roald Amundsen attained the antipode in December 1911, they were far from the coldest spot on Earth. In the Dutch university town of Leiden, another explorer had created a more frigid wonderland inside a maze of glass tubes and vessels. In Heike Kamerlingh Onnes's realm, the temperature was 200° Celsius colder than any polar explorer had ever encountered.

While the men of action pursued their geographical quest, physicists had been engaged in a competition of their own to reach the nadir of temperature. A low-temperature physicist works in a climate so severe that liquid thermometers freeze. Kamerlingh Onnes and his rivals relied on gas thermometers that not only measured their progress, but defined their goal. Held under constant pressure, a gas expands when heated and contracts when cooled. Measure the volume occupied by a certain quantity of a gas at the temperature of boiling water, 100° Celsius, and measure it again where water turns to ice, at 0°. Then it is easy to predict what volume of gas would correspond to any other temperature. Every gas would shrink to zero volume at the same temperature, where all molecular motion not required by quantum uncertainty ceases: minus 273.15° Celsius. This minimum of temperature, the absolute zero, beckoned to physicists as the poles called to the adventurers.

Real gases depart from this ideal behavior at low temperatures: they become liquids. One by one, the "permanent gases" yielded to the physicists' growing ability to produce cold. At ordinary atmospheric pressure, nitric oxide liquefied at 121°C above absolute zero, or 121 kelvins, 121 K for short. (On the absolute scale of temperature, water freezes at 273.15 K and boils at 373.15 K.) Then came "marsh gas," the methane in natural gas, at 113 K, oxygen at 90 K, carbon monoxide at 83 K, and nitrogen at 77 K. Each success presented a new liquid to be studied and eventually frozen, a new cryogen that could be used to cool samples and could serve—like the supply depots and base camps of the polar explorers—as a jumping-off point for the next step toward absolute zero. But hydrogen resisted.

One of the masters of the cold was James Dewar, the irascible Scottish inventor of the double-walled vacuum flask for storing and handling liquefied gases or conserving hot liquids—the thermos bottle. Dewar's liquefaction plant at the Royal Institution in Albemarle Street, London, seemed more like an engine room than a chemistry laboratory, filled with heavy equipment, gas engines, and compressors. To cool hydrogen below the temperature of his refrigerants, Dewar compressed the gas under a pressure of 200 atmospheres and cooled it with liquid air. When the chilled, compressed hydrogen was sprayed through a fine nozzle into a vacuum, its temperature fell. This trick, discovered in 1852 by James Prescott Joule and William Thomson, is used today to liquefy many gases on an industrial scale. It is also the basis for everyday refrigerators and air conditioners. In 1898, after more than a decade of effort—but long before the polar explorers would reach their goal—Dewar liquefied hydrogen at 20 K. He produced solid hydrogen at 14 K and was able to cool it to 13 K.

Dewar had won an important victory that opened a vast new territory. He learned that chemical reactions stop at low temperatures and that electric current ceased to flow from a battery plunged into his deep freeze. His measurements of the electrical resistance of metals and alloys at low temperatures suggested that pure metals might become perfect conductors at absolute zero. But absolute zero remained a distant goal.

As he wrote in an article on liquid gases for the tenth edition of the *Encyclopedia Britannica* in 1902, "Though ultima Thule may continue to mock the physicist's efforts, he will find ample scope for his energies in the investigation of the properties of matter at the temperatures placed at his command by liquid air and liquid and solid hydrogen. . . . [It] enormously widens the horizon of physical science, enabling the natural philosopher to study the properties and behaviour of matter under entirely novel conditions."

While Dewar and his competitors were laboring to liquefy hydrogen, a new gas was found on Earth, one that would bring absolute zero closer still. This marvelous substance was first noticed in 1868 in the geysers of hot, glowing gas that erupt from the sun's atmosphere. While daylight passing through a prism fans out from red to violet like a rainbow, the light emitted by a glowing gas is composed of specific colors that the prism separates into sharp lines. Each gas has a characteristic signature—a pattern of colored lines like a musical chord—that sets it apart from every other. To learn what the sun is made of, Pierre Jules César Janssen and Norman Lockyer observed solar flares using spectrometers, instrumented prisms that measure the wavelength of the light in each colored band. In addition to the patterns of familiar elements, they saw the mark of a new substance: it was a bright yellow line near the double stroke in the signature of sodium. For more than twenty-five years, helium, as the new solar gas was aptly named, was known only as a line in the sun's spectrum, until a pair of serendipitous discoveries brought it down to Earth. In Dmitri Mendeleev's 1889 Faraday Lecture, delivered before the Fellows of the Chemical Society in the Theatre of the Royal Institution, the inventor of the periodic table still referred to helium as "the imaginary substance."

Three years later, John William Strutt, Lord Rayleigh, would write, "I am much puzzled by some recent results as to the density of nitrogen . . ." What perplexed Rayleigh was that he measured different weights for equal volumes of nitrogen from different sources. Atmospheric nitrogen prepared by removing the oxygen, carbon dioxide, and moisture from air was a half percent heavier than a sample prepared by breaking down nitrogen-bearing compounds. The simplest way to account for the unexpected difference was to hypothesize that another ingredient remained in the air after the known constituents, other than nitrogen, had been removed. "But," said Rayleigh, "in accepting this explanation, even provisionally, we had to face the improbability that a gas surrounding us on all sides, and present in enormous quantities, could have remained so long unsuspected."

Rayleigh learned from James Dewar that the same possibility had occurred to Henry Cavendish more than a century before. Cavendish added doses of oxygen to nitrogen extracted from the air and passed sparks through the mixture to form nitrous oxide. After all the nitrogen had been consumed, a small bubble of gas—no more than one one-hundred-twentieth of the original volume of nitrogen—remained unabsorbed, an unfulfilled hint of a new constituent of the atmosphere.

Repeating Cavendish's experiment with superior Victorian technology, Rayleigh and William Ramsay isolated an unknown gas, heavier than nitrogen, that they estimated made up about 1 percent of the atmosphere. A series of elegant experiments established the properties of the new element, which Rayleigh and Ramsay called argon, after the Greek word for *idle*, because they could not make it react chemically.

Ramsay pursued the newly discovered argon everywhere—in animals, vegetables, and minerals. In March 1895 he wrote, "I am at argon again, in mice, and in a uranium mineral from Norway. I don't think that any of these sources of nitrogen contain argon." The animal and vegetable parts of the question were answered quickly: "We have settled the question of argon in the animal economy; there is absolutely no trace of it at all in peas or in mice." A mineralogist had suggested to Ramsay that he investigate a nonreactive gas released when uranium ores were boiled in acid. Ramsay obtained a small sample of Norwegian cleveite from a London mineral dealer in March 1895 for three shillings sixpence. The electric-arc spectrum of the gas he extracted displayed a brilliant golden line and several others that do not belong to the argon spectrum. Ramsay gave the unfamiliar substance the provisional name crypton and dispatched a sample for identification to William Crookes, a leading spectroscopist and a pioneer of gas-filled illuminating tubes. Crookes wired back the next morning, "Crypton is helium, come and see it." The stuff of the sun also existed on Earth.

Ramsay, Rayleigh, and others began to scour the Earth for helium. It turned up in other radioactive ores, in gases bubbling out of mineral springs, and even in the atmosphere. The hunt led Ramsay to the discovery of neon, krypton, and xenon, the chemical relatives of helium and argon. By 1896 a little helium was available in pure form for the first cryogenic studies, which showed that it remains a gas at the temperature of liquid hydrogen. Liquefying helium became the next stage in the race to approach absolute zero.

Heike Kamerlingh Onnes was superbly prepared for the competition because he had been building since 1882 the most advanced low-temperature laboratory in the world. Unlike Dewar, he understood the value of a skilled team at a time when important discoveries were still made by gentleman amateurs working in private laboratories on their estates. In 1901 he started a school of instrument makers and glassblowers to further his work. Beautifully blown glass vessels with silvered walls, nestled one inside another, reflected the standards of precision and organization of the laboratory. The Leiden Cryogenic Laboratory did not liquefy hydrogen until 1906, eight years after Dewar's triumph, but then produced it in great quantity, fit for cooling large samples of helium. Refrigerators begat further refrigerators, in a cascade that plunged to 10 K.

The technical factotum of the Leiden Laboratory was Meester Gerrit Jan Flim, a superb craftsman and gifted designer, whose title (Master) echoes the medieval system of guilds. Flim was a man of plain speech who could stand up to anyone, even his boss. A proposal he deemed ill-conceived would elicit

"I'd say, madness," and not be heard of again. The paternalistic Kamerlingh Onnes believed in a strictly ordered society. Hearing that Flim was planning to send his son to prepare for the university, Kamerlingh Onnes instructed his mechanic that this was not the lad's station in life: he should go to a good technical school. Flim simply replied, "I'd say, my boy, my money, my decision." The lad studied medicine and became a well-regarded general practitioner in Leiden.

For forty-four years, until his death in February 1926, Kamerlingh Onnes ruled as a benevolent despot, inspiring his very able staff. Pieter Zeeman, who won a share of the 1902 Nobel Prize for discovering the effect of magnetic fields on atomic spectra while an assistant in the Leiden Laboratory, wrote that Kamerlingh Onnes "ruled over the minds of his assistants as the wind urges on the clouds. He could achieve miracles with a flattering remark or witty (sometimes biting) irony." At the Great Man's funeral, the faithful technicians—in black coats and top hats—were obliged to follow the cortege on foot to a village churchyard outside Leiden. The technicians struggled to keep up with the horse-drawn hearse. Reaching the graveyard, Meester Flim cracked, "Just like the Old Man, even when he's dead he keeps us running."

To Kamerlingh Onnes, "An investigator should be almost a monomaniac, possessed of one idea only: to pursue and try to achieve but one purpose, and nothing else." The methodical approach that he took in all matters required that the task of liquefying helium be thoroughly scouted from all sides. To judge the length of the journey before him, he conducted reconnaissance experiments. By measuring helium's behavior at the lowest temperatures he could attain, he made maps that enabled him to guess the location of his goal: It appeared that helium would become liquid around 6 K.

Both Dewar and Kamerlingh Onnes figured that one hundred or two hundred liters of pure helium gas would give them a chance of success. Only a fraction of the helium would be liquefied by the sudden expansion of the chilled, compressed gas, and the liquid would occupy only one five-hundredth of the volume of the gas. Where could such quantities of gas be found? Today, you can order helium from Amazon for such simple pleasures as inflating party balloons. But in 1906, the physicist who wanted to liquefy helium had to find it and purify it for himself.

Modern calculations of stellar structure, supported by observations of helium lines in the spectra of hot stars and interstellar material, tell us that helium makes up 23 percent by mass of the visible universe. It is far less common on Earth, because helium atoms are so lightweight and so chemically inert that most of the primordial complement escaped our planet long ago. What helium we find on Earth, a few parts per billion by mass of the Earth's crust and five parts per million by volume of dry air, has been generated, atom by atom, by the escape of alpha particles—helium nuclei—from radioactive substances.

Today, some two hundred fifty billion liters of helium are available for extraction from natural gas wells each year. The helium that accompanies natural gas was found not by a scientific expedition, but by drillers prospecting for oil in Dexter, Kansas, in May 1903. At a depth of about 400 feet, the oilmen struck "a howling gasser" whose roar could be heard throughout the town. A ceremony held to light the well and celebrate Dexter's good fortune brought visitors on excursion trains from miles around. At the climax of the proceedings, a flaming bale of hay was suspended over the well—and instantly snuffed out, extinguishing a small town's dreams of untold wealth. The state geologist, Erasmus Haworth, brought a sample of the gas that would not burn to the University of Kansas, where his junior colleague, David F. McFarland determined a most anomalous composition. Typical natural gas at the well is approximately 95 percent combustible methane, but the Dexter gas contained but 15 percent methane, 72 percent nitrogen, and 12 percent "inert residue."

The high ratio of nitrogen to methane explained the relative incombustibility of the Dexter gas, but what of the inert residue? In December 1905, McFarland and Hamilton P. Cady announced that the residue contained helium—nearly 2 percent of the total—and congratulated Kansas on "the possession of an unlimited and easily available supply of what has been considered a very rare element." During the next year, Cady and McFarland found helium in forty-four well gases, but always in smaller amounts than at

Dexter. Subsequently, extensive helium-bearing gas deposits with as much as 7 percent helium were identified around Amarillo, Texas, but no effort was made to develop these deposits until the Great War raised the government's interest in helium as a safe lifting gas for dirigibles.

Dexter's windy well contained more helium by far than Dewar and Kamerlingh Onnes dreamed of, but it lay beyond their reach. Dewar had to rely on a far more modest source: bubbles in the King's Well at Bath, which contained a tiny proportion of helium, mixed with neon. At one point, assuming that his main competitor worked at the same ambitious scale as he did, Kamerlingh Onnes appealed to Dewar to supply him with helium from Bath. After years of work, Dewar had lost all his treasured store of helium through a technician's error and was struggling to resupply himself. He declined politely, but Kamerlingh Onnes had other resources.

"Fortunately," he reported, "the Office of Commercial Intelligence at Amsterdam under the direction of my brother, Mr. O[nno] Kamerlingh Onnes, succeeded in finding in the monazite sand the most suitable commercial article as material for the preparation, and in affording me an opportunity to procure large quantities on favorable terms." The monazite that the Office of Commercial Intelligence found in North Carolina is rich in thorium, the radioactive mother of radium. In its interstices, the sand holds one atom for every radioactive disintegration since the rock was formed—a milliliter (one cubic centimeter) of gas for each gram of sand. To liberate the gas, Kamerlingh Onnes's men pulverized hundreds of pounds of sand and heated the fine powder above 1000 °C. The raw gas passed through a series of chemical treatments before they cooled it to the temperature of liquid hydrogen to freeze out oxygen and other contaminants, then filtered it time and again through charcoal at the temperature of liquid air. Through perseverance and care, the Leiden team amassed 200 liters of pure helium for the experiments, with another 160 liters held in reserve—some tens of grams, all told.

Amply supplied with pure helium, Kamerlingh Onnes was ready for his assault on the recalcitrant gas. He knew, he later wrote, that the execution of the experiment bordered on the impossible. The pumps and compressors would have to work for hours without fail. Any impurity that leaked into the system might freeze and plug the tubes and valves, as Dewar once had been undone by frozen neon. Even the Leiden Cryogenic Laboratory, that early exemplar of big science, did not have enough experienced operators to look after the helium circulation and run the compressors to produce cryogens at the same time, so they had to generate vast stocks of liquid air and liquid hydrogen in advance. The production of liquid hydrogen began at half past five on July 10, 1908, and took the whole morning.

Further preparations occupied most of the afternoon. Finally, at 4:30 P.M., the helium experiment began.

Precooling gave way to cycles of compression, cooling, expansion, and recirculation. The tubes of the helium liquefier remained transparent, witness to the extraordinary purity of the Leiden helium, but for hours there was no sign of liquid helium in the collection vessel. The last bottle of the store of liquid hydrogen was connected to the liquefier, but still no liquid helium was seen. Suddenly, around seven in the evening, apprehension turned to joy when a visitor to the laboratory suggested illuminating the collection flask from below and the surface of the liquid helium showed itself. "It was a wonderful sight," Kamerlingh Onnes wrote in his Nobel lecture, "when the liquid, which looked almost unreal, was seen for the first time." The liquid helium was so transparent that no one had noticed it flowing into the vessel. Not until the container was almost full did the surface stand out sharply, like the edge of a knife.

Within hours, the *New York Times* reported, "Professor Dewar triumphantly read the communication" from Kamerlingh Onnes describing the liquefaction of helium to the British Association meeting in Dublin.

From his outpost at 4.2 K, the boiling point of helium, Kamerlingh Onnes was poised to survey the virgin territory. His chief competitor was vanquished and no one else would liquefy helium until 1923. On that first night, when the pressure in the collection chamber was reduced, sixty milliliters (one-quarter cup) of the precious liquid cooled by evaporation to 1.7 K. "Doctor Diepvries" (Dr. Deepfreeze), as Dutch cartoonists called their new celebrity, had made a new world, and he was its monarch.

Intense as the competition was to liquefy helium, low temperature was not the only important goal. Like the harpsichord maker who fashions a beautiful plane to help him shape the perfect soundboard, Kamerlingh Onnes had liquefied helium to create a wonderful tool. His motto, "*Door meten tot weten*," (Through measurement to knowledge), kept watch over the laboratory's door. Measurements he would make—with great care and revolutionary results.

To make possible thorough measurements at 4.2 K, liquid helium must be insulated extraordinarily well from the outside world of 300 K. Kamerlingh Onnes devoted more than two years to developing his helium cryostat, a vacuum bottle suspended inside two of Dewar's thermoses on which vertical strips had been left unsilvered to serve as viewing windows. The inner flask was filled with liquid hydrogen, the outer one with liquid air, both easily replenished in the Leiden laboratory. In 1911, the wonderful tool was ready.

Measurements of the densities and thermal properties of helium liquid and vapor and measurements of the electrical resistance of metals at low temperatures could finally begin.

As Stephen Gray had been first to discover, materials vary enormously in their capacity to transmit electricity. Those that block the passage of an electric current are termed insulators. Those with little resistance, like gold, copper, and silver are called conductors.

When an electric current passes through a typical conductor, heat is generated: the greater the resistance, the greater the heat. Resistance is a good thing, in its place. Without resistance to convert electrical energy to heat and light, there would be no electric blankets, toaster ovens, or incandescent light bulbs. But in power lines, the windings of motors, or the coils of strong electromagnets, resistance is an embezzler, diverting our energy resources into waste heat.

James Dewar's measurements had shown that pure metals become better conductors, that is, show less resistance, when they are cooled. There was no satisfactory theory of metals in 1911, but Kamerlingh Onnes believed that resistance would continue to decrease gradually until it vanished at some temperature a bit above absolute zero, provided the samples were pure enough. He hoped that what happens at very low temperatures, when the thermal agitation of atoms and molecules is stilled, might provide important clues about the structure of atoms and the way they cooperate to form metals and alloys. The Leiden studies of platinum and gold below 15 K had shown their resistance leveling off. Kamerlingh Onnes blamed the persistent resistance on impurities. To eliminate contamination, he moved on to quicksilver—mercury—a liquid at room temperature that boils at the temperature of a pizza oven and is readily purified by repeated distillation. Polar explorers know that mercury thermometers freeze on a three-dog night, when the temperature drops as far below the freezing point of water as human body temperature lies above it. In place of the long, fine wires of gold and platinum they had used before, the Leiden team made chains of U-shaped capillary tubes filled with mercury that froze into gossamer threads when chilled below 234 K—about 38° below zero on either the Fahrenheit or Celsius scale.

The crucial experiments began on April 8, 1911. While Kamerlingh Onnes and Flim looked after the apparatus and Cornelius Dorsman tracked the temperature, Gilles Holst repaired to a quiet room fifty meters distant from the shudders and rumbles of the cryogenic plant. There he configured a galvanometer—a sensitive voltmeter—to measure the resistance in the samples. Reference values for gold and mercury at 4.3 K were unexceptional. As the temperature reached 3 K, Holst—who would later direct one of the great industrial research laboratories, the Philips Physical Laboratories—reported no sign of resistance in mercury: "Kwik nagenoeg nul," (Mercury essentially zero) reads the lab notebook.

With only two data points, the loss of resistance seemed at first to be what Kamerlingh Onnes had anticipated. Subsequent experiments with refined apparatus showed that the resistance did not glide gently to zero as the temperature fell. On October 26, the Leiden team established that in the space of a few hundredths of a degree the resistance plummets to less than one one-hundred-thousandth of its room-temperature value and becomes unmeasurably small. The electrical nature of the mercury had changed in some fundamental way that neither Kamerlingh Onnes nor anyone else had foreseen or could explain. On one side of the border, mercury was a normal metal; on the other, it became a miraculous "supraconductive" substance in which electricity flowed without resistance, as different from ordinary mercury as water is from ice. The discovery was all the more striking because it occurred just a few hundredths of a degree below the temperature at which helium liquefies. The new technology of liquid helium—which itself rested on the theoretical understanding of gases and on experimental studies of helium's properties—was making new science possible.

"Dr. Deepfreeze" pursued the new phenomenon of superconductivity with a single-mindedness reminiscent of Stephen Gray. He began to test every plausible material. Pure platinum and gold retained their resistance, contrary to his prejudices. Mercury contaminated with gold or cadmium became superconducting. Even the mercury amalgam used to silver mirrors suddenly lost all resistance at the temperature of liquid helium, so purity had been a false lead and the time spent distilling mercury could have been saved. Tin became superconducting at 3.7 K, indium at 3.4 K, a tin-mercury foil at 4.2 K, thallium at 2.36 K, and lead at 7.2 K. All this was happening just below the temperatures accessible before the liquefaction of helium. Dewar—the pioneer of low temperatures and resistance measurements—was frozen out of the new science.

Kamerlingh Onnes found that very fine superconducting wires could carry impressively large currents, but their superconductivity vanished when a certain critical current was surpassed. One fine lead wire cooled by liquid helium remained superconducting up to a current of six amperes and melted when the current exceeded thirteen. At a lower temperature, superconductivity is more robust: a wire can carry larger currents before resistance returns.

The discovery of superconductivity in tin and lead, which can be formed easily into any desired shape at room temperature, made possible all manner of new experiments and raised hopes for immediate practical applications. In his 1913 Nobel lecture, Kamerlingh Onnes asked whether the absence of resistive heating "makes feasible the production of strong magnetic fields using coils without iron, for a current of very great density can be sent through very fine, closely wound wire spirals." But when he tried to make a strong electromagnet by passing a large current through a coil of superconducting wire, he discovered another important limit to the magic. A coil of a thousand turns of lead wire wound onto a small bobbin ceased to be superconducting when the current reached only a tenth of what a short sample of the wire had carried without resistance. The coil's own magnetic field reduced the critical current in much the same way that warming the wire did. Subsequent experiments showed that the influence of a small permanent magnet is enough to kill lead's superconductivity entirely. High magnetic fields would have to wait.

Kamerlingh Onnes designed an ingenious experiment to verify that once started, an electric current should continue running around a superconducting loop forever. He joined the ends of the coil of lead wire, placed it in a magnetic field produced by an ordinary electromagnet, and cooled it to 1.8 K. Then the electromagnet was switched off and removed, inducing in the superconducting coil a modest current that could be detected by its magnetic influence on a small compass. At room temperature, the induced current would die out in a fraction of a second. The current continued to flow until the liquid helium coolant evaporated away and the wire warmed above 7.2 K and regained its resistance. Kamerlingh Onnes exulted, "A coil cooled in liquid helium and provided with current at Leiden might . . . be conveyed a considerable distance and there be used to demonstrate the permanent magnetic action of a superconductor carrying a current." Years later, in 1932, Meester Flim was chosen to fly to London for a demonstration at the Royal Institution, bearing a liquid helium cryostat containing a lead ring in which a current had been set flowing. A modern version of the persistent-current experiment, in which a current flowed undiminished for two years, ended only when a strike by truck drivers cut off deliveries of liquid helium.

Superconducting magnetic energy storage systems, which rely on persistent currents in superconducting magnets, hold promise for load leveling in power distribution systems.

The Great War interrupted cryogenic research in Leiden and London alike. In 1915, when Dewar was a vigorous seventy-two years old, the Royal Institution volunteered for war work. He developed sturdy versions of his famous flask to transport liquid oxygen to field hospitals where the oxygen was used to treat victims of German gas attacks, and to supply oxygen gas to airmen at high altitudes. For his own research, Dewar turned to "something cheap—soap," to study the properties of thin films and bubbles. Kamerlingh Onnes, who became professor emeritus in 1915, would not liquefy helium again until 1923. In that same year, liquefiers were also built in Toronto and Berkeley. The Leiden monopoly was broken, but an exclusive oligarchy still ruled the lowest temperatures.

Today, liquid helium is a commercial commodity, available for a few tens of dollars per liter in quantity. The coldest spot on Earth is still in physics laboratories where record low temperatures below one nanokelvin—a billionth of a degree above absolute zero—are now routinely achieved in Bose-Einstein condensates, clouds of atoms brought nearly to rest by deft manipulation with laser beams. NASA's Cold Atom Laboratory aims to cool small collections of atoms to approximately one picokelvin, just a trillionth of a degree above absolute zero, in the microgravity environment of the International Space Station.

12

UNCLE TED

Nearly fifty years passed between the discovery of superconductivity in Leiden and the construction of the first strong superconducting magnet, a forebear of the magnets now used in particle accelerators and magnetic resonance imaging machines. Ted Geballe was one of those who contributed the most to realizing Kamerlingh Onnes's dream. He stopped co-teaching a Stanford seminar on energy at the age of ninety-nine but continued research into his second century.

As Ted described the San Francisco of his youth: "Back then a boy could roam the city freely by bicycle, cable car, or streetcar." Although his high school was named for Galileo, its most illustrious alumni were the DiMaggio brothers, Vince, Dom, and Joe. The main interest was baseball, not telescopes. "I wasn't someone who was driven to science as a youngster," he said. "I was interested in sports like baseball and football.

An interest in science came to the aspiring athlete by osmosis through his older brother Ron and the summer-vacation visits of his Aunt Pauline, whose claim to family fame was that she had been Linus Pauling's first science teacher. "My aunt was a great teacher. She was childlike in the sense that she dealt with children as equals. She loved teaching and never talked down to anyone. Whatever she did, she related it to chemistry." While cooking, she would explain why it takes longer to boil an egg in the mountains than in San Francisco. Coming upon a pump on a jaunt in the country, she would teach her young companions how it lifted water. More important than any fact or detail was Aunt Pauline's infectious enthusiasm for understanding how things work. Naturally, Ted and Ron had a chemistry set. "We almost set fire to the house and the usual kinds of things." Pauline Geballe was still feisty and energetic a generation later when she showed her brother's grandson, Bob Cahn, how air pressure could crush a can and taught him about negative numbers. Her first edition of *The Discovery of the Elements*, by Mary Elvira Weeks, published in 1933, is one of Bob's treasures.

Ted went "East" for college, all the way across San Francisco Bay to Berkeley, not as a center fielder but as a chemistry undergraduate. There he began working in William Francis Giauque's laboratory, which had succeeded the Leiden Cryogenic Laboratory as the coldest spot on Earth in 1933. Ted's road to superconductivity began in a junior-year physical chemistry laboratory, where the teaching assistant was one of Giauque's graduate students. With the innocent certitude of youth, the TA told Ted that nobody in Giauque's bailiwick, particularly Giauque, could understand modern ideas about electrons in metals. He refused to believe that you could think of the ten billion trillion conduction electrons in a cubic centimeter of metal as a gas when they were buffeted by electromagnetic forces. Ted and his young mentor tried to do an experiment in the junior lab to measure properties of the electron fluid. The experiment failed, but Ted learned that one can also learn about electrons in metals from failures. That was precisely the spirit that Gilbert Newton Lewis, who headed Berkeley's small College of Chemistry, cultivated: learn by experimenting rather than just from formal courses. Professors and textbooks don't have all the answers.

Just before his Cal graduation, Ted was called up by the Army, sent to Fort Warren in Cheyenne, Wyoming. He was able to obtain a five-day leave to drive back to San Mateo, get married to Frances "Sissy" Koshland, his wife for the next seventy-eight years, and report back to Fort Warren. A few months after Pearl Harbor, he was shipped overseas and spent the next three years maintaining tracking devices for antiaircraft guns in Australia, New Guinea, and the Philippines. "Near the end of the war," he recalled, "Bell Labs sent us black boxes containing vacuum tubes that were supposed to help the gunners aim. That was the first time I heard of Bell Labs. If only they had sent instruction manuals, we might have won the war sooner."

When he got out of the service, Geballe hesitated between joining his father's wholesale shoe business and going to graduate school. At twenty-seven, "I felt old . . . maybe I was already past my peak for creativity." He decided to give science a try, because "my whole time in the army I never had as much fun in a single day as I'd had in the worst day at Berkeley as an undergraduate. The idea that I could go back and get in that atmosphere seemed like such a dream, and my wife Sissy urged me to give it a try." He returned to Berkeley to work with Giauque, and right away, the fun returned.

In the spirit of Lewis's instructional ideal, Ted had to present a seminar on a subject outside his field as part of the qualifying exam. Browsing through journals in the library, he stumbled on a paper that talked about transporting electrons around a superconductor without changing their energy. From his junior-lab experience, Ted knew enough to be amazed. "Superconductivity was so fascinating that I started reading up on it. I soon found out 'super' is not a Hollywood exaggeration. If you started a current flowing in a coil of good metal like copper, it might last a second or so. In a superconductor, if you started a current flowing at the beginning of the universe, it would still be flowing today provided you kept its cold enough, below the transition temperature where it loses its 'super' property. That's what we mean by perfect conductivity, but that's not all! A superconductor has magnetic properties, just as remarkable: it doesn't allow a magnetic field to exist inside the superconductor."

The remarkable magnetic property of superconductors was discovered by Walther Meissner and Robert Ochsenfeld in Berlin in 1933. Meissner and Ochsenfeld placed a warm sample of tin in a magnetic field so the lines of magnetic force permeated the material. When they cooled their sample below tin's critical temperature of 3.7 K, the magnetic field was expelled, penetrating no more than a few millionths of a centimeter below its surface, so that the lines of force flowed around the sample. This was not just a consequence of perfect conductivity, but a completely new aspect of the phenomenon of superconductivity.

Here was the key to the disappointing performance of Kamerlingh Onnes's first superconducting magnet. A material cooled below its critical temperature becomes superconducting because it requires less energy to be superconducting than to be normal, that is, resistive. But Meissner's discovery showed that the material must expel magnetic field to become a superconductor—and that means doing work. If the energy price for pushing out the magnetic field is more than the energy benefit from being superconducting, the material loses its superconductivity and becomes normal. Right at the critical temperature, even a tiny magnetic field is enough to destroy superconductivity. If the sample is cooled further, more magnetic field can be tolerated. No matter how cold the sample is made, there is a limit to the field that can coexist with superconductivity. For Kamerlingh Onnes's lead wire, that field is far too small to make a really useful magnet.

"I admire Meissner a great deal for what he did," Ted Geballe mused. "He did physical measurements that changed the way we think about superconductivity completely, and he found whole new families of superconductors—elements such as niobium and tantalum, and their alloys like niobium carbide." Niobium supplanted lead as the substance with the highest critical temperature, at 9.25 K. Then came niobium carbide, superconducting at 11 K and, in 1941, niobium nitride, at 15 K.

Giauque believed that careful measurement made on well-prepared samples would achieve a certain kind of immortality while the theories that explained them might come and go. For his thesis research, Geballe chose to determine the low-temperature thermodynamic and magnetic properties of a large crystal of copper sulfate pentahydrate, a common salt that had been used for centuries as a fungicide. Its structure had just been determined in England. Giauque thought that a large sample would give the best results because the surface-to-volume corrections would be small. Geballe's first challenge was to grow a large single crystal. But how? Giauque suggested he write to Mervin Kelly, president of Bell Labs, where large crystals of lithium niobiate had been grown to make transducers needed for the war effort. The trade secret needed to grow them turned out to be a home remedy. A five-page reply from Alan Holden, a Bell research scientist, explained that to maintain the proper concentration ratio at the interface between a crystal and the solution from which it grows, you have to stir it. How to do that without damaging the crystal? Salvage the agitator from an old washing machine that sloshes back and forth, keeping the crystal seed in contact with a slowly cooling saturated solution. Holden later published his prescription in the *Transactions of the Faraday Society*. After a year and a half, Geballe had mastered this decidedly mundane technique to grow the crystal and grind it into an ellipsoid in preparation for his thesis measurements.

In 1952, Ted took his new PhD to the source of the undocumented black boxes and crystal-growing expertise, Bell Labs in Murray Hill, New Jersey. It was a place where anything seemed possible. The invention of the transistor a few years earlier had justified for all time the Bell Telephone Company's investment in its research program, but that was only the beginning.

For Ted, "The Bell atmosphere was overwhelming. It seemed like half the solid-state physics in the world was being born there. It was Mount Olympus. I hardly believed it when the managers told me I could do whatever I wanted, but it was absolutely true. They wanted us to be making research decisions on our own. Later on, I could see that 90 percent of Bell Labs was still designing vacuum tubes and things like that, but the game was changing: the solid-state era was underway. Transistors and other faster, smaller, and less energy-consuming solid-state devices were being invented by scientists recruited to Bell from both sides of the Iron Curtain after World War II."

Geballe was attracted to two exciting projects. Should he study semiconductors—materials halfway between metals and insulators, from which transistors are made? Or should he work with Bernd Matthias, who was leading an effort to find new superconductors? As he was deliberating and reading and trying to decide which way to go, the Bell management gently suggested that he go into semiconductors—especially germanium. In contrast with the more than a year he spent growing a single crystal for his thesis, he was handed a beautiful single crystal of germanium purified to a level far beyond anything available commercially and the opportunity to study its low-temperature properties. No one had ever looked at good semiconductors at low temperatures. There was a joke that people had been making good measurements on bad materials and bad measurements on good materials. ("Which is which?" was part of the joke.) Here was a chance to work with good—pure—materials and to make good—imaginative—measurements. How could any Giauque student resist?

Superconductivity remained an enigma for the forty-six years following Onnes's discovery until in 1957 a trio—John Bardeen, Leon Cooper, and Robert Schrieffer—came up with the explanation, inevitably called BCS theory. But this was still years in the future when Ted finally made the switch to superconductivity. He described how this came to pass.

Around 1950, when the world's most famous theorists had been unable to explain superconductivity, Enrico Fermi sought clues that might be exposed by finding new superconductors. "Fermi convinced two young visitors at Chicago, John Hulm and Bernd Matthias, to search for new, unexpected superconductors in the hope that they might provide the needed clues. Matthias and Hulm constructed a simple setup for rapid synthesis and mea-surement. Shortly before leaving Chicago, John discovered a compound of three atoms of vanadium and one of silicon that set a record for the highest temperature superconductor—17 kelvins. When Bernd returned to Bell, I joined him."

Bernd was passionately involved in creating new materials by substituting one element for another. He lived and breathed the elements. Colleagues revered him as the modern successor to Mendeleev. Geballe recalled, "Bernd had all this marvelous intuition based upon his deep knowledge of the peri-odic table and his imagination. He wanted to be unusually sensitive and inspirational and a pipeline to up above. His own special genius would be just knowing the right thing to do."

Matthias cultivated an air of the supernatural. Asked by his Swiss Federal Institute of Technology (ETH/Zürich) classmate Rolf Steffen to explain what lay behind a particularly outlandish, but typically successful, speculation—that cesium, the most reactive of alkali metals, would become superconducting under pressure—Bernd replied, "Rolfli, I felt it in my urine!"

Together, Matthias and Geballe and their team at Bell became the dominant inventors of new superconductors: Ted recalled, "Some of my happiest days with Matthias were Sunday mornings, when we went looking through all the journals to get ideas. We took any unexpected result as a signal that something was worth investigating. We would go up to the Bell library, which was organized much better than Stanford or any other place I know of for this sort of thing. The technical journals from all over the world were displayed alphabetically in the room. We would look at a title and guess 'That's a metal,' and 'That's going to be an insulator.' When it wasn't, that was a clue that something new might be going on and was worth investigating, and we made a list.

"What made it really good was that we had trusted technicians who knew how to make things and were resourceful, so our list was turned into samples in record time. The test for superconductivity was sensitive and simple and required a sample only a few millimeters long. There were a lot of new things to be discovered." In a year, perhaps ten or twenty new substances would turn out to be superconducting, enough to keep the team excited about the search and busy with detailed studies of the most promising materials.

Besides creating new materials, Matthias and Geballe carried out experiments to help them understand the nature of superconductivity. They shared a healthy disrespect for theories. Geballe described his own approach this way: "I always wanted to pick things that were a little too hard for theorists. I want to discover and then understand rather than understand and then verify something that is already understood. When you're looking at how an ensemble of atoms or molecules collectively behave new phenomena emerge and you have to find that experimentally. That's why condensed-matter science is a good subject.

"With excellent support from our staff we discovered a great many new superconductors. A compound of niobium and tin looked like it might be much better than vanadium-silicon. We didn't know what the transition temperature was going to be. We thought it might even be above liquid nitrogen at 77 kelvins. We found it was actually 18 kelvins. When we applied a magnetic field of a tenth of a tesla, the superconductivity remained, but we didn't make a careful measurement. When Bernd reported on the niobium-tin in a conference it was regarded as 'schmutz physics,' uninteresting because the high transition temperature was expected not to survive at high current or in a high magnetic field according to the established wisdom. Niobium-tin remained a useless oddity because we had missed its remarkable properties."

One day in 1960, Ted's friend Rudi Kompfner challenged him to do something useful for Bell Labs by producing a small superconducting magnet for a device to track communications satellites. A few superconducting magnets had been wound from niobium wire, but they achieved feeble fields compared to standard copper-and-iron electromagnets. Matthias and Geballe regarded building a magnet as a distraction from their search for new superconductors. Their taste was for the fundamental, rather than the applied. The task of building the magnet fell to Bell's metallurgy department, to a team led by Gene Kunzler, himself a student of Giauque. Geballe and Matthias took the role of advisors, proposing promising materials. After some encouraging results had been obtained with a tiny magnet wound with gold-plated molybdenum-rhenium wire, attention turned to niobium-tin. The gold plating was a Geballe touch: it insulated one turn of wire in the magnet coil from the next. Gold may be a good conductor by normal standards, but it is infinitely more resistive than a superconductor at liquid-helium temperatures. If an unexpected surge of current killed the superconductivity, the gold would take over the task of carrying the current and prevent overheating and a meltdown.

Switching to niobium-tin was easier said than done. While niobium and tin like to combine in the right proportion—three niobiums for each tin atom—the resulting compound is not ductile; it is too brittle to be drawn into wire. In one way to work around this obstacle a mixture of powdered niobium and tin was poured into a niobium straw. After the filled tube was squeezed down to wire-like dimensions and bent into the desired shape, it had to be baked for sixteen hours at 1000°C to make the niobium and tin react into the alloy.

When Kunzler and his colleagues put a sample of the ersatz wire, cooled with liquid helium, inside the giant magnet in December 1960, it remained superconducting in the strongest field the magnet could produce. "No one expected the result," Ted remembered.

But that was only the first surprise. The niobium-tin wire continued to carry a large current—a thousand amperes per square millimeter of cross section—although the magnetic field would easily have destroyed super-conductivity in lead. "This was totally unexpected. I was stupefied. We were surprised by the critical field, but we were totally amazed by the critical current." Within a few months of the discovery, the fabrication of niobium-tin test magnets was fairly routine. Less than a year later, Kunzler and his team had operated an experimental seven-tesla superconducting magnet.

"At first I didn't understand the implications for technology," said Geballe. "Then I realized what Kamerlingh Onnes knew, that the cost of refrigeration for superconductivity was nothing compared to the savings in electrical power for running a magnet. We were thrilled because, here was niobium-tin, which most people weren't interested in and disregarded as too messy. Yet it had these marvelous properties. The high-energy community just flipped when they heard about it."

Remarkably, the distinctive—miraculous—properties of superconducting alloys had been encountered a quarter century earlier by Lev Shubnikov's team at the Ukrainian Physical-Technical Institute in Kharkov. Perhaps the discovery was simply too far ahead of its time to find immediate practical application, but the arrest and execution of Shubnikov in Stalin's Great Terror of 1937 and the German invasion of the Soviet Union in 1941 terminated cryogenic research in Kharkov.

Why did niobium-tin's superconductivity survive in the enormous magnetic field? Previously, the only trick known for avoiding the energy penalty for expelling the magnetic field was to use a sample so thin that the field can penetrate across it without being expelled. Of course, a thin film is not able to carry a large current.

Not until the mid-1950s were the leading Russian-language physics journals systematically translated into English, so most western scientists were slow to appreciate developments made in the Soviet Union. The disconnect was especially marked in the case of superconductivity; the pioneering work published in 1950 by Vitaly Ginzburg and Lev Landau remained little known in the West for the better part of a decade. As we shall see in the following chapter, the Ginzburg-Landau picture is a precursor to a key element of our theory of weak and electromagnetic interactions.

Although as Kharkov colleagues they were familiar with Shubnikov's discovery, they regarded alloys as "unsavory business," in the words of Ginzburg's Nobel lecture, and did not attempt an explanation.

Building on their work, and with Shubnikov's results in mind, Aleksei Alekseevich Abrikosov reasoned over the period 1953–1957 that a second type of superconductor, including many alloys, would only partially exclude magnetic fields. Abrikosov found it theoretically possible to impose microscopic zoning regulations that would permit some regions to remain normal while neighboring regions become superconducting. Magnetic field could be corralled in the normal regions, instead of being expelled completely, which would reduce the energy cost of going superconducting. The gain had to be balanced against the additional cost of all the normal regions that had more energy than the superconducting ones. Maybe it was worth paying for the normal regions and maybe it wasn't. It would depend on the material. But if a material could be found that would benefit from this arrangement, superconductivity might coexist with a high magnetic field. Niobium-tin turned out to be such a material. Abrikosov's insight only gained wide recognition after the fact.

At the victory party Ted Geballe threw to celebrate the niobium-tin discovery, an alert Bell patent attorney asked about the other superconductors.

"Shouldn't we patent them all?" With a characteristic touch of theater, Bernd rattled off a list that included the ductile solid solutions niobium-zirconium and niobium-titanium and wrote them all down on the inside cover of a matchbook. That's how Bell got the patent.

Kunzler's discovery of niobium-tin's remarkable robustness gave even greater impetus to the search for new superconductors. Niobium-tin had a high transition temperature and could tolerate large magnetic fields, but it was a troublesome material. Matthias and Geballe wanted to find something better. The stakes were high and even bizarre possibilities were worth pursuing. In 1953, John Daunt and James Cobble had reported that the radioactive man-made element number forty-three, technetium, was superconducting above 11 K—the highest critical temperature of any pure element. Since there could be no technetium mines on Earth, the material had to be prepared from Oak Ridge reactor wastes. To overcome this inconvenience, Matthias created an alloy with the same crystal structure out of molybdenum and ruthenium, the naturally occurring elements 42 and 44 that lie before and after technetium in the periodic table. The faux-technetium became superconducting at 10.6 K. Refined measurements pushed the critical temperature of technetium below 8 K, making Matthias's imitation better than the real thing.

John Hulm's alloy of molybdenum and rhenium, a stable element in the same chemical family as technetium, had a transition temperature of about 12.5 K. "Bernd thought that molybdenum-technetium would have a good transition temperature and could be drawn into wire." This time, the inconvenience of technetium's radioactive instability could not stand in the way. Matthias simply returned from Los Alamos, where he was a consultant at the time, with half a gram of technetium smuggled out of the weapons laboratory. The results were modestly positive. "It had good properties if it weren't radioactive and weren't expensive," Geballe remembered with a wry grin. Neither molybdenum-ruthenium nor molybdenum-technetium became a practical superconductor.

Today, niobium-tin is reserved for special applications that require the very strong magnetic fields only it can sustain. The workhorse of high-field, high-current superconductivity is niobium-titanium, one of those ductile type-II superconductors that Bernd Matthias wrote on a matchbook. Although it has a lower critical temperature and lower critical field than niobium-tin, niobium-titanium can be drawn into flexible filaments finer than a human hair and, with a copper matrix for support, can be made into a cable that is one of the true technological wonders of our time. It was the essential component of magnets in Fermilab's Tevatron, the pioneer high-energy superconducting synchrotron.

Niobium-titanium magnets steer beams around the LHC, the proton-proton collider that looms so large in the present and future of particle physics. By the mid-1980s, magnetic resonance imaging machines using niobium-titanium cable at the boiling point of liquid helium brought superconductivity out of the laboratory and into medical diagnosis and people's lives. Superconducting magnets now operate so reliably that they are used in remote, hostile environments at mines where they have revolutionized the old technique of the magnetic separation of materials. In the clay-processing industry, superconducting technology has produced an entirely new grade of materials by removing mineral impurities to increase the whiteness of clay used in papermaking.

Early in 1986, Georg Bednorz and Alex Müller, two physicists working at IBM's Zürich Research Laboratory, found a ceramic compound of lanthanum, barium, copper, and oxygen that became superconducting at 30 K, above the boiling point of liquid neon. The search for high-temperature superconductors had known many false leads, so the copper oxides were not an overnight sensation—not yet.

At the Materials Research Society meeting in Boston in December, Ted learned that Shoji Tanaka's University of Tokyo group had duplicated the Zürich experiment and observed the superconducting transition magnetically. Paul Chu, who had earned his spurs under Matthias, told Ted that he too had confirmed the result at the University of Houston. The news took Geballe's breath away. "The minute I saw that it was real . . . This was the thing I had been looking for a long time and we got started on it before I got back. We started making the compound in the laboratory at Stanford immediately."

When strontium replaced barium, the superconducting transition temperature approached 40 K. By Christmas, Chu pushed the onset of superconductivity above 50 K by squeezing the barium compound. It seemed that every condensed-matter group in the universe dropped what they were doing to join the chase for new superconductors.

Techniques ranged from the traditional—substitutions of one element for another, controlled variations in proportions, and the application of high pressure—to shake-and-bake: grind up any materials you can get your hands on, mix them together, and pop them in the oven. An astonishing range of the new substances showed signs of superconductivity.

The excitement and competition were intense.

In February, Paul Chu announced that he had found a material—yttrium-barium-copper oxide—with a transition temperature of 94 K, beyond the liquid nitrogen boundary. "To me the discovery of high-temperature superconductivity is a gift of nature almost too good to be true," Ted reflected. "I had never expected to see superconductors above 77 K. I thought we might get to 25 or 26 K."

Four weeks later, with discoveries being made daily, a meeting of the usually sedate American Physical Society turned into a superconductivity happening. On the evening of March 18, 1987, a standing-room-only crowd overflowed the New York Hilton's 1,140-seat Rendezvous Trianon Ballroom for a hastily arranged session on the new superconductors. After introductory remarks by two invited panels, five-minute summaries of the latest news from one low-temperature group after another continued until 3:15 in the morning.

Participants compared the atmosphere to a rock concert and dubbed the event "the Woodstock of physics." [It's fair to ask how many of them had ever been to a rock concert, let alone Woodstock.] Room-temperature superconductivity seemed around the corner.

Rumors circulated that speculators were trying to corner the yttrium market. Patent applications flew like confetti. Inevitably, some reports were mistaken and some too enthusiastic, but the phenomenon was real: A New Age had dawned.

Normally reasonable people lost their composure. Nobel laureate Phil Anderson proclaimed that the new materials could make magnets ten times more powerful, making gigantic particle accelerators unnecessary. Marvin Cohen, another prominent condensed-matter theorist, told the national media we could look forward to going to the hardware store to buy some electrical wire and being asked, "Normal or superconducting?" They were not the first to be dazzled by a superconducting future. When soon-to-be Nobel laureate Leon Lederman was director of Fermilab in 1979, he rhapsodized about superconducting light bulbs on the *Phil Donahue* television show, forgetting for just a moment that without resistance a light bulb won't emit the faintest glimmer, let alone glow white hot.

There is still no superconducting wire at your local hardware store and probably few requests for it. The record-high critical temperature of 133 K, held by a mercury-barium-calcium-copper oxide, is still far short of room temperature, 293 K. (Evidence for superconductivity in a compound of lanthanum and hydrogen under 1.9 million atmospheres pressure has been found at 260 K by a team including Zachary Geballe, Ted's grandson.) Moreover, like niobium-tin before, the new high-temperature superconductors are awkward materials. They are generally brittle or fragile, not yet the sort of stuff you want to use to wire your doorbell. But they are fascinating substances that hold great promise for many applications, from wireless communication to motors and energy storage—and perhaps, in the long term, accelerator magnets.

For now, the best option to produce the magnetic fields needed for the next stage of the Large Hadron Collider remains niobium-tin, which scientists and engineers are working hard to domesticate. The immediate task is to build accelerator magnets that reach eleven to twelve teslas, eventually approaching sixteen teslas. One of the leading fabrication techniques is an evolution of the method used at Bell Labs in 1960. A stacked assembly of powder-filled niobium tubes is drawn or extruded down to wires of the desired diameter, which are then made into cable that is wound into coils of the desired configuration. Once the coils are formed, insulated, and stabilized, they are reacted for forty-eight hours in an inert atmosphere of argon at temperatures reaching 640° Celsius. A far cry from winding doorbell wire around an iron spike!

13

AN ELECTROWEAK THEORY

W e named chapter 10 "Beta Version" to signal that Fermi's theory of beta decay, extended to incorporate parity violation, is highly serviceable yet cannot be the final word on the weak interaction. We cited two related shortcomings: the probability that neutrinos interact with matter grows with energy, eventually surpassing 100 percent; and the theory makes sense only as a one-act play. Both soft spots can be traced to Fermi's idealization that the weak interaction occurs at a point—in contrast to electromagnetism and the strong interaction, which are propagated by force particles, the photons and gluons.

If the weak interaction were also carried by a force particle, this time with a large mass—more than a half-billion electron volts, say—the range of the weak interaction would be limited to a fraction of the size of a proton. [The range of a force is measured by the inverse of the force carrier's mass.] All the successes of Fermi's theory would persist, but the impossible infinitely rising interaction rate would be deferred to neutrino energies far above those of the most energetic cosmic rays ever observed. Introducing a force carrier in this way provides symptomatic relief, but it falls well short of a cure. Just as in the Fermi theory, quantum corrections—multiple acts—increase crazily, instead of contributing ever smaller refinements. In addition, the problem of interaction probabilities that rise without limit returns in a new form.

We name the force carrier that mediates beta decay W, for weak. Following the example of electromagnetism, and to reproduce what was known about beta decay, muon decay, and other well-studied phenomena of the late 1950s and early 1960s, W would be a boson that carries one unit of spin and one unit—positive or negative—of electric charge. Unsuccessful attempts to produce weak bosons through neutrino interactions in the early 1960s showed that if the W did exist, its mass had to exceed a few billion electron volts. Only in 1983 did the conjectured W boson become real, with a mass near eighty billion electron volts, in experiments at CERN. Today, a few W bosons are produced every second of operation at CERN's Large Hadron Collider. The Particle Data Group's W-boson dossier runs to forty-seven pages.

Thought experiments enable physicists to spot problems in a theory or to anticipate new phenomena. Our consideration at the end of chapter ten, "Beta Version," of a muon neutrino colliding with an electron to produce a muon and an electron neutrino is one example. That exercise showed us that the Fermi theory could at best be a low-energy approximation to a real theory of weak interactions.

Now let us imagine beams of neutrinos and antineutrinos meeting head-on to produce a pair of weak bosons, one carrying charge plus-1 and the other charge minus-1. In the graphical language of Feynman diagrams, we can picture a neutrino splitting into an electron and a positive W^+; the electron combines with the antineutrino to form a negative W^-. To this day, no one has witnessed neutrino-antineutrino collisions, neither with colliding neutrino beams nor in collisions of a neutrino beam with fossil neutrinos produced in the early universe. But our imaginations can perform feats that our laboratories cannot. In this case, the probability of interactions grows rapidly with energy and would eventually exceed 100 percent. If W bosons are to be the salvation of our theory of the weak interactions, we must find a way to impose discipline on the way they behave.

What could be better behaved than a gauge theory, in which symmetry begets the interactions? But which symmetry? We haven't found a principle that dictates which gauge symmetry corresponds to a force law observed in nature. What we have learned to do is to identify patterns that suggest some gauge symmetry, to work out the consequences, and to see whether the resulting mathematical construction faithfully describes nature. Sometimes we find the alternate realities instructive, even when they differ from the real world.

First to give a detailed account of what we now know to be the right symmetry behind the weak interaction was Sheldon Glashow, an American postdoctoral fellow working at Niels Bohr's Institute for Theoretical Physics in Copenhagen. Enrico Fermi had modeled the structure of his 1933 "Attempt at a theory of the emission of beta rays" on the quantum theory of electromagnetic radiation. With the evidence before him in 1960, Glashow saw a deeper connection. "At first sight," he wrote, "there may be little or no similarity between electromagnetic effects and the phenomena associated with weak interactions. Yet certain remarkable parallels emerge with the supposition that the weak interactions are mediated by unstable bosons." In each case, a single coupling strength suffices to describe a wide class of phenomena. In that sense, both electromagnetism and the weak interaction can be called universal. The Fermi theory describes both beta decay and muon decay, for example. By hypothesis, both electromagnetism and the weak interaction are transmitted by spin-one fields.

There is one not-so-little problem. The charged W bosons must be massive, to account for the short range of the weak interaction, but the photon mass is zero. "[S]urely," Glashow acknowledged, "this is the principal stumbling block in any pursuit of the analogy between hypothetical [weak bosons] and photons. It is a stumbling block we must overlook." It is an inspired evasion: acknowledge the problem, set it aside, but never forget it. Glashow put forward his candidate for the gauge symmetry and showed where it might lead. That symmetry, as we will make explicit later in this chapter, has two components: one resembles the isospin symmetry of protons and neutrons that gives rise to

Yang-Mills theory, the other is similar to the phase symmetry that underlies quantum electrodynamics. Writing six decades after Glashow, we will address the stumbling block first and then make explicit the symmetries.

Luckily, particle physics can learn from other scientific specialties; that is why the best physics departments urge—or even compel—their students to master at least a beginner's knowledge of many areas and why department colloquia survey recent developments across physics and astronomy and beyond.

In this case, the critical insight is found in the second miracle of superconductivity, the exclusion of magnetic field from a superconductor that we met as the Meissner effect in chapter 12. If a magnetic field can penetrate no more than a hair's breadth into a superconducting medium, that means that the electromagnetic force carrier—the photon—must become massive inside the superconductor. Under ordinary circumstances, quantum electrodynamics requires the photon to be massless, but within the special environment of a superconducting medium, the photon acts as a massive force carrier. The larger the photon mass, the shallower the depth to which magnetic field penetrates.

Might it make sense to imagine that our entire universe is an environment in which the weak gauge boson is massive—something analogous to a superconducting medium? The answer is yes! That verdict is possible because of an important general lesson: not every symmetry that we discern in nature's laws is openly displayed in the expressions of those laws. A few examples will help us explore this idea.

A liquid is a disordered collection of atoms or molecules held together by electromagnetism. It has the same appearance from every vantage point, manifesting the fact that the theory of electromagnetism is indifferent to direction and location. In disorder lies great symmetry, contrary to what we may infer from the simplest examples of Western decoration. A crystal that forms when a liquid is cooled is a regimented collection of the same atoms or molecules that make up the liquid, held together by the same electromagnetic interaction. It exhibits regularly spaced atoms arrayed in ranks and files and columns that single out preferred locations and directions. The crystal's orderly structure conceals the translational and rotational symmetries of electromagnetism.

Many snowflakes are flat ice crystals radiating six branches that are very nearly identical in appearance. They are held up as iconic illustrations of symmetry. Rotate a snowflake by one-sixth of a turn and its appearance is unchanged. That sixfold symmetry is captivating, but it is the merest vestige of the symmetry displayed by an unfrozen droplet of water. The renowned astronomer Johannes Kepler, mathematician to the court of Holy Roman Emperor Rudolf II, was so taken with the six-cornered starlets alighting on his coat in the Prague winter that he began to ponder how, if every one was different from all the others, each was hexagonal in form. Kepler documented his musings in a booklet of two dozen pages titled *Strena seu De Nive Sexangula*, as the year 1610 turned to 1611. He presented the New Year's gift, *On the Six-Cornered Snowflake*, to court counselor Johann Matthäus Wacker von Wackenfels, his friend and occasional patron. Kepler reasoned that there must be a cause of the six-legged pattern and argued that it would not be found in the shapeless water, but in some agent. That is not the whole truth, but it marks the beginning of a scientific inquiry.

Physicists in search of insight have learned the value of idealization. We neglect minor effects so we can refine our understanding of major ones. We reduce a problem to its essence by isolating a system or by treating real-world complications as controllable corrections. "Neglect air resistance," or "neglect friction," we tell beginning students as they study the classical mechanics of Newton's laws: projectile motion, billiard-ball collisions, or those infernal blocks sliding down inclined planes. A favored turn of phrase is to "turn off" some interaction, a token for concentrating on big effects before dealing with the subsidiary. We have done just that (more or less) in chapter 8 by noting that, in the absence of electromagnetism, what anyone chooses to call a proton or a neutron is purely arbitrary—a matter of convention.

In that spirit, we can imagine turning off gravitation so that a drop of water is a perfect sphere suspended in space. [In the microgravity environment of the International Space Station, astronauts have displayed spherical drops that are not deformed by gravity and do not fall to the floor. Not that there was any doubt!] Turn the drop by any amount around any line that passes through its center, and it looks the same. That sameness—symmetry—makes manifest the fact that electromagnetism, which holds the water molecules together, picks out no favored direction in space. And yet, in the frozen form of a snowflake, the water crystal only looks the same when rotated about one line through its center, and only by one-sixth of a turn at a time. The full three-dimensional symmetry of electromagnetism is hidden. We might say that the form of the snowflake makes the true symmetry a secret!

Kepler's surmise that the sixfold symmetry of a snowflake is not a matter of chance is correct—though the orientation of the arms and the detailed shape of the flake are randomly determined. But it turns out that the "six-cornered" shape is not the work of an outside agent; it is a consequence of the shape of water, not as a formless vapor or a flowing liquid, but as a molecule. Only in the 1930s did X-ray diffraction studies reveal that a water molecule is bent in a way that encourages the formation of hexagonal crystals. The bonds that join two hydrogen atoms to the oxygen atom form an angle of 104.45°.

Now, instead of inspecting a crystal from the outside, let us dive inside. Imagine a very tiny physicist, a millionth of a millionth of normal size, living and working in the recesses of a magnetic crystal of iron. The laws of electromagnetism that, together with quantum mechanics, determine the shape of the crystal world, favor no direction over any other. But this picophysicist would have a hard time discovering nature's rotational symmetry, because on every street corner of the crystal world stands a compass, and the needle of every compass points the same way. The compasses are monuments to the way things were at a time that no one remembers, just after this world came out of the fiery furnace.

Together, the compass needles establish a magnetic field that pervades the whole of the picoworld, including the picophysicist's laboratory and instruments. If the experimental equipment were not much affected by the magnetic field that the regimented compass needles impose, our tiny colleague might find that the laws of nature look approximately the same from every direction. But if magnetism strongly influenced the picoapparatus, the picoinvestigator might miss the idea of full rotational symmetry altogether. In any event, pico-experiments would be hard-pressed to reveal that three-dimensional rotational symmetry is exact for the fundamental laws of nature.

Nor could the picophysicist demonstrate directly that the orientation of the compasses, the emblem of stability and order in the crystal world, is determined by happenstance. (Imagine the cultural significance that would have accumulated over time for so many compass needles all indicating the same direction!) Too small by far to reorient all the compass needles at the same time, the picophysicist would find it an unachievable task to show that any other alignment is possible, let alone equivalent. The picoscholar could at best conjecture that a hidden symmetry lies behind the order of the world and test that idea through its consequences.

At some considerable peril to our diminutive colleague, we macroscopic outsiders could be of assistance. We could raise the crystal's temperature above 1,043 kelvins, until heat's random motion caused the needles to flutter about in a highly symmetrical state of disorder. Then, as the crystal cooled, we could watch all the street corner semaphores settle into a new pattern, equally regimented but differently aligned. The chosen direction—the new true way—would change on every cycle of heating and cooling, retaining no memory of what had gone before. Over many trials, the changing orientation would reveal by its aleatory nature the secret rotational symmetry of the laws that govern the magnetic crystal picoworld.

As a final example of how symmetries may be hidden by circumstances, we offer a student favorite that you might want to try for yourself. Feel free at any moment to switch from a real experiment to a thought experiment, which sounds more profound in Einstein's formulation—*Gedankenexperiment*. Procure a good bottle of wine, one with a perfectly formed dimple, or punt, in its base, along with a perfectly spherical pearl. Empty the bottle by a method of your devising. Set the bottle upright and, as carefully as you can, insert the pearl and balance it at the center of the punt. View the bottle from above. Gravity draws the pearl toward the center of the Earth, but neither east, nor west, nor any other direction is preferred. In classical mechanics, the pearl's location is called a point of unstable equilibrium because the slightest displacement will cause it to tumble. But in the classical world—at least in a vibration-free environment—the pearl, once deftly placed, will stay forever atop the dimple; nothing tells it which way to go.

The quantum world is different: quantum-mechanical jitters jostle the pearl infinitesimally this way and that. Nature is patient; eventually, the pearl will yield to gravity's tug and it will fall, coming to rest at some location in the groove at the bottle's base. Its resting place appears to designate a preferred direction, which seems at odds with the cylindrical symmetry of our experimental apparatus. By repeating the experiment afresh, we can test whether the chosen direction reveals some bias in the laws of nature. To minimize systematic effects, it is good practice to carry out each trial with a fresh bottle of wine. After each repetition, the pearl will find a new resting place. Every direction is equally probable, and each choice of direction falls to chance. An ensemble of many experiments exhibits the two-dimensional rotation symmetry displayed by the experimental setup.

What physicists call the vacuum state is the configuration of lowest energy for a physical system. The structure of a magnetic crystal arises from the propensity of compass needles representing individual iron atoms to align, minimizing the total energy. The result is not one unique vacuum state, but an infinity of equivalent vacua corresponding to all the directions in which the aligned compass needles might point. These are all connected by symmetry—any orientation can be rotated into any other. We refer to the equivalent vacuum states that share the same energy as degenerate vacua. Similarly, any location at which the pearl should come to rest in the groove at the base of a wine bottle has the same energy, the lowest energy possible for that system. Rotations about the symmetry axis of the wine bottle can tune any of these outcomes into another.

When circumstances compel the existence of degenerate vacua and chance chooses one of them, we say that the symmetry is spontaneously broken. This is not to say that the symmetry is only approximate, or otherwise defective. What might have been an overt symmetry of the theory or the apparatus is now a hidden, or secret, symmetry of the outcome, because the vacuum state does not exhibit the symmetry.

The phenomenon of spontaneous symmetry breaking is not confined to the symmetries of space and time, such as rotation and translation; it shows up as well for the abstract symmetries that dictate interactions. The first example of this kind, explored in 1950 by Vitaly Ginzburg and Lev Landau of the Lebedev Physical Institute in Moscow, has much in common with our pearl in the wine bottle. The Russian theorists sought to understand how the two remarkable features of superconductors—the absence of electrical resistance and the exclusion of magnetic field—might be linked. They imagined two kinds of actors for their superconductivity script: charge carriers of the familiar sort, carrying current with energy-dissipating resistance, and exotic newcomers—let's call them supercarriers—with the astonishing power to transport charge without any resistance at all. Like Dirac's script for quantum electrodynamics, the play that Ginzburg and Landau wrote for these characters is based on quantum-mechanical phase symmetry. The supercarriers change the outcome.

To learn how the system behaves, we identify the state of lowest energy, which will depend on how many supercarriers are active. At temperatures above the critical temperature for superconductivity, activating the supercarriers must cost energy, and so be disfavored.

Ginzburg and Landau offered a visual aid, representing the energy of the would-be superconducting substance as a shape with cylindrical symmetry. The distance from the central axis measures how many supercarriers are active, and the position around the circumference indicates the quantum phase. The elevation represents the energy. If it costs some energy to activate the supercarriers, and the more supercarriers the more energy, the energy surface will be shaped like a wine glass. The state of lowest energy corresponds to no supercarriers—it is like placing a pearl at the bottom of the glass—so quantum electrodynamics behaves normally. The photon is massless as usual, magnetic flux comes and goes as it pleases, and current flow experiences resistance. No direction is selected, and so the quantum phase symmetry is not called into question. We say that a symmetry is manifest—overt—if its action leaves the vacuum state unchanged.

Below the critical temperature, Ginzburg and Landau observed, it must be beneficial to activate supercarriers; doing so must tend to lower the system's energy. Taking into account the requirements of a well-behaved theory, they asserted that the energy surface should have a shape much like the wine bottle we have described. (Others visualize a sombrero.) Now, like the pearl in our student exercise, the system will settle in a state of minimum energy that is governed by its population of active supercarriers and a randomly chosen phase. When a symmetry operation changes one vacuum state into another, the symmetry is said to be hidden. The system is now superconducting, thanks to the active supercarriers, and, as a little algebra shows, the photon has acquired a mass within the superconductor. The lesson is general: a gauge boson related to a hidden symmetry is massive.

In outline, that is all we need to know to construct a generalization of Fermi's theory in which the W boson has a mass. However, a hint is not a comprehensive framework; many details needed to be learned before all the elements were in place and understood. By 1957, John Bardeen, Leon Cooper, and Robert Schrieffer at the University of Illinois presented their microscopic theory of superconductivity, which has far greater explanatory power than Ginzburg and Landau's instructive short subject. It took nearly a decade for theorists to make complete sense of the Bardeen-Cooper-Schrieffer theory and to learn in detail what spontaneous symmetry breaking means for either global symmetries or gauge symmetries.

Progress was driven both by condensed-matter physicists and by particle physicists investigating relativistic quantum field theories. Curiously, overcoming Glashow's stumbling block to a theory of the weak interactions was the primary motivation of no one. The particle physicists were more concerned with questions of principle or hopes of rescuing Yang-Mills theory as a description of strong nuclear forces by giving masses to its force particles.

All these efforts culminated in 1964, when three groups, working independently, showed just how hidden symmetries lead to massive force carriers in gauge theories. Writing in distinct styles that reflect their individual scientific backgrounds, François Englert and Robert Brout of the Free University of Brussels, Peter Higgs of the University of Edinburgh, and Gerald Guralnik, Richard Hagen, and Tom Kibble, working at London's Imperial College, brought clarity to what had been a perplexing subject.

Their work showed that if any symmetry operation changes the vacuum state into another, the corresponding gauge boson—force particle—acquires mass. Because the gauge symmetry is only hidden, not eliminated, it enforces mathematical relationships that govern how the particles in the theory interact. These are beautiful to see; the intricate interplay that we witness while carrying out a calculation is akin to observing irregularly shaped forms assemble themselves into a flawless and elegant shape. These insights open the possibility that massive force particles are compatible with gauge-theory discipline that enforces good high-energy behavior.

We look back on this work as a great turning point in physicists' quest to understand the fundamental interactions, but at the time no one rushed to exploit the marvelous tool that the six theorists had created, not even the toolmakers themselves. Since we were not quite conscious of contemporary research at the time, we can only surmise—examining the literature—that other interesting developments in experiment and theory saturated the community's attention. Some of these were indeed attention grabbing: the discovery of the Omega-minus particle, which we have already discussed, and then in the years to come, the detection of the microwave background radiation, a relic from the early universe, and the observation of a subtle difference in the behavior of matter and antimatter.

The lull ended in 1967, when Steven Weinberg put forward "A Model of Leptons." As Glashow had done, he opened by citing a motivation and an obstacle. "Leptons interact only with photons, and with the intermediate bosons that presumably mediate weak interactions. What could be more natural than to unite these spin-one bosons into a multiplet of gauge fields? Standing in the way of this synthesis are the obvious differences in the masses of the photon and intermediate [boson], and in their couplings." Weinberg

bet that he could understand these differences by imagining that the symmetries relating the weak and electromagnetic interactions are hidden by the
character of the vacuum state.

He did not address beta decay—the first indication of weak interactions—and
other processes involving protons and neutrons because he didn't see how to
deal with their weak interactions. Quarks were still hypothetical, neither
established nor regarded as an entirely respectable hypothesis. Quantum
chromodynamics, the theory of strong interactions based on quarks and
gluons that we described in chapter nine, "A New Law of Nature," would not
emerge for another six years. Five decades on, we know more about the basic
constituents, so we will describe the theory in today's context.

Following Weinberg, we begin our dramatis personae with the leptons. The
leptons that carry electric charge –1 (electron, muon, and tau) exist as both
left-handed and right-handed entities. Only the left-handed neutrinos (one for
each lepton flavor) are included in the model. All the charged leptons—left-
handed and right-handed—experience electromagnetic interactions. The
three observed flavors of neutrinos are only left-handed; they are linked with
their charged counterparts by the weak interaction. We do not know whether
right-handed neutrinos exist in nature, nor what properties they might have.
If they do exist, they must experience neither electromagnetism nor the
beta-decay interaction, so we haven't devised a way to detect them directly.
Neutrino practitioners call these hypothetical neutrinos "sterile" because they
are indifferent to the known interactions.

Putting aside the strong interactions but admitting quarks to the cast of
characters, we can amend Weinberg's opening to read, "Leptons and quarks
interact with photons and with the intermediate bosons that presumably
mediate weak interactions." Matching the six flavors of leptons are six flavors
of strongly interacting constituents, all present in left-handed and right-
handed versions. Like the left-handed leptons, the left-handed quarks can
be transformed from one flavor to another by the weak interaction. To very
good approximation, the up quark is paired with down, charm with strange,
and top with bottom. Both left-handed and right-handed quarks experience

electromagnetism; the strength is governed by the quark charges: +2/3 for up, charm, and top, and −1/3 for down, strange, and bottom. No charge-changing weak interaction has yet been detected among the right-handed quarks, which is to say that no right-handed partner of the W boson is known.

The pairs connected by the weak interaction—electron neutrino with left-handed electron, left-handed up quark with left-handed down quark, and the rest—suggest one element of a gauge symmetry. It is called (left-handed) weak isospin for its similarity to the strong-interaction isospin that links proton and neutron. Just as in Yang-Mills theory, this gauge symmetry implies three spin-one gauge bosons, carrying charges +1, 0, −1. The charged ones are candidates for the W bosons; the role of the neutral one is to be determined. To incorporate electromagnetism, a quantum phase symmetry is called for. As Glashow had done, Weinberg chose a symmetry based on a quantum number called weak hypercharge, which combines electric charge and weak isospin. That introduces another electrically neutral gauge boson, this one coupled to weak hypercharge.

All this is very elegant, but it makes for a rather denatured representation of the real world. To this point, the construction implies the existence of four massless weak and electromagnetic force particles, whereas nature has but one, the photon. There is another shortcoming. Paul Dirac's work showed that the electron's mass joins the left-handed and right-handed pieces together. That is no problem in quantum electrodynamics; they both carry the same electric charge, and so are affected identically by the electromagnetic phase symmetry. But in the case at hand, left-handed and right-handed components behave differently under the symmetries. Only the left-handed bits are affected by the weak-isospin symmetry, and the weak hypercharges are also different for the left-handed and right-handed pieces. Quark and lepton masses are therefore incompatible with the gauge symmetries, so the constituent particles all must be massless. This is a point worth emphasizing: because the weak interaction treats left-handed and right-handed particles differently, we need to find a mechanism that gives mass to the pointlike, seemingly elementary, fermions. Parity violation has a new implication. This is both a burden and an opportunity.

To set up a situation in which the full symmetry can be hidden, it is necessary to introduce a new element: Weinberg chose a family of two pairs of spinless particles with charges +1 and 0, 0 and −1. These couple to weak isospin and weak hypercharge, respecting those symmetries. They also interact among themselves to contrive the familiar wine-bottle (or sombrero) energy surface and to select at random a vacuum state that hides the weak-isospin and weak-hypercharge symmetries but respects electromagnetic phase symmetry. As we have learned to expect from the Meissner effect and subsequent developments, force particles that correspond to hidden symmetries become massive. The two charged gauge bosons become the W^+ and W^- particles, as desired. The neutral weak-isospin gauge boson and the weak-hypercharge gauge boson rearrange themselves into a massless photon and a massive particle named Z^0. This new force particle mediates a previously unknown weak interaction that changes neither charge nor flavor. The "neutral-current" weak interaction (to differentiate from the "charged-current" beta-decay interaction) acts on both left-handed and right-handed particles, but differently. The existence of this new interaction stood out as the first prediction on which the electroweak theory, as this synthesis is called, would stand or fall.

This is heady stuff, but Weinberg's model was not an immediate sensation. As the late Harvard theorist, Sidney Coleman, commented, "Rarely has so great an accomplishment been so widely ignored." According to the INSPIRE database of the high-energy physics literature, only twelve of the more than 14,000 citations to Weinberg's 1967 paper appeared in the first four years after publication. Weinberg himself returned to other matters. In 1968, Abdus Salam of Imperial College and the International Center for Theoretical Physics in Trieste announced in a lecture that he had independently developed the same theory, elaborating his earlier work with John Ward. He, too, did not immediately pursue the electroweak theory.

Then came two important developments. In 1970, Glashow, along with John Iliopoulos and Luciano Maiani, showed how to incorporate the strongly interacting particles. At the time, when quarks were still regarded with suspicion by many, only three quark flavors were indicated: up, down, and strange, along with two pairs of leptons. Glashow, Iliopoulos, and Maiani showed that quarks must come in pairs to match the leptons and insisted that a fourth quark, which we now know as charm, had to exist. A year later, Gerard 't Hooft, a graduate student working with Martinus (Tini) Veltman at the University of Utrecht in the Netherlands showed how a gauge theory with hidden symmetry could be a consistent—and calculable—many-act theory. By 1973, Benjamin Lee and Jean Zinn-Justin at Stony Brook University, as well as 't Hooft and Veltman, had constructed rigorous proofs that the electroweak theory developed from the ideas of Glashow, Salam, and Weinberg made mathematical sense. How would experiment respond?

Before the electroweak theory took shape, no one had written a one-act play in which a muon neutrino or antineutrino bounces off an electron and the two particles emerge—like billiard balls—with their identities unchanged, but with energy and momentum altered. In Weinberg's model, a muon-(anti) neutrino and electron can exchange a Z boson. This is the simplest example of a neutral-current interaction that might be studied in the laboratory. In July of 1973, experimenters at CERN, led by André Lagarrigue, presented the first specimen of a putative neutral-current event. It was a photograph taken in the Gargamelle bubble chamber that showed, with high probability, an electron recoiling from an encounter with a muon-antineutrino. Named for the mother of the giant Gargantua in François Rabelais's epic, Gargamelle was a cylinder nearly five meters long and two meters in diameter, filled with the refrigerant bromotrifluoromethane that provided both a large target mass and high sensitivity to electrons and photons. Since the neutrino interacts only weakly, it leaves no ionization contrail in the chamber.

Even a flawless specimen does not necessarily prove a discovery; alternative interpretations, however improbable, are difficult to rule out. But the Gargamelle team also presented evidence that neutrinos scattered from protons and neutrons in the heavy liquid target without changing into muons. Experimenters at Fermilab seemed at first to support the discovery, then withdrew their positive evidence, and finally came out in favor of neutral-current interactions. Their struggle was parodied as "alternating neutral currents."

Within a year, no doubts remained; Gargamelle had indeed discovered a new fundamental weak force. The phenomenon was the focus of neutrino-scattering experiments at CERN, Fermilab, Argonne National Laboratory, and Brookhaven National Laboratory. Other experimental approaches followed in short order; many experimenters wanted a piece of the action.

Once neutral-current interactions were observed, the search for charm became urgent. The paired quarks guaranteed that Z-boson interactions would not change quark flavor—an essential point because meson decay experiments had shown that transitions from strange quarks to down quarks were either absent or greatly suppressed. Glashow was highly confident. At an April 1974 Conference on Experimental Meson Spectroscopy in Boston, he presented "Charm: An Invention Awaits Discovery," setting out the case. When he and "B. J. Bjørken" had speculated about a fourth quark flavor a decade before, their justification was a simple notion of symmetry. There were at the time four known leptons (electron, muon, and their neutrinos), so why shouldn't there also be four kinds of quarks? Now, citing his work with Iliopoulos and Maiani as better motivation for charm, he laid out search strategies. Ever ebullient, he offered three possibilities for the next conference: "1. Charm is not found, and I eat my hat. 2. Charm is found by hadron spectroscopers, and we celebrate. 3. Charm is found by outlanders, and you eat your hats." The rest, thanks to the November Revolution, is history. An electroweak theory began to seem inevitable. Was the 1971 version really the answer?

As new experimental techniques came into play over the next five years, the theory's reputation fluctuated. Some experiments confirmed the predictions, then another new observation deviated. Ambulance-chasing theorists had the time of their lives, contriving new variations to accommodate every experimental twitch. At length, the evidence stabilized in support of the 1971 edition of the theory. An elegant electron-scattering experiment carried out at the Stanford Linear Accelerator Center by Charles Prescott and collaborators exploited an interference between the weak and electromagnetic interactions to amplify the Z boson's effect. Their observations showed that the

Z boson mediated parity-violating interactions between electrons and nucleons just as Weinberg and Salam—supplemented by Glashow, Iliopoulos, and Maiani—predicted. Like the charged-current interaction, the neutral-current interaction could differentiate left from right! Hope grew that this electroweak theory might indeed be a new law of nature.

The unification of electromagnetism and the weak interaction is not complete, because the theory is built on two separate symmetries, each with its own interaction strength. The strengths of electromagnetism and the charged-current interaction are fixed in advance; details of the neutral-current interaction, including its strength, depend on how the neutral bosons corresponding to the weak isospin and weak hypercharge symmetries rearrange to make the photon and Z. Until we have a more complete theory, this information must be inferred from experiment. The common origin of the photon, W^+, W^-, and Z means that the intrinsic strengths of the interactions they mediate are not very different. Why, then, does the weak interaction seem so weak? It is because the exchange of a massive particle suppresses reaction rates.

If the theory is correct, a single number, called the weak mixing parameter, determines a host of experimental observables. It governs the detailed properties of the neutral-current interaction, as well as the masses of the weak force particles. Within the electroweak theory, the mass of the W boson must be at least forty times the proton mass; the predicted Z-boson mass is no less than the W mass. The force particles would be unstable, decaying into a lepton plus antilepton or quark plus antiquark pairs. A W^+ might decay into a positron and an electron neutrino or into an up quark and an anti-down quark. The Z boson would decay into a particle-antiparticle pair: a positive muon plus a negative muon, a charm quark and an anti-charm, or even a neutrino and an antineutrino.

The massless photon, which corresponds to the surviving electromagnetic symmetry of our theory, illustrates one lesson of the work of Englert and Brout, Higgs, and Guralnik, Hagen, and Kibble: For any element of the gauge symmetry that does not alter the vacuum state, the corresponding force particle remains massless. That is hardly a surprise, but when we are dealing with hidden symmetries, there is a kicker: a new massive scalar (spinless) particle appears. That particle is an excitation of a scalar field, in the same way that the force particles are excitations of gauge fields. In its interactions with the W and Z, the scalar particle respects the gauge symmetry, but the gauge symmetry does not dictate interactions among the four scalars introduced to hide the symmetry. Three of the four were swallowed by the fields that became the massive W^+, W^-, and Z; the fourth emerges as the new massive scalar. Peter Higgs gave what most particle physicists have regarded as the easiest-to-parse statement of the existence of massive spin-zero particles. For that reason, as well as some accidents of citation history, such particles are popularly known as Higgs bosons. If it existed, the Higgs boson would be an avatar of the mechanism—akin to the Ginzburg-Landau scheme—by which the electroweak symmetry is hidden.

We owe to Weinberg the realization that the scalar might have another function of great conceptual importance. Interactions of the Higgs field with the constituents can generate masses for the quarks and charged leptons. (The combination of left-handed and right-handed fermions joined to Higgs respects the gauge symmetry.) If that is realized in nature, it helps us understand how those bits of nothing acquire mass—through their interaction with the Higgs field. By itself, the electroweak theory does not reveal why the quark and charged-lepton masses assume the values they do. Nor does the electroweak theory predict the mass of the scalar boson; it is another unknown quantity, alongside the weak mixing parameter, that must be drawn from observations. Once the mass is specified, the Higgs boson's decay signatures follow at once.

By the late 1970s, the electroweak theory had passed many tests, but only at the one-act level. Would it turn out to be, like the Fermi-inspired theory, only a reliable low-energy guide, or would it live up to its promise as a precision tool? Many predictions remained unexplored, and existing measurements needed to be taken to a level of precision that would be sensitive to quantum corrections. And, of course, it is one thing to have circumstantial evidence for the force particles W and Z, quite another to produce and study them in detail. If all this could be done, would the Higgs boson really be there? Buoyed by the great wave of discoveries of the 1970s, physicists began to devise ways to reach the high energies that would be required to discover the electroweak bosons and to test the theories. The path to the essential new instruments needed to do so is our next topic.

14

THE STUFF THAT BEAMS ARE MADE OF

At 2:00 in the afternoon on Friday, January 29, 1954, Enrico Fermi took the stage of Columbia University's McMillin Theatre to deliver his retiring presidential address to the American Physical Society. At fifty-two, Fermi was one of the Immortals of modern physics, revered for his accomplishments in research and admired for his ability to present complicated subjects with disarming clarity. On this day, in unhurried, reassuring diction redolent of his Roman origins, he introduced to his audience the emerging field of particle physics.

Fermi recounted the successes of nuclear physics since the discovery of the pion, the agent that binds protons and neutrons into the nuclei of the elements. Hideki Yukawa had imagined the pion as the nuclear force carrier in 1935, but the search for it had been full of twists and turns. In 1947, physicists finally found what Yukawa asked for, "but," Fermi said with a gleam in his eye, "to our dismay, we got a lot more." There was the muon, at first mistaken for the pion, which provoked I. I. Rabi's legendary "Who ordered *that*?" Then came a "stupendously great" number of so-called elementary particles—enough to test the classification mettle of a botanist. Nuclear physics was evolving from the study of the nucleus, radioactivity, and nuclear chemistry—the simple rearrangement of protons and neutrons—toward the investigation of pions, strange mesons, hyperons, and even antinucleons. Nuclear physics was giving birth to particle physics. To make sense of this "tantalizing vista," Fermi said, physicists would need a host of new data at higher and higher energies.

Early experiments in particle physics were carried out using the found beams of cosmic rays that shower our planet from every direction and the alpha particles that emanate from a speck of radium. To Fermi, the energy of the cosmic rays—high-energy protons and iron nuclei that he speculated were accelerated by wandering magnetic fields in the interstellar medium—seemed almost limitless, but there was a catch. Earth's atmosphere filters the primary cosmic rays, so physicists had constructed cosmic-ray observatories on the Jungfraujoch, the Pic du Midi, and other picturesque mountaintops. Even at altitude, the flux was frustratingly small: for the studies Fermi had in mind, the rain of primary cosmic rays was a gentle mist—about one per square inch per hour. (For the experiments we are carrying out today, the flux is about one *per square mile* per hour!) Physicists were clamoring for larger and more powerful particle accelerators for the same reason we create artificial sources of illumination to supplement the sun, moon, and stars: to make beams with the qualities they needed available on demand.

The July 1950 issue of *Physics Today* had heralded "the springtime of Big Physics." The Massachusetts Institute of Technology alone boasted seven accelerators exceeding one million electron volts, with two more on the way. Machines exceeding 300 MeV were popping up at major research institutions, including Columbia, Berkeley, Illinois, and the General Electric Company. By 1952, the Cosmotron at Brookhaven National Laboratory surpassed the billion-electron-volt mark, boosting protons to 3.5 GeV the following year, at a cost of nearly nine million dollars. A proton synchrotron at Birmingham University in England joined the electron-volt-billionaires' club in 1953 at 1 GeV, and the Berkeley Bevatron reached 6 GeV two months after Fermi's talk to the Physical Society. These frontier accelerators were growing in size, as well as cost and energy: The Cosmotron ring was about seventy meters (228 feet) around, and the Bevatron 120 meters (394 feet).

Projecting the energy, size, and cost of particle accelerators forty years ahead to 1994, Fermi dreamed of an instrument that would reach far into the unknown, to an energy of five million billion electron volts—5,000 TeV. He displayed a fanciful "preliminary design" for a gargantuan machine more than 31,000 miles around, encircling the globe at an altitude of a thousand miles, well above the lanes plied today by the International Space Station. Fermi estimated that his dream accelerator would cost $170 billion. A machine of similar design to answer our generation's burning questions would be more than twenty times as large—a loop about two-fifths as large as the moon's orbit.

According to the charming testimony of his disciple, Vincenzo Viviani, Galileo Galilei conducted the first accelerator experiments near the end of the 16th century to study the motion of falling bodies. Trudging up the 294 marble steps of the Leaning Tower, worn and polished by the passage of myriad feet, the modern visitor to Pisa can imagine walking in the Maestro's footsteps. He carried different weights of the same material to the top, dropped them off at the same instant, and heard them impact the ground together, about three and a half seconds later. Galileo called his experiments *cimenti*—ordeals, for the unyielding way he put hypotheses to the test.

Earth's gravitation—the acceleration of massive bodies toward the center of the Earth—is the same for us as it was for Galileo. Near Earth's surface, speed increases by twenty-two miles per hour during each second of fall. Jump from a sixteen-foot perch, and you will hit the ground in one second at twenty-two miles per hour: *oooff*! Jump from a sixty-four-foot perch, and you will hit the ground in two seconds at forty-four miles per hour: *splat*! Jump from the highest structure and air resistance will hold your speed under the terminal velocity of 125 miles per hour, but that is small comfort on impact.

Because we cannot change the strength of Earth's gravity, we can extend Galileo's method only by eliminating air resistance and climbing higher or falling further. Three centuries before Enrico Fermi described his fantasy machine, Galileo analyzed the ultimate terrestrial gravitational accelerator in his *Dialogue Concerning the Two Chief World Systems*. A cannonball dropped into an evacuated column bored along Earth's axis would accelerate all the way to the center, flashing by about twenty-one minutes later at nearly 18,000 miles per hour. Once past the center, the projectile would slow under the tug of gravity.

Approaching the other side of the Earth like a ball tossed into the air, it would come to momentary rest at the surface, fall back, then shuttle forever between the antipodes. Eighteen thousand miles per hour is a big step from the seventy-five miles per hour of Galileo's legendary test masses, but—at five miles per second—it is a tiny fraction of the speed of light, paltry even compared with the 8,000 miles per second of the alpha particles emitted by radium.

Actually using a gravitational accelerator for experiments would not be terribly practical. It would be inconvenient to build detectors in the molten core of the Earth, though lab directors might delight in telling certain experimenters to go to Hell.

Since we call the big machines accelerators, it is natural to wonder just how fast the protons are traveling. "Really fast," and "nearly at the speed of light" are not completely satisfying responses, and to be told that the 980-GeV protons and antiprotons in the Fermilab Tevatron moved at 99.999 954 167 percent of the speed of light, whereas the 6.8-TeV protons in CERN's Large Hadron Collider move at 99.999 999 048 percent doesn't engage our intuition, even if it is geeky fun. It comes a little closer to human terms to know that the Tevatron's protons passed Chris's office window at Fermilab 47,700 times each second, and that protons zip around the LHC 11,245 times a second. We find it a bit more tangible to note that the Tevatron beams moved 307 miles per hour (495 kilometers per hour) slower than the speed of light—which is more than 670 million miles per hour, nearly 1,080 million kilometers per hour—while the LHC beams lag that limiting speed by just 6.4 miles per hour (10 kilometers per hour). An Extra-Large Hadron Collider with two times, or ten times, or even a thousand times the energy of the LHC would increase the speed of protons only by a jogger's pace. According to Einstein's special theory of relativity, it would require literally an infinite amount of energy to boost protons to the speed of light.

The Great Galilean Accelerator and Fermi's extrapolation of the technology of his day dramatize the need for new ideas and new technologies to propel experiments toward higher energies. Questions we now ask demand energies many times greater than Fermi envisioned, to look closely at quarks and leptons—today's basic constituents—and shake them up, to learn whether they can be excited or taken apart. We need probes of short wavelength, or high energy, to examine small structures. We want to make very heavy particles from great amounts of energy. We want to investigate new phenomena by going—metaphorically—where no one has gone before. Like Fermi and his contemporaries, we climb the mountain of high energy not merely because it is there, but because it matters.

What we learn at high energy shapes our understanding of everyday phenomena and our appreciation of why the world is as we observe it to be. By learning the laws that prevail at high energies, we hope to understand the evolution of the universe from newborn to its present state.

Six decades after Fermi, every particle physicist can sketch a design vastly superior to the master's. Since 2011, the Large Hadron Collider, an accelerator ring only twenty-seven kilometers (seventeen miles) around has delivered proton-proton collisions at nearly three times the energy he foresaw. The contrast between Fermi's dream machine and the LHC invites two seemingly contrary questions: "Why must accelerators be so big?" and "How can accelerators be so small?"

Modern particle accelerators rely on a few simple ideas, judiciously combined. There is no better way to see the logic behind an instrument to provide the highest-energy collisions ever available and unlock the secrets of the origin of mass than to invent it. That's just what we're going to do in the next few pages.

As a rule, particle physicists choose simple projectiles that do not needlessly complicate the interpretation of experiments. Intense beams are best made of particles that do not decay. Electrically charged particles can be accelerated and manipulated by electric and magnetic fields. The simplest of the stable, charged particles that nature supplies in abundance are the electron and the proton.

Electrons boil off a metal plate heated to about 1,700° Celsius (3,000° Fahrenheit) in a high vacuum. They are easily tugged free with the electric field supplied by a small positive voltage. Protons, the positively charged nuclei of hydrogen atoms, can be made to shed their orbiting electrons simply by flowing a stream of hydrogen gas through a spark in a vacuum, where the famous explosive potential of hydrogen doesn't come into play.

Collisions involving electrons are the simplest to interpret, because—as closely as we have been able to look, beyond a billionth of a billionth of a meter—an electron has no internal structure. A proton is composite, a collection of quarks and antiquarks and gluons, so a beam of protons of definite energy can be considered a mixed beam of quarks, antiquarks, and gluons with a distribution of energies. It is richer, but less precisely defined, than a beam of electrons. With our current technology, far higher energy is attainable with protons than with electrons.

Antiprotons and antielectrons, or positrons, are also stable, charged particles, but they must be made to order in the laboratory. Pairs of electrons and positrons are made copiously when a high-energy electron hits a metal target. Antiprotons are made less frequently in high-energy collisions, but Fermilab achieved a useful production rate of 250 antiprotons per million protons on target. That meant that Batavia, Illinois, for many years had the greatest store of antiprotons on Earth, reaching a peak of 7.59 trillion in May 2010. To underscore the exquisite rarity of antiprotons, let us note that the number of *protons* in a teaspoon of water is 200 billion times greater! Muons, with a mean life at rest of 2.2 microseconds, show promise for future machines, provided they can be gathered efficiently and accelerated quickly. The mean life of an energetic particle is extended by the ratio of its energy to its rest energy, or mass.

The obvious way to accelerate charged particles to high energies is to apply a very large voltage, but—as the experiments on Monte Generoso illustrated—it is no easy feat to produce, control, and apply millions of volts, let alone the trillions of volts we require today. Repeated application of small voltages is a far easier path to higher energies.

Ernest Lawrence's great insight was that a magnetic field could bend a beam of charged particles so it passed the same accelerating station again and again. To visualize the idea behind Lawrence's cyclotron, think of a can of tuna cut in two so that each half is shaped like a capital D. Remove the tuna and put the two halves back together with a small gap between them. Place the dees inside a magnet so the magnetic field passes in the top and out the bottom. A charged particle shot across the gap near the center of the can wheels in the magnetic field like an airplane in a holding pattern. The radius of the bend is proportional to the momentum of the particle. Suppose now that every time the particle crosses the gap it receives a small push, in the form of an accelerating voltage measured in thousands, not millions, of volts. As the particle gains energy, it spirals out toward the edges of the chamber, like Yeats's falcon turning and turning in the widening gyre, its path bent ever less sharply by the fixed magnetic field. When the beam reaches the desired energy, in an orbit of predetermined size, it can escape through a strategically placed vacuum window (a hole with a thin covering) in the side of the can or hit a target within the cyclotron. The first cyclotron, four and one-half inches in diameter, reached an energy of 80,000 electron volts in January 1931.

Armed with sealing wax, cyclotron builders waged epic struggles to maintain a good vacuum inside their machines so that collisions with air molecules would not stay the particles from their appointed rounds. When the vacuum held and the cyclotron worked, the stopping power of air outside the accelerator presented a visible measure of an atom smasher's might. Lawrence's team exulted over the five-foot-long purple streak of ionized air produced when they shot the sixty-inch cyclotron's 16-MeV deuteron beam into the laboratory. Visiting scientists, reporters, and patrons were duly impressed, too. The February 5, 1940, issue of *Life* magazine featured a centerfold of the purple glow issuing from the cyclotron over the headline, WITH HIS ATOM-BUSTING CYCLOTRON, LAWRENCE OF CALIFORNIA WINS THE NOBEL PRIZE IN PHYSICS.

The great volume of the sixty-inch cyclotron was a source of pride to its builders when they assembled inside its 220-ton magnet for a group portrait, but the spiral orbit makes the cyclotron a clumsy instrument for producing beams of extremely high energy. Enlarging the magnet and vacuum chamber by even a hundred times is a staggering thought. Vancouver's TRIUMF laboratory is home to the largest cyclotron ever built, nearly sixty feet in diameter. It accelerates prodigious numbers of protons—a million billion per second—to three-quarters of the speed of light, in one three-thousandth of a second.

A better solution for high energies is to inject the particle into an initially weak magnetic field that increases in strength as the particle gains energy, so the beam traces out the same orbit throughout its acceleration. Since the beam does not spiral outward, most of the center of the accelerator can be eliminated. A magnetic field is needed only in a narrow circular band around the orbit. The beam is contained in an evacuated doughnut that is mostly hole. One or more accelerating stations can be placed around the ring. The strength of the magnetic field that steers particles around the ring must be carefully synchronized with the energy of particles in the beam. If the magnetic field is too weak, the beam will skid into the outer wall of the beam pipe. If too strong, the beam will curl into the inner wall of the tube. The frequency of the accelerating voltage may also need to be adjusted, in synchrony with the circulating beam. Particles not exactly in time with the oscillating voltage will be accelerated in a way that brings them into step. Those moving a little faster than the nominal rate receive less than the nominal boost, while laggards receive a stronger push.

This *phase stability principle*, discovered in the mid-1940s by Vladimir Veksler at the Lebedev Institute in Moscow and Edwin McMillan at Berkeley, is a key insight. All of our high-energy proton machines are *synchrotrons*, built on this concept.

Just how small the beam pipe and magnet aperture can be made depends on how well the beam is collimated and how closely it follows the ideal path. A minor error in aiming the beam when it is injected, mis-steering caused by the inevitable imperfections of the confining magnetic field, and the mutual electrical repulsion of protons all cause particles to wander from the prescribed orbit. Keeping the beam on track in a tight bunch is a bit like shaping bread dough made with particularly lively yeast. Left alone, the loaf will balloon in all directions, but you can keep it to the desired dimensions by squeezing it side to side and top to bottom in turn. During the summer of 1952, Ernest Courant, cyclotron pioneer Stanley Livingston, and Hartland Snyder at Brookhaven National Laboratory on Long Island showed that a very small, dense beam can be produced by alternately squeezing top to bottom and side to side with magnetic quadrupole lenses.

The quadrupole lens has four magnetic poles arrayed around the beam pipe at 1:30, 4:30, 7:30, and 10:30 o'clock. In the same language, a bending magnet is a dipole, with two magnetic poles, north and south, at 12:00 and 6:00 o'clock. In one orientation, say north, south, north, south, the quadrupole magnet focuses particles vertically (squeezes top to bottom) while defocusing horizontally (spreading out side to side). Rotated by a quarter turn, or three hours, to south, north, south, north, the quadrupole magnet focuses particles horizontally (squeezes side to side) while defocusing vertically. Physicists call this technique alternating gradient focusing, or strong focusing. Remarkably, after the Brookhaven scientists published their idea, the American-born engineer Nicholas Christofilos, who was working on his own in Athens, returned to the United States and demonstrated through patent applications and an unpublished report that he had invented the same concept in 1950. Strong focusing gives a remarkable degree of control over the beam. Nowhere in the Large Hadron Collider, the grandest machine we have, is the beam profile broader than a millimeter.

The idea of the synchrotron and the invention of strong focusing have enabled accelerators to shrink even as they have grown. As physicists reach for higher energies, the outer dimensions of accelerators grow ever larger, but what is on the inside, the profile of the beam pipe under high vacuum, shrinks. The vacuum chamber of the Berkeley Bevatron, a six-GeV proton accelerator built without strong focusing in 1954 to discover the antiproton, is one foot high by four feet wide. Fermilab's Main Ring, built in the early 1970s to reach four hundred billion electron volts, had an elliptical beam pipe only about two inches by four inches, so the volume under vacuum was smaller than the Bevatron's. Bevatron magnets are twenty and one-half feet wide and nine and a half feet high; Main Ring magnets are twenty-five inches by fourteen inches. Although at 2π kilometers (3.9 miles) in circumference the Main Ring is more than sixty times as long as the Bevatron, there is more steel in the Bevatron magnets. The reduction in the size of magnets and vacuum system dramatically decreased the cost of accelerators per million electron volts, a trend that must be maintained by other innovations if we are to achieve still higher energies.

A device that is smaller on the inside but larger on the outside brings its own challenges. No power on Earth can ensure that a beam tube miles long and a few inches across will survive installation free of dents and clear of foreign objects. Just as a distracted surgeon will leave an instrument or a sponge inside a patient, a technician who temporarily packs the beam pipe with paper tissues during a messy procedure will forget to remove them. Alien lunchbags—even beer bottles—will materialize. As beam pipes shrink in aperture, fewer bizarre objects can find their way where they don't belong, but the problem of removing obstacles from the beam pipe becomes more challenging.

When long stretches of Fermilab's Main Ring vacuum doughnut had to be swept of debris in 1971, physicist Bob Sheldon thought of the ferrets used to flush rabbits from their burrows in his native England. The grand sum of $35 brought a black-footed ferret named Felicia—a pound and a half of weasel curiosity—from a game farm in Minnesota. Wearing a leather harness that trailed a strong, lightweight cord, the masked prairie-dog predator scampered a hundred meters through pipes that would transport the proton beam to experimental halls. After Felicia emerged from the stainless-steel burrow, technicians attached a parachute-like swab to the string and pulled it back through the pipe while Felicia celebrated with a meal fit for a mink—chicken livers and fish heads. Felicia became a media darling, but the half-mile-long stretches of vacuum pipe in the accelerator itself were too much for her. She was replaced by a spear powered by compressed air, an elegant but far less adorable solution.

The first experiment to take data at Fermilab, as the machine was being commissioned in the early hours of February 12, 1972, measured particles produced as the circulating protons collided with a rotating foil inside the beam tube. It was symbolic of the international character of particle physics that the experimenters represented the University of Rochester, Rockefeller University, the National Accelerator Laboratory (as Fermilab was then known), and the Joint Institute for Nuclear Research at Dubna in the Soviet Union. In our great accelerators, internal-target experiments are limited by the requirement that they not disturb the circulating beam appreciably. Experiments in which the primary beam is extracted from the accelerator at top energy and either used directly to study collisions with a variety of targets or to produce beams of rare, unstable, or neutral particles that cannot be accelerated efficiently themselves account for much of the richness and productivity of the proton and electron accelerator laboratories.

No advance in the technique of particle physics has been more consequential than the development of colliding beams. Smashing an energetic beam into a target at rest is a highly inefficient way to concentrate great energies in a tiny volume. According to special relativity's laws of motion, only one-sixteenth of 1 percent of the energy in a 5,000-TeV beam that collides with a stationary proton can go into particle production. All the rest is taken up by the forward motion of the products. To double the energy available for producing new particles would mean quadrupling the beam energy.

Head-on collisions of oppositely directed beams of the same energy are far more efficient for producing new particles. In this case, the target is another beam, and essentially all the energy of the two beams can be devoted to the creation of new particles. Doubling the energy in each beam doubles the energy available for particle production. The first written evidence of this insight is in a patent application filed in 1943 by the same Rolf Wideröe whose earlier publication had led Ernest Lawrence to conceive the cyclotron back in 1929. Wideröe observed that "head-on collisions must be avoided on the road, but collisions of protons could be of great benefit to research." In practice, making two beams of particles collide is like hitting a swarm of birdshot with a blast from a second shotgun.

Exquisite timing and deadly aim are needed, of course, but hits would be rare indeed unless the shotgun fired many pellets in a tight bunch.

The dream that colliding beams might be practical instruments for research took a big step toward reality with the publication early in 1956 of a letter to the editor of *The Physical Review* by Donald Kerst and colleagues at the Midwestern Universities Research Association, a hotbed of accelerator innovation. They wrote, "It is well appreciated that the energy which will be available for interactions . . . will increase only as the square root of the energy of the accelerator. The possibility of producing interactions . . . by directing beams against each other has often been considered, but the intensities of beams so far available have made the idea impractical. [New developments] offer the possibility of obtaining sufficiently intense beams . . ." In their short note—just over two columns in the journal—they estimated the number of circulating protons needed for experiments, considered the vacuum that would ensure tolerable backgrounds and losses from scattering on residual gas, and identified important issues in beam dynamics.

Gerard K. O'Neill of Princeton University took up the cause, noting that beams accelerated to high energy in a synchrotron could be transferred into two storage rings with a common straight section in which the opposing beams would interact. Head-on collisions of 25-billion-electron-volt (25-GeV) protons, for which accelerators were being built at Brookhaven and CERN, would be equivalent to a single beam of 1.3 trillion electron volts (1.3 TeV) hitting a stationary proton.

Colliding beams were first realized at Frascati (Italy), Novosibirsk (USSR), and Stanford in the early 1960s. The Princeton-Stanford Colliding Beams Experiment (CBX), carried out by O'Neill, Carl Barber, Bernard Gittelman, and Burton Richter, entered Chris's student consciousness through a story in the *New York Times* reporting the first collisions of 300 million-electron-volt electron beams. If that seems puny, consider that to achieve the same result in a fixed-target setting would require an electron beam of about 350 billion electron volts (350 GeV), which was only attained a quarter-century later at Fermilab! The *Times* reported that "two electrons come close enough for a collision only once every fifteen or twenty minutes." Soon thereafter, Chris read that the scientists did not know what would happen when they made the high-energy electrons collide head-on. "What," he thought, "could be more exciting than not knowing the answer?" Thus began his fascination with high-energy colliders, and with the potency of *We do not know . . .*

Over the following decade, a sequence of machines in Novosibirsk, Frascati, Cambridge (Massachusetts), Orsay (France), and Stanford developed the collider art and made pioneering measurements of particles created in electron-positron annihilations. Any lingering suspicion that they were merely niche machines evaporated in 1974 when Richter's team at SPEAR discovered the narrow charmonium resonances, launching the November Revolution. During the discovery shift, psi-particles came at the rate of one per second. By the year 2023, nearly twenty electron-positron colliders in eight countries have contributed decisively to the study of heavy quarks (charm and beauty), and to searches for dark matter and other new particles. Beginning in 1989, the Stanford Linear Collider—a Z factory—and CERN's Large Electron-Positron Collider, LEP, which reached approximately 105 GeV per beam—enabled definitive tests of the electroweak theory that helped establish it as a law of nature. Running at 45.6 GeV per beam, the energy for peak Z production, LEP produced one of those electroweak bosons every second. Proposed electron-positron machines would make possible high-precision studies of the Higgs boson and provide a complementary look at the energy domain of the Large Hadron Collider.

Electrons have no discernible structure, so the full energy of an electron is available in a collision, whereas a proton's energy is shared among its constituents. The simplicity of an electron also means that electron-positron collisions are relatively free of clutter.

However, electrons are lightweight particles that radiate electromagnetic energy readily when they are accelerated—when a force changes their speed or direction. The electrons and positrons circulating at peak energy in LEP, for example, lost about 3 percent of their energy on every turn around the machine to what is called *synchrotron radiation*, and the energy loss grows as the fourth power—the square of the square—of the beam energy. This example shows that there is a practical limit to the sustainable electron energy, for a machine of a given radius, and that we have been operating close to that limit. To push electrons significantly higher in energy, we will have to give up the circular machine's advantage of passing the same accelerating stations many times, and employ a linear accelerator with accelerating stations along its whole length. For a proton—1,836 times the mass of the electron—at the same energy, in a machine of the same size, synchrotron radiation dissipates ten trillion times less energy per turn.

Could a proton-proton collider really open a window to a future of much higher-energy collisions? There were many unknowns—and many doubts. In 1965 CERN, led by its director general, Victor Weisskopf, took the plunge; Brookhaven, lamentably, did not. CERN began operating the first proton-proton collider, the Intersecting Storage Rings (ISR), in 1971—two interlaced rings approximately one kilometer in circumference, crossing at eight interaction regions. The top energy of 31 GeV per beam was equivalent to collisions of a 2000-GeV (2-TeV) beam with a stationary proton target. Over the first few years of running, the rate of collisions moved to tens of thousands, and then millions, per second. ISR experiments overturned a longstanding expectation that the total interaction rate would settle at a constant value in a high-energy "asymptopia"—the rate increases with energy—and suggested the importance of quark-quark scattering in producing particles at large angles to the direction of the colliding beams. Some major discoveries escaped: the charmonium states found at Brookhaven and SLAC in 1974 and the upsilon family discovered in 1977 at Fermilab were produced in great numbers at the ISR, but the early ISR detectors could not see them.

Accelerator science is a practical art that brings together idealized design and real devices. It is a craft propagated through apprenticeship and osmosis, more than formal courses. Nearly all accelerator physicists learn their trade in a few great accelerator laboratories—no more than a dozen—around the world. As experimenters were learning to study proton-proton collisions in the new environment, the machine builders had what project leader Kjell Johnsen called "the finest instrument one can imagine for research in accelerator physics." The ISR experience built confidence that the performance of future proton-proton colliders could be predicted—and attained. The path to high-sensitivity experiments at extremely high energies was now clear.

Physicists built two great proton synchrotrons during the 1970s, both single-beam machines designed to accelerate protons to higher energies than ever before, supplying not only protons but derived beams of pions, kaons, muons, neutrinos, electrons, photons, and other particles for a wide variety of experiments. The Main Ring at the Enrico Fermi National Accelerator Laboratory in Batavia, Illinois, 6.3 kilometers (3.9 miles) around, attained its design energy of 200 GeV in March of 1972 and reached 400 GeV later that year. The machine is housed in a tunnel seven meters below the prairie, but an ingenious composition of earthworks—a berm over the tunnel and a canal to circulate cooling water—contrives a play of light and shadow that is a dramatic sight for air travelers. The Main Ring reached its peak energy of 500 GeV in May 1976. At CERN, the Super Proton Synchrotron reached 400 GeV in June 1976, and has operated for decades at 450 GeV. The SPS is a bit larger than the Main Ring, at 6.9 kilometers around, and more robustly designed. The first accelerator to cross national boundaries, with two-thirds of its circumference in France, it lies forty meters below the surface.

Advances in accelerator technology enable new discoveries, while physics imperatives prod the development of new accelerators. The discovery at CERN in 1973 of the new weak force called neutral currents—a prediction of the nascent electroweak theory—gave urgency to the search for the carriers of the weak interactions, W and Z. At the time, we could estimate the masses to lie between 60 and 100 GeV (about 60 to 100 times the proton mass), too high to be created at any accelerator then in existence.

The Main Ring at Fermilab and the Super Proton Synchrotron at the European Laboratory for Particle Physics each consisted of a single accelerator ring built to deliver high-energy protons to stationary targets. Adding a twin ring to either machine to enable head-on proton-proton collisions seemed impractical. Making collisions with protons in a modest, lower-energy, auxiliary ring would not produce the required energy. If only we had on hand a bottle of antihydrogen, we might create an antiproton beam circulating in the opposite sense to the protons, within the same ring. Antiprotons are rare in our world, but studies by Gersh Budker in Novosibirsk and Simon van der Meer at CERN showed that it might be practical to accumulate intense beams of antiprotons produced in proton-nucleus collisions. Their work inspired Carlo Rubbia, David Cline, and Peter McIntyre to propose in 1976 that either the Main Ring or the SPS might be refitted as a single-ring proton-antiproton collider that would enable a decisive search for the hypothetical electroweak bosons. The sense that the electroweak bosons might be within reach moved Chris to compose a comprehensive primer on how they might be produced and detected.

In a conventional accelerator magnet, current flowing through a coil of copper cable produces a magnetic field that partially magnetizes a carefully formed iron yoke. The magnetic field shaped by the iron guides particles around the accelerator ring. When a low-energy beam is first injected into a large synchrotron, the magnetic field must be small, vulnerable to tiny fluctuations in persistent currents and residual magnetism that would be utterly negligible at high field. If the magnetic field is inconsistent around the ring, the beam may be thrown off its intended track. A synchrotron that increases beam energy too much can be an unruly beast. In the 1950s, Bob Wilson, later the master builder of Fermilab, and Matt Sands of Caltech independently put forward the idea of a cascade of accelerators, each one feeding the next. This way, the minimum magnetic field in each machine is large enough to be regulated precisely. Each of the machines is used to accelerate particles in the energy range for which it is best suited, as an automobile transmission shifts through several gears to attain top speed.

A chain of smaller accelerators fed 8-GeV protons into the Main Ring at Fermilab. In one straight section of the tunnel, a battery of eighteen radio-frequency cavities delivered a nudge to the passing protons, raising the energy by 2.8 MeV per turn, about 125 GeV per second, as the magnetic field strength kept pace. In three seconds, the protons reached the peak energy of 400 GeV. Then, over one second, the protons were extracted from the ring and directed to experiments, after which the magnets relaxed to injection strength. The cycle repeated every ten seconds. The SPS accelerator chain followed a similar cadence.

Conventional magnets are costly to operate because resistive heating in the cables dissipates vast amounts of electrical energy. In collider mode, the counter-rotating proton and antiproton beams would be stored for hours at the highest energy at which the magnets do not overheat. Fermilab scientists judged the Main Ring deficient for conversion to a collider, but CERN, where the SPS could be held for long periods at 270 GeV, took up the challenge. Additional innovations by van der Meer meant that the SPS Collider could attain both the energy and the collision rate the search required. By 1981, more than ten billion antiprotons at a time could be accumulated and concentrated into a fine beam to be accelerated in the SPS along with protons circulating in the opposite direction. In a small fraction of these collisions, a quark contained in a proton would fuse with an antiquark contained in an antiproton to form a W or Z.

The W boson is highly ephemeral; it lives on average a third of a yoctosecond (a trillionth of a trillionth of a second), so experimenters must piece it together from its decay products. Most convenient for the discovery experiments was the disintegration into an electron and a neutrino, the fate of about one W in nine. Roughly one SPS collision in one hundred million would lead to a detectable W production and decay, so if the machine could provide a thousand collisions per second, a few examples might be detected in a month of running. The production and decay of a Z boson into an electron and a positron would be ten times more challenging. Two Underground Area detectors, named UA1 (a 135-person team led by Rubbia) and UA2 (fifty-nine collaborators led by Pierre Darriulat), viewed and analyzed the proton-antiproton collisions. The UA1 detector was gigantic for its day at ten by six by six meters (thirty-three by twenty by twenty feet), and daunting in its complexity.

UA2 was more modest in scope, but was optimized for the W and Z searches. UA1 and UA2 could identify electrons and muons and measure their energies and directions, and could infer the presence of neutrinos through undetected "missing" energy.

Early in 1983, both experiments announced the discovery of the W boson. Only a few months later, first UA1 and then UA2 announced that they had found the Z boson as well. The masses, production rates, and other characteristics aligned with theoretical expectations. It was a stunning triumph for experimental technique, accelerator technology, and the electroweak theory. For their "decisive contributions to the large project which led to the discovery of the field particles W and Z, communicators of the weak interaction," Carlo Rubbia and Simon van der Meer shared the 1984 Nobel Prize for Physics.

The SPS alternated between fixed-target and collider modes until 1991, when collider operation ceased. The antiproton source, which has been progressively improved, enables antimatter experiments to this day, including the production and detailed study of antihydrogen atoms. So far, the properties of the antiatoms precisely match those of their matter counterparts.

While CERN was building the SPS Collider and reaping its rewards, Fermilab launched its own innovative accelerator project, the first large-scale use of superconducting magnets. Magnet coils made of superconducting cable that loses all resistance to electricity when cooled within a few degrees of absolute zero have decided advantages over copper coils. Because there is no resistive heating, very large currents can pass through superconducting coils. Moreover, the strength of a superconducting magnet is set by the current in the coils themselves, not by the magnetic properties of iron. A superconducting synchrotron had been a gleam in Bob Wilson's eye from Fermilab's earliest days, and work to develop superconducting magnets suitable for a high-energy accelerator began in 1973, shortly after the Main Ring was commissioned. Although individual superconducting magnets had been in use for years in other applications, an accelerator calls for mass production of high-quality magnets with reproducible properties. A new accelerator in the Main Ring tunnel would require approximately one thousand bending and focusing magnets that would operate near 4 kelvins—4 degrees above absolute zero.

The superconducting dipoles, constructed out of niobium-titanium alloy, produced twice the bending strength of the conventional magnets. The new machine, initially called the "Energy Doubler," reached the world-record energy of 512 GeV on July 3, 1983. By 1984, when the machine attained 800 GeV, within range of 1000 GeV or 1 TeV, it was renamed the Tevatron. A superb antiproton source that built on the experience at CERN enabled operation as a proton-antiproton collider, with the first collisions on October 13, 1985. Over time, the collision energy increased to 900 GeV and then 980 GeV per beam, close enough to 1 TeV to justify the Tevatron moniker. Like the SPS, the Tevatron alternated between fixed-target and collider modes. As the Tevatron and its detectors, CDF and DZero, surpassed the capabilities of the SPS Collider, it became for a quarter century the dominant high-energy collider in the world.

The legacy of the Tevatron experiments includes many results for which the high energy of a hadron collider was decisive. Chief among these is the discovery of the top quark, which for fifteen years could only be studied at the Tevatron. Exacting measurements of the top-quark mass and W-boson mass punctured the myth that hadron collisions—sometimes derided as smashing garbage cans together—are too complicated to be exploited for precision measurements. By the end of the Tevatron's reign in 2011, nearly a thousand experimenters were active there. Coming to terms with the attendant many-body problem helped lay the foundation for the several-thousand-member teams at the Large Hadron Collider.

Even before its scientific achievements began to mount, the successful operation of the Tevatron gave physicists confidence to conceive still greater machines. The United States launched the Superconducting Super Collider, or SSC, a machine eighty-seven kilometers around, that would collide proton beams of 20 TeV each. The choice of proton-proton (rather than proton-antiproton) collisions was driven by the desire for very high collision rates. To our everlasting sorrow, the SSC fell victim to a convulsion of "fiscal responsibility" in 1993, after the expenditure of two billion dollars. Back when the Tevatron was turning on, CERN began construction of a ring four times larger than the SPS to house the Large Electron-Positron Collider, LEP. Eventually, that large tunnel could be reused for a superconducting proton-proton collider—the LHC.

The Large Hadron Collider is installed in a tunnel twenty-seven (26.7) kilometers in circumference astride the Swiss-French border near Geneva. The tunnel lies 100 meters (~330 feet) below the surface on average, tilted very slightly to take advantage of sound rock and minimize the depth of shafts. Under the Jura Mountains, the tunnel is as deep as 175 meters (575 feet) below the surface; toward Lake Geneva, it is only 50 meters (~165 feet) deep. The deep tunnel minimizes the intrusion on surface activities and provides good shielding for radiation. The LHC tunnel is a stretched circle, consisting of eight arcs, each fitted with 154 bending magnets, and eight straight sections in which beams are injected, accelerated to 6.8 TeV (6800 GeV), conditioned, brought into collision, and eventually dumped.

Many exotic technologies come together to make the Large Hadron Collider. Beams typically circulate in the machine for ten hours or more. A proton must travel at least ten light hours, or ten billion kilometers—roughly the distance to the edge of the solar system—without encountering a stray gas molecule. In ordinary air, which holds twenty-five billion billion molecules per cubic centimeter, a stored proton would collide with a molecule within a few microseconds! The ultrahigh vacuum in the beam tube is ten trillion times smaller than normal atmospheric pressure, both to keep the beam circulating and to minimize the background to experiments from beam-gas collisions. In addition, high vacuum must be maintained to insulate the superconducting magnets and liquid-helium transfer lines enclosed in a cylindrical cryostat 914 millimeters (thirty-six inches) in diameter, in the style of a Dewar flask or thermos bottle.

As in the pioneering Tevatron, the 1,232 dipole magnets that bend particles around the ring use niobium-titanium cables that become superconducting at temperatures below about 10 kelvins. Each fifteen-meter-long LHC dipole contains two 50-millimeter-diameter beam tubes, side-by-side with centers offset by 194 millimeters. The superconducting coils are configured so the magnetic field points up in one tube, down in the other, so that proton beams moving clockwise and counterclockwise bend around a common center. The magnets are gently curved to match the protons' paths. Cooled by superfluid helium to 1.9 kelvins—colder than the cosmic microwave background radiation, whose 2.725-kelvin temperature is a measure of the chill of outer space—the LHC dipoles produce a magnetic field of 8.33 teslas designed to steer protons of up to 7 TeV around the ring. The superconducting cable is a true engineering marvel.

Thousands of individual filaments as small as 7 micrometers in diameter (about one-seventh the diameter of the average human hair) are twisted into strands that are braided into cable—a design pioneered at the Rutherford Laboratory in the United Kingdom. The cable that went into the LHC dipoles would reach from CERN to Fermilab, and a little beyond! Maintaining the magnets at such a low temperature requires the largest cryogenic plant in the world.

Many other magnets—the total is 9,600—define the LHC: the 520 quadrupoles needed for strong focusing, a host of correction elements that optimize the proton's trajectory, and special quadrupoles that focus the beam down to a spot twenty micrometers in diameter at the collision points—tinier by far than the period at the end of this sentence.

Superconductivity also enables the radio-frequency cavities that accelerate protons and maintain the 2,556 bunches of protons in dense packets to ensure a high collision rate for the experiments. Each bunch in the LHC is a few inches long and contains some 100 billion protons. Successive bunches are separated in time by twenty-five nanoseconds, which corresponds to a spacing of 7.5 meters, about twenty-five feet. The detectors must therefore be ready for bunch crossings at the rate of forty million per second, although large gaps inserted in the pattern of bunches to allow time for kicker magnets to come on to inject or dump beams reduce the average crossing rate to a mere thirty million per second. Eight cavities, each delivering an accelerating kick of two million volts, are installed on each beam in a dedicated straight section.

A proton destined for the Large Hadron Collider emerges in a puff of hydrogen molecules from a small gas bottle into an electric field that separates the constituent protons and electrons and accelerates the protons through approximately 100,000 volts. A second element, called a radio-frequency quadrupole, focuses the proton beam and accelerates it to 750,000 electron volts (750 keV). Next, a linear accelerator raises the proton energy to 50 million electron volts (50 MeV), with the protons moving at 31.4 percent of the speed of light. The first circular accelerator, 157 meters in circumference, is the four-ring Proton Synchrotron Booster (PSB), which accelerates approximately ten trillion protons per ring to 1.4 billion electron volts (1.4 GeV) and 91.6 percent of the speed of light in about half a second. Next, the proton synchrotron (PS), 628 meters around, raises the energy to 25 billion electron volts (25 GeV) and the speed to 99.93 percent of lightspeed in one to two seconds.

Three trillion protons emerge from the PS in eighty-one bunches separated by twenty-five nanoseconds, the structure required for the LHC. Two hundred forty-three bunches are loaded into the Super Proton Synchrotron (SPS), which is 6.9 km in circumference, and accelerated to 450 GeV, reaching 99.9998 percent of the speed of light in 4.3 seconds. Twelve of these loads fill the LHC's dual superconducting rings over four minutes and twenty seconds per ring. Then, the counter-rotating protons—300 trillion per ring—are raised over twenty minutes to 6.8 TeV. After a period of cleanup and collimation, the beams—having traversed much of the history of CERN accelerators—are ready to be brought into collision. The LHC's mammoth versatile detectors, called ATLAS (for A Toroidal LHC ApparatuS) and CMS (Compact Muon Solenoid), witness more than a billion head-on proton-proton collisions per second, the raw material of discovery.

The innovation of colliding beams, the mastery of superconductivity, the development of strong focusing, and the refinement of supporting technologies that turn good ideas into practical devices make possible accelerators that surpass Enrico Fermi's wildest dreams—at a small fraction of the size and expense.

15

OMMATIDIA

Cuttlefish—giant squid—in the genus *Architeuthis* have single eyes a foot or more in diameter, complete with eyelid, iris, lens, and retina. These cycloptic organs testify to the evolutionary advantage that sight affords. Evolution favors solutions that suit the circumstances. An eye like that of a vertebrate (or cuttlefish!) wouldn't work if reduced to the scale of a fly; such a small eye would produce a blurry image. Rather than a single lens, a fly has hundreds or thousands of separate eyes—ommatidia—through which light must pass before being detected by a retina-like tissue. The dragonfly's 60,000 ommatidia make it the most eagle-eyed of insects, but its sight cannot rival that of vertebrates, whose retinas have roughly half a million receptors in each square millimeter, and thus millions of receptors altogether.

Our human eyes have two kinds of light receptors: rods and cones. The cones are responsible for color vision, while the rods are especially effective when there is little illumination. Eyes form in the embryo when an epithelial extension of the forebrain reaches the exterior covering of the head. The epithelium—a boundary layer that contains no blood vessels—folds back on itself, forming a cup, the surface of which becomes the retina. A surprising feature of this architecture is that the photosensitive cells do not lie at the surface where the light arrives, but behind the nervous system's circuitry, as if the components had all been assembled backward.

The signals generated when light hits the rods and cones are not sent directly through the optic nerve, but instead are processed locally. Indeed, as the embryology suggests, we can consider the interconnections of the nerves at the retina to be a part of the brain itself. Their output—about ten megabits of information per second—is forwarded to the central brain, where it is processed further. The information transmitted is not simply a pattern of excited rods and cones, but a partial interpretation, sensitive to contrast and shape.

For many of us, especially in our later years, nature's generous gift of vision needs some supplement. It seems that some seven centuries ago, spectacles came into use in Italy to restore clear vision to elderly presbyopics. Eyeglasses quickly became emblematic of scholarship and erudition. Before his death in 1313, Friar Alessandro della Spina of Pisa passed along the secret he had learned to restoring clear vision to those in need. A fresco in the small 8th-century church of Santa Lucia in Treviso, a day's walk north of Venice, painted in 1352 by Brother Tommaso da Modena, depicts the 13th-century biblical commentator Cardinal Ugo di Provenza wearing spectacles, although he died a half century before they were available.

The first eyeglasses enabled the old to see once more what was visible to the young but did not confer new powers of sight. Not until the beginning of the 17th century did Dutch lens makers learn to combine lenses to create instruments that magnified small objects and made distant objects seem nearer. The harnessing of optics begun in northern Italy three centuries before rebounded from Holland to Venice where in May 1609, Galileo refined the telescope and turned it toward the heavens. In Galileo's hands, the twenty-fold magnification of his spyglass was an instrument of revelation—and revolution. In his *Starry Messenger* of 1610, illustrations drawn to record his observations revealed that the lunar surface is not perfectly spherical, featureless, and plain, but is sculpted with valleys, mountains, hills, and rough places—just as Earth is. An instrument that enhanced human vision undermined the ancient belief in the perfection of heavenly bodies.

Even more astonishing was his discovery of cosmic stars (✷) close to Jupiter (◯), forming a line near the plane of the solar system. On January 7, 1610, they appeared as ✷   ✷ ◯   ✷ but the next night as ◯ ✷ ✷ ✷ . On January 10 he saw ✷ ✷ ◯ , and on the 11th, ✷ ✷   ◯ . January 12 revealed ✷ ✷◯ ✷ and the 13th, ✷ ◯✷✷✷ . Meticulous observations through March 2 left Galileo with no room for doubt that the little bright spots were not distant stars, they were bodies in orbit about Jupiter. He called them Medicean stars to honor the noble family he hoped to attract as patrons. In Galileo's telescope Venus displayed phases like the moon's, phases that directly contradicted the geocentric, Ptolemaic model. For those who would deny the Copernican model and insist that everything revolved around the Earth, all this was anathema.

At very nearly the same time, Simon Marius, court astronomer to the Margrave of Ansbach in what is today the region of Middle Franconia in Bavaria, carried out observations similar to Galileo's. His publication in 1614 of *The Jovian World, discovered in 1609 by means of the Dutch telescope,* provoked Galileo's ire, because the advertised date of discovery appeared to assert priority, in spite of the later publication date. In fact, Marius followed the Julian calendar favored by Lutherans, whereas Galileo used the Gregorian calendar adopted by the Roman Catholic authorities. The ten-day offset between the two calendars suggests that Galileo and Marius perhaps began to record their decisive observations on the same night.

The Jovian World reports the first tables of the orbital motion of Jupiter's moons. And even if his reputation has been eclipsed by Galileo's, Marius did attain a bit of immortality: we know the Medicean stars by the names he proposed: Io, Europa, Ganymede, and Callisto.

Seventy-nine satellites of Jupiter are known today, fifty-three bearing official names. Just as the technological innovation of the spyglass enabled the discovery of the four largest moons in the 17th century, new tools—telescopes equipped with photographic plates, space probes, and digital astrophotography—made possible subsequent waves of discovery.

The magic of lenses opened a microcosm as well. What could be more interesting than life itself? With their microscopes, Robert Hooke in London and Antony van Leeuwenhoek in Delft found microorganisms and made the fundamental discovery that the creatures that emerged from standing water and decaying plant-life were not spontaneously generated. Hooke asked in his 1665 masterpiece, *Micrographia*, "Whether all those things that we suppose to be bred from corruption and putrification, may not be rationally suppos'd to have their origination as natural as these Gnats, who, 'tis very probable, were first dropt into this Water, in the form of Eggs."

These early explorations beyond the natural range of human senses demonstrated the nature of the solar system and showed that life was a quality that did not arise spontaneously again and again. Once our vision was extended to the very large and the very small, our perception of our place in the universe changed.

The microscope and telescope provided artificial means of extending the evolution of human sight and opened new worlds, but certainly no optical microscope could open atoms to view. The alpha and beta particles that were Ernest Rutherford's subatomic quarry were far too small to be seen with ordinary light. To extend scientific observation beyond the microscopic to the atomic required entirely new ways of seeing.

In his early days at the University of Manchester, Rutherford's standard trick for observing rays emitted by radioactive substances was one he learned from Sir William Crookes: the phosphor-coated screen that we encountered in the quest to split the atom. To make visible the rays emitted by radium, Crookes devised a simple apparatus that he called the spinthariscope, from the Greek word for *spark*. A needlepoint dusted with a trace of radium is placed near a zinc-sulfide screen set at one end of a short tube and viewed through a lens at the other end of the tube. "The surface of the screen is seen as a dark background," Rutherford testified in his 1904 book, *Radio-Activity*, "dotted with brilliant points of light which come and go with great rapidity. . . . The experiment is extremely beautiful, and brings vividly before the observer the fact that radium is shooting out a stream of particles." The impact of each particle on the screen is marked by a flash of light.

Bob recalls: "The A. C. Gilbert chemistry set I had as a youngster featured a spinthariscope. The silvery steel device was no more than an inch long, cylindrical, with two flattened ends, one of which held a small lens. Slipping into my bedroom closet and shutting the door behind me, I waited a few moments for my eyes to adapt to the darkness. My spinthariscope never produced quite the spectacle that Rutherford promised, but I could make out occasional flashes of light, which provided immediate evidence for radioactive decay." Even the most elaborate chemistry sets available in today's safety-conscious society contain neither radiation sources nor detectors, but the inquisitive can find instructions online for constructing a spinthariscope using an americium-241 source liberated from a home smoke detector.

A decade later, Bob had a more professional encounter with phosphorescence. As a twenty-year-old, he had managed to get a summer job at the Stanford Linear Accelerator Center (SLAC) while it was still under construction. His boss was Dick Taylor, who would share the 1990 Nobel Prize as one of the leaders of a decisive experiment that probed protons with high-energy electrons, yielding evidence for tiny, charged constituents of the proton subsequently identified as quarks. Arriving at SLAC in the summer of 1965, Bob met with Taylor expecting someone in the mold of Edward Purcell, the buttoned-down Harvard Nobel laureate who had taught him quantum mechanics. The contrast between Purcell's elegance in manner, expression, and approach to physics and Taylor's gruff, profane style was stunning. This bear of a man from the steppes of Medicine Hat, Alberta, seemed scarcely to belong to the same species as Purcell.

"I learned that my assignment was to devise a way to measure the position of the electron beam as it made its way to the target. I was to use Rutherford's tried and true detection medium: zinc sulfide. A yellowish slurry of the phosphor could be sprayed on thin mylar sheets. The portion of the screen through which the beam passed would emit the same greenish glow as the numbers on a radium watch dial (which used radium mixed together with zinc sulfide) and could be viewed remotely with a television camera. One of my designs had zinc sulfide sprayed on a conical surface, mounted so it could rotate about its axis. Part of the cone was cut away to allow free passage of the beam between profile measurements. To rotate the viewing area in and out of the beam required some device that could push or pull a plunger attached to the cone.

"Technician Sid Frenkel was assigned to me (or I to him), at my service as long as he wasn't busy tending to Taylor's temperamental sports car. When I explained my designs, Sid immediately knew what it would take to move the cone remotely. Off we went to a used appliance store where he scavenged the actuator for the water-inlet valve from an old washing machine. The motion of a piston that had opened and closed the valve served admirably to push and pull the plunger and rotate the phosphor-painted cone, a solution that would have pleased the frugal Rutherford."

Although the phosphor-screen telescope extended human senses, its performance was still limited by the capabilities of the human observer. Rutherford's colleagues found that they could not count with certainty more than ninety scintillations per minute nor fewer than about five. Eye fatigue—or perhaps simple boredom—made it difficult to continue counting for more than two minutes at a time.

Ernest Rutherford and his assistant, Hans Geiger, knew that an alpha or beta particle passing through a gas would leave a trail of charged ions and electrons in its wake, but they knew as well that the charge along a track would be too small to measure conveniently. What they needed was a way to multiply the ionization. In 1908, they applied a high voltage—not so high as to cause sparking—to a wire running down the center of a brass tube filled with air at low pressure. Electrons left behind after the passage of an alpha particle would then rush toward the center wire, crashing into atoms on the way, freeing more charged particles and generating a cascade that multiplied the small current through the gas initiated by the alpha particle itself. The new device extended our vision, this time with electronic signals rather than flashes of light.

Rutherford and Geiger's counting tube—a precursor of the Geiger-Müller counter—is of historical interest for two reasons. It enabled the first detection of a single bit of matter in swift motion by its electrical influence, setting the stage for the widespread development of electronic detectors. Of more immediate consequence for our conception of nature, the counter behaved erratically, recording a spread of readings for alpha particles expected to trigger identical responses. It seemed that some alphas were taking longer paths than foreseen through the tube, encountering more air molecules, and generating a larger current than they should. Might they be knocked off course traversing the mica window through which they entered the tube? Rutherford asked Geiger to investigate how much thin layers of matter could scatter alpha particles, which he had previously thought too heavy to be deflected much. That assignment prompted the series of investigations that led Geiger and Rutherford's student, Ernest Marsden, to discover evidence for the atomic nucleus. Recall that their stunning observation, first made in 1909, was that roughly one alpha particle in 8,000 was reflected from a gold foil back toward the source.

The phosphor-coated screen and the Geiger-Rutherford counter tube paved the way for future detector components. The zinc sulfide emits light when a charged particle strikes it. The electronic counter amplifies, then measures, the ionization produced when a charged particle zips through matter, knocking electrons out of atoms along its path. These two basic functions—light emission and ionization—are the foundations for most of the components that our experimental colleagues use today.

Particle detectors have advanced dramatically over the years since Rutherford's boys first measured alpha scattering from nuclei. Both the gas-filled counting tube and the small zinc sulfide screen viewed with a microscope were capable of measuring the passage of a single particle. Geiger and Marsden mapped out the scattering of alpha particles by rotating the zinc-sulfide screen and its attached viewing telescope through a range of angles around the target.

We can imagine an improved apparatus with many screens deployed in a spherical array around the alpha-particle source and target, with an observer stationed at each screen, viewing it with a microscope. This would have speeded the work of Rutherford's team, but there are limits to how many observers could work harmoniously in close quarters—not to mention the problem of eye fatigue and limitations on the number of counts per minute.

The counting tube of Rutherford and Geiger provides an electrical output, which can be recorded without an observer's participation. Finding electrical replacements for Geiger's and Marsden's eyes is more difficult, but an invention called the photomultiplier tube does the trick. A photon entering a photomultiplier tube strikes a metal surface coated with a cocktail of elements often including the alkali metals sodium, potassium, and cesium. The photon's energy liberates an electron, which escapes from the surface by the photoelectric effect. Now the challenge is similar to the one that Rutherford and Geiger faced: how to multiply one electron into many. The solution again is to apply a large voltage that attracts the electron and causes it to crash into atoms, liberating more electrons. In a photomultiplier tube this process is repeated up to ten times. At each stage, the electrons are multiplied by a factor of three or more, turning the initial electron into a hundred thousand or even a million.

Scintillators and gas counters were not the only means of detecting high-energy charged particles. Cloud chambers and, later, photographic emulsions became attractive alternatives in many applications. Indeed, even when freed from the need for observers, the gas-filled counter and scintillator seem out-matched by these competitors. Counters and scintillators record the passage of a single particle through a small locale, while emulsions and cloud chambers provide full three-dimensional pictures of many simultaneous tracks.

For cosmic-ray experiments it was pointless to sit by a zinc-sulfide screen in the hope that some particle might happen by. Even if it did, little would be learned. An isolated signal couldn't tell you where a particle was coming from or whether it was accompanied by other particles. Throughout the cosmic-ray era and into the beginning of the age of accelerators, electronic counters played a subsidiary role to the dominant optical devices.

Once accelerators overtook cosmic rays as the preferred source of high-energy particles, counters became more competitive, but the subsequent invention of the bubble chamber pushed the counters into the background once again.

The signal reported by a counter can serve as input to an electronic circuit that analyzes it or combines it with other inputs to reach a logical decision—to compute the momentum of a charged particle from the curvature of its track in a magnetic field, and to use that result to make a decision about whether to record the event, for example. The science historian Peter Galison classifies the two ways of seeing as Image and Logic.

Even if we value eyewitness testimony, the likes of Geiger and Marsden seem overqualified to be taking turns counting scintillations in the dark. It doesn't take an atomic physicist to count flashes—even at the rate of ninety a minute. All the intelligence required could be supplied by a logic circuit called an adder. The first examples were made of a few vacuum tubes, the sort found in old radios and television sets. Adders are cheaper and more reliable than human observers; they don't ask for a salary, and only need to be fed some electricity. Unfortunately, they heat up in operation, so they are prone to fail like incandescent lightbulbs and need to be replaced. The first digital computer, ENIAC, contained 17,468 vacuum tubes that failed at a rate of one every day or two. Heat output and limited reliability impose practical limits to how complex an intelligence you can build out of vacuum tubes.

Like many budding scientists of our generation, we got our early forensic training when the family TV stopped working. We would carefully remove all the vacuum tubes from their sockets, take them to the local electronics store, and test each tube in a kiosk-sized device with a satisfying array of sockets, switches, analogue meters, and indicator lamps. Having identified the defective device, we would buy a replacement and return home to insert each tube back in its proper place and turn the set on. In memory, the diagnosis and repair always succeeded on the first try!

To compete with the bubble chamber and its three-dimensional view of an entire event, electronic counters needed to progress from individual ommatidia and compound eyes to cooperative multisensory arrays. The transistor, still a novelty when we took our undergraduate lab courses, enabled this evolution.

Transistors increased reliability and speed, reduced size, and cut heat production. At first, the gains were modest: substitute one transistor for one vacuum tube. Then came integrated circuits—the computer on a silicon chip. Electrical signals could be routed through cables containing many smaller wires to create a nervous system for the detector. It was this advance that allowed electronic detectors to evolve rapidly, supplanting even the discerning power of bubble chambers. The miniaturization continues apace; the A15 Bionic processor that powers an iPhone 14 packs fifteen billion transistors onto a one-square-centimeter surface.

A particle detector is a very specialized organism. This is a metaphor, but our colleagues who design, build, and commission detectors do sometimes speak of breathing life into their creations. A general-purpose detector has three primary senses, one sensitive to the tracks of charged particles, another that measures the energy particles carry, and a third that identifies muons. Networks link the arrays of sensors to electronic brains while complex circulatory systems supply electrical power, refrigerants, and the exotic gases some sensors require. All the elements are supported by a skeletal structure that must ideally be both insubstantial and sturdy.

The winning strategy in the evolution of particle detectors has been the same as in insect eyes: the repetition thousands or even millions of times of the same elemental sensor. Like an ommatidium, each element has limited capabilities, but combining enormous numbers of simple sensors creates a powerful device. Over time, the sensors themselves have become more capable, their performance profiting from—and also driving—technological innovation.

Many of today's detectors at high-energy accelerators measure and identify the particles created in collisions between two beams aimed directly at each other rather than between a beam and a stationary target. The detector surrounds the region in which the collisions occur and is generally a series of cylindrical shells nestled one inside another, each looking with its myriad electronic ommatidia inward toward the collision point. Endcap detectors help complete the picture from fore and aft.

Since the cloud-chamber era it has been common practice to immerse the detector in a magnetic field to bend the paths of the charged particles into circles or helices, and so to measure their momenta. The lower a particle's momentum, the tighter the bend. In a two-tesla magnetic field, a charged particle with momentum 3 TeV bends with a radius of 5 kilometers—nearly a straight track. The bending radius of a 3-GeV particle is more detector scale, 5 meters, and of a 300-MeV particle, half a meter. Particles of still lower momentum loop around and around in tight gyres, never reaching the outer layers of the detector.

The tracking device—usually several nested layers of sensors—is like a fingertip, where the high density of nerve endings provides our sense of touch with an exquisite sensitivity to position. The tracker is engineered to report the paths of particles with great accuracy, but to contain little material so that nearly all the particles created in a collision pass through without suffering much deflection or loss of energy. In today's collider detectors, the inner layers are made entirely of semiconductor—silicon—devices. Pixels, like those in a digital camera, lie at the core of the detector. They are well adapted to dealing with the highest density of particle tracks. Next come layers of silicon microstrip detectors, perhaps supplemented by wire chambers—modern versions of the counter tube.

Beyond the tracking layers lies the calorimeter, more akin to our human heat receptors—more diffused, less localized. It is in the calorimeter that particle personalities reveal themselves. The calorimeter's function is to stop particles within a short distance and capture all their energy. Basically, the particles smack into the calorimeter and make a three-dimensional splash, a shower of subatomic debris. The more energy a particle carries into the calorimeter, the bigger the splash.

Almost from the first contact with a layer of material, electrons and photons provoke an electromagnetic shower of photons, electrons, and positrons, which develops quickly and is confined to a narrow cone. In contrast, pions, protons, and other strongly interacting particles initiate a hadronic shower, which consists mostly of pions and nuclear fragments, is slower to develop than an electromagnetic shower, and spreads more from side to side. Many modern experiments employ separate electromagnetic and hadronic calorimeters, so the response of each can be optimized. In both cases, the calorimeter's task is to measure the extent and nature of the shower.

The calorimeter has its origins in the early 1780s, when Pierre-Simon Laplace and Antoine Lavoisier in Paris constructed a device to determine the amount of heat liberated by combustion or respiration. They confined a guinea pig in their calorimeter and measured the amounts of carbon dioxide and heat it emitted. They replaced the guinea pig with just enough carbon to produce the same volume of carbon dioxide as the animal had exhaled. The heat released by the burning carbon matched the heat the guinea pig had generated. The similarity led Laplace and Lavoisier to conclude that metabolism is a slow combustion process. The calorimeter is just a calorie counter—not for food intake, but for heat output.

The calorimeter in a particle detector measures deposited energy. While that energy is ultimately degraded into heat, the particle calorimeter doesn't measure a rise in temperature, but instead measures ionization and radiation, on their way to materializing as heat.

Some crystals, such as table salt's cousin sodium iodide (with a little dose of thallium added), convert the ionization and excitation produced by an electromagnetic shower into a light pulse or scintillation. The crystal is transparent so the scintillation light, in an amount proportional to the energy deposited, can be transmitted to photomultiplier tubes where it is amplified and measured. Even better-performing scintillating crystals—cesium iodide, bismuth germanate, lead tungstate—are available if you are willing to pay the price. For less exacting applications, bargain-priced plastic scintillators suffice.

The alternative is to measure ionization directly. The bigger the shower, the more ionizing particles are created. The best way to capture ionization is to alternate layers of dense material, which initiates the shower, with layers that sense and amplify the ionization.

Why should the calorimetry be needed at all in a particle detector? Isn't it enough to measure the tracks of the charged particles? First, only charged particles can be tracked. Some particles, including photons, neutrons, and neutrinos, are invisible in the tracker. Sometimes charged-particle tracks are too close to each other to be separated. More significantly, the individual particles may not be particularly important. Processes that are due to interactions of quarks still never produce free quarks. An outgoing quark materializes as a collection of ordinary particles. To measure the energy of the quark, we cannot do better than to measure the energy of all the particles that it became.

The muon—the heavier sibling of the electron—does not feel the strong force of the nucleus, so cannot initiate hadronic showers by blasting nuclei apart. When a high-energy muon is jostled by the electric fields inside atoms, it does not spray out energetic photons as an electron would, so does not trigger an electromagnetic shower, either. Being charged particles, muons do register in a tracker and typically leave a light trail of ionization in a calorimeter, but they almost never deposit much energy. Traversing a meter of iron, a muon loses perhaps 1.5 GeV of energy, so even a 10-GeV muon—soft on the scale of Large Hadron Collider products—can pierce six meters of iron, to be registered in tracking layers that make up the outer shell of the detector. We may think of the muon as a phlegmatic personality, expending minimal amounts of energy and leaving its surroundings little affected by its passage. Its signature is a trace in the inner tracking layers, little activity in the calorimeter, and counts in the muon chambers behind thick iron shielding.

Neutrinos, which carry no electric charge and do not feel the strong force, leave not even a wisp in the tracker and pass through the calorimeter as if it weren't there. Indeed, for neutrinos, a slab of iron is very nearly transparent. A pion speeding toward the same block of iron will see each nucleus and each electron in it as if it were an object of a certain size, what physicists call its cross section. A charged pion sees the nuclei as big—on the nanonanoscale of the particle physicist—because the likelihood of their colliding is large. To a pion with momentum 1 TeV, a nucleus looks like an obstacle a few millionths of a billionth of a meter in diameter. Atomic electrons appear smaller to the pion because the electromagnetic interaction is weaker and makes a collision less probable. A high-energy pion could make it through a thin block of iron with high probability, but if the slab is fifteen centimeters thick or more, interactions with iron nuclei will block its path.

The weakly interacting neutrinos see the nuclei and electrons as extremely tiny, easy to breeze right by. To a 1-TeV neutrino, roughly the highest energy to be expected at the LHC, a nucleus looks like a spot less than a ten-billionth of a billionth of a meter in diameter. Confronting a slab of steel three meters long, such a neutrino sees so little obstruction from the nuclei and electrons that the chance it will actually collide with one of them is about one in a hundred million. For a typical collider detector, the neutrino is not merely phlegmatic, it is subliminal—so much so that we infer the presence of a neutrino by the absence of detected momentum that was expected in a certain direction.

The complete collider detector is the ultimate in introspection, an organism with successive layers of oculi, all looking inward, the ultimate surveillance state. The innermost eyes of the tracker have the finest resolution. The energy sensors—the calorimeters—surround these, remeasuring the charged particles and catching the neutral particles that eluded the tracker. Finally, the muon detectors stand watch on the outer reaches to register the most penetrating kind of charged particle. An individual experimenter's attentive eye has been succeeded by tens of millions of silicon devices and sense wires.

Detectors do not, of course, spring spontaneously into being. The people who design them or perfect new and improved detector elements are a special breed. One of these wizards is David Nygren, who spent the longest stretch of his career in Berkeley. Tall, rangy, and tidy, Nygren conveys a quiet competence. He grew up in the Pacific Northwest, graduated from Whitman College in Walla Walla, and earned his PhD at the University of Washington.

Nygren is revered for inventing in 1974 the Time Projection Chamber, which only *sounds* like a machine that would send you back to the future. The TPC is a large vessel that holds a sensitive volume of gas or liquid. It uses a combination of electric and magnetic fields to perform three-dimensional reconstructions of particle trajectories. A charged particle curving through the TPC's magnetic field leaves a trail of ions and electrons in its wake. The electric field guides the electrons to an array of wires and gives a picture of the initial particle's path as viewed from the end of the detector. By measuring the arrival time of the electrons, you can figure out for how long and thus how far they drifted, which tells you where they were produced.

The TPC is a sort of electronic cloud chamber or bubble chamber, but able to operate at vastly higher rates. It has turned out to be a miraculous detector. It has been used to disentangle stupendously complex events involving hundreds or even thousands of overlapping tracks. Each detector module for the next-generation neutrino experiment based at Fermilab is a rectangular prism fourteen meters square by sixty-two meters long that will hold 17,500 tons of liquid argon. The TPC is also the device of choice in detectors that may look for years without ever seeing the kind of event they seek—a signal for dark matter, for example. For all this, the Time Projection Chamber is not quite the miracle that representatives of the US military hoped for when they asked Dr. Nygren if was really true that he had invented a machine for time travel.

Most particle physicists attach specific personality traits to denizens of the subatomic world: everyday electrons, workaday pions, strange kaons, exotic mesons composed of heavy quarks, and the rest. The instrument makers among us have a heightened feel for objects and materials. Some have cultivated an uncanny intuition for the atomic realm itself—the nanoworld in which showers are born and tracks are left. The trace of a fast-moving particle is born of a profusion of collisions and exchanges of photons and electrons between atoms, details of which depend on the properties of the gas, liquid, or solid being traversed. To most particle physicists, these minutiae are akin to the fearsome intricacies of organic chemistry or metabolic cycles. To an instrument maker, the variety opens possibilities.

Combining just the right materials with electric and magnetic fields you can work magic. Looking at a familiar object with fresh eyes you may crack a vexing design puzzle. Some years ago, Bob had the good fortune to work with the wizard. Nygren was looking at a silicon-strip detector, but he wasn't looking at it the way everyone else did. Instead of seeing a rectangle perhaps one centimeter across and a few centimeters long, he looked at the thin edge, a fraction of a millimeter across, and divided into strips each just fifty micrometers—a twentieth of a millimeter—wide. From this perspective, the edge became a line of very tiny pixels, each comparable to the breadth of a human hair. What good was that?

The silicon device could detect high-energy charged particles, of course; it could also detect X rays. If an X ray smacked into one of the tiny end-on pixels, it would reveal with very great precision exactly where it had landed. Nygren saw that by turning the silicon-strip detector so the thin edge became the business end, instead of the long strip, he might make an X-ray imager with superior position resolution.

But there was more. The silicon detector could count individual X-ray photons one by one. The contrast between places where there were many X-ray photons and other places where there were fewer would be measured exactly. Altogether, this would make a nearly perfect X-ray imager. A high-resolution picture with excellent clarity seems extravagant for imaging broken bones, but might enable early detection of incipient breast cancer by revealing a very small calcification.

There was one hitch: if the thickness of the silicon detector was one-tenth of a millimeter, you would need a hundred measurements to make an image just one centimeter in extent. The solution was to move the target past the detector, making a movie instead of a still image. Now the challenge was to make a mammogram in minutes, not hours or days.

If it were possible to focus more X rays into the narrow channel that was the edge of the silicon detector, the patient wouldn't have to be too extraordinarily patient. In contrast to a convex lens that focuses light, an X-ray lens must be thick at the edges and thin at the center. After experimenting with upper and lower rows of teeth in the shape of parabolas, which was costly and complicated to manufacture, the team sought simplicity. A young Swedish member of the team, Björn Cederström, thought it would be more practical to use teeth with straight edges in a sawtooth pattern. This would be easier to produce and thus less expensive. To get the right shape he drew a plan with big teeth in front, followed by successively smaller teeth, producing the required form, thick at the edge, thin at the center. Bob realized that the same effect could be achieved by using teeth all of the same size and tilting the upper and lower jaws. The alligator lens produced just the needed shape and would be simpler to make, since the teeth were straight and all were the same size.

It was at this point that the wizard stepped in again, this time with a new take on an everyday object. The solution could be found about a mile from the laboratory, where a shop offered vintage vinyl records, doggedly preserving the legacy of acoustic recordings against the seemingly implacable tide of digital recording. Record grooves would stand in for the teeth and the material, polyvinyl chloride, wouldn't attenuate the X rays too much. The team could cut out two radial strips to become the tilted jaws. For the grand sum of $4.95 the group procured *Autobahn*, a 1974 album from the German electronic band, Kraftwerk.

Side 1 contained the title song, twenty-two minutes and forty-three seconds of uninterrupted grooves. The depth and spacing of the grooves were not ideal, so they commissioned a limited-edition silent disc that yielded 333 upper and lower teeth of the required dimensions.

In the end the sawtooth X-ray lens neither saved the vinyl record business—audiophiles accomplished that—nor turned out to provide strong enough focusing to speed up the mammography adequately, but it proved an important principle, reported in the journal *Nature*. Applying the lessons, Cederström and his Stockholm supervisor, Mats Danielsson, stacked a number of perfectly aligned edge-on silicon detectors to simultaneously collect separate pieces of the X-ray image. About a thousand microdose mammography instruments of their design have been deployed to image patients. Work continues to extend the edge-on technology beyond mammography to computerized tomography.

As it happens, Berkeley was home to another encounter between silicon detectors and vinyl records. In the year 2000, Berkeley physicist Carl Haber had been using light and imaging in an ultraclean environment to make very precise measurements of silicon detectors destined for the ATLAS experiment at CERN's Large Hadron Collider. Carl heard a National Public Radio interview in which musicologist and Grateful Dead percussionist Mickey Hart spoke of vintage audio recordings so fragile that to play them would destroy them.

Before the onset of digital recordings, electromechanical storage came in a variety of formats, including wax-coated discs and cylinders and an evolving line of phonograph records. Most familiar during our college years were "long-playing" vinyl discs twelve inches in diameter that were played back on a turntable spinning at 33⅓ revolutions per minute. An "LP" stored sound in the form of excursions left and right, up and down, along a spiral microgroove about 460 meters long that was laid down over more than 700 turns. These undulations were detected by a phonograph needle riding the groove and then decoded into sound—music and speech.

A monograph on acoustical engineering convinced Haber that the instrument that he and his postdoc, Vitaliy Fadeyev, used to measure fine details of silicon detectors could be adapted as an optical scanner to map with great fidelity the features carved into the grooves on a phonograph record. These measurements could then be processed into an electronic file, which in turn could be rendered into sound using standard computer programs.

They commissioned their new sound-reconstructing device—sort of a digital microscope fitted to a turntable—on a 78-rpm shellac disc bearing a 1950 Weavers recording of "Goodnight, Irene," the signature song of the American blues and folk singer known as Lead Belly. Perhaps to immortalize this early step, Carl called the production version of his device IRENE, invoking the acronymic alibi, "Image, Reconstruct, Erase Noise, Etc."

The IRENE team could now play a record without ever touching it—or, to be scrupulous, by touching it only with light. In fact, they could hope to recover the sound with greater clarity than any phonograph could, because their software could launder the digital map, identifying and expunging scratches and dirt that would have produced hisses and pops in a mechanical reproduction, while compensating for speed variations and warped surfaces.

The mature IRENE is in everyday use around the globe, recovering our cultural heritage from treasures of music and the spoken word to anthropologists' field recordings of declining languages. For opening this new ear on our past, Carl was named a MacArthur Fellow in 2013.

Among the audio treasures unlocked by IRENE are some lovely novelties. For a century and a half, Édouard-Léon Scott de Martinville's phonautogram of "Au claire de la lune," preserved in 1860 in scratches recorded on soot-coated paper pulled past a recording stylus formed from a pig's bristle, was an example of write-only memory. IRENE's reconstruction was aided by Scott's methodical approach: he encoded the speed at which the paper moved past the stylus by recording along with the folksong the ring of a tuning fork as a reference tone.

And if our ears do not deceive us, a March 11, 1885, glass disc from Alexander Graham Bell's laboratory contains the first known recording—now recovered by IRENE—of that most potent of Anglo-Saxon expletives, uttered reflexively when an attempt to record "Mary Had a Little Lamb" went awry.

The pursuit of fundamental physics makes unprecedented demands on imagination and technology. The modern-day wizards who answer our challenges not only enable us to uncover new laws of nature, they create the unexpected: better medical imaging, reclaimed audio recordings from the past, and more.

High-energy physics is pursued by international collaborations of scientists. They need to exchange tables, figures, designs, photos, and ideas quickly. The demand for the rapid exchange of all sorts of data was addressed in 1989 by the English computer scientist, Tim Berners-Lee, then a CERN fellow. His World Wide Web changed not only particle physics, it changed the world. His solution was just what the particle physicists needed, and, as it turned out, just what everyone else needed, too. Making the technology freely available was a most enlightened decision.

16

A WORLD IN A GRAIN OF SAND

The human mind seems programmed to acquire language at the age of one to two years. It is optimized at eight or nine to memorize baseball statistics. What evolutionary advantage this peculiar form of numeracy gave the developing species has not been explained, not even by the late paleontologist and baseball maven, Stephen Jay Gould. Perhaps the compartment of Bob's brain devoted to .367, Ty Cobb's lifetime batting average, evolved to store the location of a stash of nuts or the best route for evading an ill-tempered mastodon. Or maybe specialized pleasure centers hold numbers that resonate with the mythic past of baseball: Babe Ruth's 714 home runs; Joe DiMaggio's 56-game hitting streak; Rogers Hornsby's .424 batting average in 1924, the greatest of his three seasons over .400.

Whatever the sociobiological rationale, we were at the peak of our devotion to baseball—and baseball statistics—in 1954. In homage to his Brooklyn birthplace, Bob rooted for the Dodgers. Chris favored the Philadelphia Phillies over the hapless A's. We both spent the days of summer playing pickup ball games, following the exploits of our favorite teams, and playing *All-Star Baseball*.

In "the first scientific baseball game," the lifetime batting records of current heroes and storied figures of the past are represented on pie charts—actually, bagel charts, because the inside is missing—about the diameter of a baseball. The border of a stiff paper disc is partitioned into bands for strikeouts, singles, home runs . . . every possible outcome. A swath labeled "1" covering twenty-five of the 360 degrees of the circle proclaims that on 7 percent of his trips to the plate, the Great Babe Ruth, as he is called in the rules of *All-Star Baseball*, hits a home run. A yawning band 44° wide warns that the Babe strikes out one chance in eight. Two slices covering 69° reveal that he draws a walk almost one time in five.

When it is a player's turn to bat, he mounts his disc on a spinner, and where the spinner comes to rest determines the result. Because the discs are based on career statistics, or statistics for a given year, each batter performs in *All-Star Baseball* just as he would in real life. Small red pegs track the progress of runners around the bases. Just as in real baseball, long stretches of routine at-bats are punctuated by moments of high drama when one swing of the bat, one spin of the pointer, could change the course of human events—or at least decide the game.

Of course, the simulation was incomplete. There was no way to represent the performance of pitchers, so the Babe batted just as well against the immortal Walter "Big Train" Johnson as he did against some mediocrity who was included in the 1954 edition of *All-Star Baseball* because he had one good year in 1953. Nor was fielding represented in detail, so no twirl of the spinner would ever make Willie Mays turn his back to the plate to chase down a Vic Wertz drive in the deepest recesses of the Polo Grounds. Still, *All-Star Baseball* was true to life as far as it went—proceeding by chance compounded on chance, building an intricate plot out of simple events, just as baseball does. And nothing stopped a boy from imagining for himself the heroics on his cardboard diamond.

During his tenth summer, Chris merely loved baseball above anything else; Bob was absolutely fanatical. He and his friends organized an *All-Star Baseball* league that ran all summer long, with teams selected in a draft at the beginning of the season, a formal schedule, box scores, and official statistics. The young Dodgers fan learned a lot of mathematics from the sports pages of the *San Francisco Chronicle*. Like many American boys of his generation, Bob perfected his long division by calculating batting averages. It was perhaps less typical that he treasured—in addition to his baseball glove—a protractor and a slide rule. He had explored the wonders of these tools, along with fractions, algebra, and geometry, with his father, a mechanical engineer with a fine practical sense for mathematics. To Bob, mathematics was not just something to learn in school, it was part of life. If the reality simulated by *All-Star Baseball* was not to his liking, he knew how to change it.

No major league player has hit over .400 since the Second World War—not in an official sense. But when Ted Williams returned to the Red Sox from military service late in the 1953 season, he batted an astonishing .407 in thirty-seven games. The Splendid Splinter knocked out thirteen home runs and collected nineteen bases on balls while striking out only ten times in ninety-one at-bats. His slugging percentage was a phenomenal .901. This stupendous feat moved Bob and his protractor to enshrine that magical short season on an *All-Star Baseball* disc. At an early age, he already knew that part of the fun of doing simulations is making the rules.

When we were whiling away our summer days playing, or, in Bob's case, improving, *All-Star Baseball*, we had no idea that many scientists had begun to devise and play similar games. Nor did it ever occur to us, any more than it does to young animals at play, that we might be rehearsing for "serious" situations we would confront as adults. How did it happen that the experiments that shape the development of particle physics came to depend on the simulation of events even more complicated than baseball games? The catalytic event, fittingly enough, was a game.

When the Manhattan Project scientists dispersed from Los Alamos after the victory over Japan, Stanisław Ulam became an associate professor of mathematics at the University of Southern California. With his impish grin and expressive eyebrows, Ulam was an original—even among the colony of originals at Los Alamos. Animated, disinclined to sit still, Stan Ulam was always fidgeting, his fingers in perpetual nervous motion. If he was not smoking a cigarette, he was fondling his penknife, dealing cards, or jotting notes on tiny scraps of paper in a seismographic hand. And talking. He was a link with the legendary Polish school of mathematics in Lwów, Poland's "Little Vienna." For Ulam's student circle, mathematical research meant talking, disputing, laughing, and joking, while scribbling equations, conjectures, proofs, and counterexamples on the Café Szkocka's marble-topped tables.

When they posed a problem or found a solution worth recording, they summoned a waiter to bring their communal notebook, later published as *The Scottish Book*, from its place of honor behind the bar.

Stan was a thoroughly verbal man. At Los Alamos, he skittered from one colleague's office to the next to kibitz, interrupt, and think out loud. Even more than most scientists, Ulam at work seemed indistinguishable from Ulam at play. With a touch of false modesty, he claimed to have little technical skill in mathematics. The gift he would have chosen for himself was a feeling for what he called the gist of the gist of questions, the essential insights that opened new lines. Admiring colleagues marveled that Stan never seemed to do a hard day's work in his life, yet left a trail of wonderful results wherever he took his apparently idle conversation.

Early in his second semester in Los Angeles, Ulam was felled by debilitating headaches that brought on disorientation and aphasia. His condition became so precarious that surgeons had to open his skull to relieve the pressure on his swollen brain and treat the infection by sprinkling penicillin powder directly on his inflamed gray matter. After the crisis had passed, Ulam filled his convalescence with brief walks by the ocean and countless games of solitaire in the waterfront cottage he had rented for his family on Balboa Island. His head swathed in bandages, he stood at a long, narrow table and dealt solitaires one after another, for hours at a time. He played mechanically, pausing only to record the success or failure with a pencil on one of his trademark tiny pieces of paper. His favorite game was named for Richard Canfield's turn-of-the-century gambling houses in Saratoga Springs and New York City.

Canfield is a pure game of chance that requires no strategic decisions and presents no opportunity for the player to demonstrate skill. The outcome is determined by the order of the cards. Whatever the ebb and flow scripted by Fortune, a game with no element of skill might seem pointless to any but the most fatalistic player. It is a perfect challenge for a mathematician.

Stan asked himself what was the probability of a winning Canfield. Some probabilities—especially those that refer to a single event or a short chain of events—are easy to calculate. The chance of drawing an ace from a shuffled deck of cards is one in thirteen, because there are four favorable outcomes in fifty-two cards. Drawing four consecutive aces is much rarer, once in 270,725 chances. And drawing four cards in a specified order will happen only one time in six and a half million. Knowing the probability of simple events, it is possible in principle to calculate the probability of more complex chains of events. But where there are many cases to consider, or many paths that lead to the same outcome, the calculation may be too horrific to contemplate. Although he seemed to have all the time in the world on his hands, it was inconceivable for Stan to consider every possible case. There are more than 8×10^{67} possible orders of a deck of fifty-two cards, but the age of the universe—13.8 billion years—is only a bit over 4×10^{17} seconds. There is no chance of having one deck in every order or playing fast enough to get through all the possibilities.

Then, in the blur of Ulam's Canfield marathon, it came to him: he cared about a practical answer, not a mathematical formula. In a sufficiently complicated problem, sampling may be better than examining all the chains of possibilities. He could simply play out a few hundred hands—ensuring a random sample by shuffling the cards thoroughly—tally the wins and losses, and know the probability of success—not exactly, but well enough. Why couldn't sampling help solve complicated physics problems, too? Ulam's question came from the same sort of common sense that had inspired Chicago Cubs outfielder Ethan Allen to invent *All-Star Baseball* while he watched teammates play cards during the 1936 season.

Ulam described his next inspired step as common sense applied to mathematical formulations of physical laws and processes.

As chance would have it, during the fall of 1945 Stan had been discussing combinatorial problems for computing machines with another "somewhat practical mathematician," the Hungarian émigré John von Neumann. Von Neumann had been Ulam's first host in America, at the Institute for Advanced Study in Princeton, and recruited him in 1943 to the Manhattan Project. As a consultant to the Army's Ballistic Research Laboratory, he had access to the only electronic digital computer in the world, the Electronic Numerical Integrator and Computer, or ENIAC. The ENIAC was a ponderous beast: forty ten-foot-high racks of electronic components, including 18,000 vacuum tubes, that stretched for a hundred feet around three walls of a specially prepared room. But it could perform 333 multiplications a second. (This is far faster than the rate at which any human can calculate, but trifling compared to the 3.6 *trillion* operations executed each second by a MacBook Pro in the year 2023!) If the ENIAC could be instructed to shuffle an imaginary deck or whirl a numerical spinner, it could be programmed to do a surrogate experiment—a simulation. The advantages—in time, cost, and the increased number of cases considered—would be obvious. Some desirable experiments may be too costly, too risky, or literally impossible to do; they might instead be simulated. Every baseball fan would like to see a World Series between the 1927 Yankees and the 1955 Dodgers, but that will never happen on the diamond. It can be simulated any time in *All-Star Baseball*!

When Ulam returned to Los Alamos in the spring, his poker partners transformed the Canfield epiphany into a practical tool. It was an unusual group. Von Neumann, for example, frequently had to be reminded what game they were playing, because his mind was elsewhere. He began to formulate a simulation of neutron chain reactions that the ENIAC could carry out. Reactors or nuclear weapons release neutrons, which can slip through an atom's veil of electrons and approach the nucleus without being deflected by electromagnetism. If by chance they strike a nucleus directly, they can bounce off, giving up some of their energy to send the nucleus recoiling away. Or they can have more violent encounters, all according to definite probabilities. Some nuclei gobble up neutrons voraciously, removing them from the swarm. Others go to pieces when pierced by a neutron, emitting more neutrons as they split, opening the possibility of chain reactions.

Though every collision of a neutron with a nucleus can be described by mathematical equations, von Neumann realized that it would be simpler to follow a number of neutrons as they rattle around the material, pausing at every tick of a clock to ask where each neutron is and decide what it does next. The neutron's fate is determined by drawing a random number, which is the digital equivalent of the spinner in *All-Star Baseball*. Each neutron—with all its offspring—is followed until it escapes the device or has so little remaining energy that it is no longer of interest. This method makes it easy to consider many alternative designs. Changing materials or shapes doesn't mean rebuilding a prototype, it is just a matter of changing the table of probabilities. A Los Alamos code descended from von Neumann's outline has been applied to tens of thousands of problems by scientists around the globe.

During one of those animated discussions that Stan thrived on, Nicholas Metropolis—a legendary pioneer of scientific computing—gave the method the perfect name: Monte Carlo, in honor of the casino that was the cash cow of the Principality of Monaco. Stan had an uncle who was forever borrowing money from relatives to go gambling in Monte Carlo, and eventually was buried there. Stan himself loved the Monte Carlo Casino and claimed that its roulette wheels always loved him. The wonderfully evocative name contributed to the rapid spread of the method. Many particle physicists today feel certain that they understood the Monte Carlo method as soon as they heard its name.

While the Los Alamos scientists and von Neumann's East Coast collaborators used the Monte Carlo method to study neutron propagation on the ENIAC, they developed schemes for selecting numbers at random and analyzed how closely a Monte Carlo sample of given size estimates the true answer. Many other scientists applied the spirit of Monte Carlo—the spirit of play—using the tools available to them: hand calculators and mechanical devices.

∞

Robert Rathbun Wilson sat at his laboratory bench spinning a small drum—a round Quaker Oats box on a stick, if truth be told. Each time the cylinder came to rest, Wilson peered along an index to read from a graph pasted to the side of the box. He wrote a number into a table in his notebook, ran his finger down the scale to pick off another number, made a notation in his notebook, then spun again. At two o'clock on an Ithaca morning, his tie still neatly knotted, Wilson was intent, focused.

Richard Feynman loitered back-lit in the doorway, angular yet casual, relaxed yet ready to pounce—watching, watching. Looking with those raptorial eyes of scientific conscience, the virtuoso calculator of the Manhattan Project was nature's surrogate, on the lookout for cheating. *What was Wilson doing?*

"Right after the War," Wilson reminisced, "when I went to Cornell after being at Harvard for a year, I could see that building new kinds of detectors was important. Knowing how electrons and gamma rays interacted with matter was the key to building those detectors. I also knew enough about the theory that people were using to know it was wrong. So I thought I'd just do this with the Monte Carlo method, following a gamma ray or an electron going through a millimeter of lead at a time and then seeing what happens."

In addition to the curves on his graphs, Wilson had marked a band of numbers, like a ruler, along one edge. When Feynman had watched for a while, he began to write down the numbers as they came up. A few minutes later, he said, "It won't work. Your wheel is crooked." Now, Wilson was an artist, a sculptor, a proud craftsman—not to mention, years later, the founding director of Fermilab. Though his wheel of fortune was made of humble materials, it was to him a beautiful machine. It spun smoothly, without the slightest wobble.

He shot back, "It's not either crooked!"

He had kept track of these numbers from the beginning, long before he felt Feynman's presence in his doorway, to be sure that they were all equally likely in the long run. He also had verified that his imaginary gamma rays were being absorbed at the proper rate as they went through simulated matter.

He was sure that Feynman had not taken enough numbers. Feynman arched a skeptical eyebrow.

"Okay, Dick!" Wilson said, "Let's bet!"

Betting and banter made the simulation go much faster. The competition kept them up most of the night, and the stakes grew to what seemed like big money to a couple of young professors—up to a dollar a spin.

"Of course, Dick being human (haha) . . ." Wilson's eyes sparkled. "Well, he won twenty or thirty dollars from me. I could have said that the wheel was no good, but I gave him a lesson in statistics and pointed out that he was just lucky." Wilson's wheel was honest, but honest fluctuations emptied his pockets.

Chris has never been a gambler, but he learned an important lesson in statistics when he bought his children their first personal computer, an Atari 800. "The first exercise I did on the Atari was to simulate the behavior of a magnetic substance in a two-dimensional world—in *Flatland*. It's a very physicist thing to do. I really did it as play, to learn about computer graphics and sound effects, not because there was some deep question I wanted to study. (Of course, once I got started, I couldn't help myself.) I calculated what should happen to check my simulation. I spent hours and hours, many late nights, watching the system evolve when I changed the temperature or disturbed it in some other way. What astonished me was how dramatic the fluctuations can be, how far the system can depart from average behavior. I think it's a general human characteristic—it was certainly true for me—to underestimate the fluctuations, to expect every event to be more or less typical. I knew better, intellectually, from my calculations, but doing some experiments and seeing the rare events occur made me understand the limits of my intuition." As George Eliot echoed Aristotle's *Poetics* in *Daniel Deronda*, "It is a part of probability that many improbable things will happen."

Before personal computers, an awareness of what statistics really meant was a major distinguishing difference between experimentalists and theorists—even theorists like Chris, who spend most of their time thinking about experiments. Experimentalists got their education in statistics by doing experiments and finding out how often even careful measurements differ from the expected average. Now everyone can get that experience—at least vicariously through simulations—and gain a more intuitive understanding of what statistical fluctuations mean.

Wilson's detector simulations paid off better than his gambling. He built a much more elaborate Monte Carlo device that started and stopped on the aleatory passage of cosmic rays. This refinement eliminated any unconscious bias of the operator's hand and ensured that the spin was highly random. Student assistants took over production, spinning the wheel for hours on end to accumulate large samples of electrons with different energies traversing many kinds of detectors. (We do not know whether Feynman lightened the students' wallets, too.) Wilson concluded that the way people had described the passage of photons and electrons through matter was indeed unreliable, as he had believed from the start. His Monte Carlo calculations paved the way for better detectors. Wilson's wheel of fortune was a great advance, until computers came along and put it out of business.

Simulations of high-energy physics experiments have grown from a curiosity to an indispensable tool since those homespun origins in oatmeal boxes and hand-drawn graphs. Today, nearly all particle and nuclear physics experiments, as well as many space, medical, and security applications, rely on software toolkits such as Geant4, more than a million lines of object-oriented C++ code written and maintained by an international team of about 135 physicists and computer professionals. Geant4 is more than a software engine to propagate particles through a geometric representation of a detector. It is a comprehensive repository of models—tuned to measurements—that simulate the interactions of particles with matter. Those models are continually refined through comparison with new measurements testing detector components and the observed behavior of complete detectors that often explore new realms of energy and data rates.

Not only are entire detectors simulated—instead of individual components—but the sources of background noise and physics signals are modeled, too. Physicists think in idealized terms of the quarks and gluons and such of their Feynman diagrams, but detectors respond to the end products—hadrons, leptons, photons—that materialize by the hundreds in high-energy collisions. Monte Carlo simulations of physics bridge the gap between theoretical abstractions and what detectors can actually observe. Modern experiments routinely generate as many simulated collisions as they record. In a recent year, the ATLAS experiment at the Large Hadron Collider simulated fifteen billion events representing 26,000 distinct processes! Grains of sand transformed into silicon microprocessors give us a new perspective on the world—indeed, on the nanonanoworld!

A computer can run through many possible chess moves to determine the best (by some metric) play or consider many routes between two airports to optimize speed, comfort, and cost. Similarly, experimentalists can train computers to devise the most favorable criteria for selecting events of interest. These methods are known as boosted decision trees or neural networks.

A simulation can turn a quark into its material manifestation, a shower of charged and neutral particles colliding with every piece of the detector they encounter. Can a computer reverse the process, run the spray of particles backward and infer what quark, lepton, or force particle was the source of the myriad signals in the detector? We cannot run the Monte Carlo method backward, but we can train a computer just the same way that we, ourselves, learn, by studying examples. Feed the computer many displays of events (generated by Monte Carlo or chosen from carefully selected data sets) started by top quarks, many examples of those started by W bosons, and so on. Let it analyze the differences so that it can make judgments about a stream of new events: a top-quark event? A W-boson event? Electronic particle detectors evolved to supplant the human observers Hans Geiger and Ernest Marsden staring at screens. Now electronics allied with artificial intelligence can create super-Geiger-Marsdens to evaluate events at today's colliders.

Each time that we apply artificial intelligence and machine learning in a new domain, we need to ask how much we should trust our electronic partners and in what circumstances should we be comfortable turning decisions over to them. Chatbots raise questions of the same sort for the public at large.

∞

Kevin Einsweiler is one of the planetarians—scientists who travel the globe in search of the richest opportunities for discovery. A Minnesota native, he took his PhD in experimental particle physics at Stanford studying electron-positron collisions in 1984, before moving to CERN, the European Laboratory for Particle Physics in Geneva, as a postdoctoral associate. There he joined the UA2 experiment investigating proton-antiproton interactions at the SPS Collider, leading an upgrade of its data-acquisition electronics and then driving the first precise measurement of the W-boson mass.

In the spring of 1990, Kevin returned to the United States to work on a detector for the Superconducting Super Collider, the American project that would have reached more than sixty times the energy of the SPS Collider, and three times the ultimate energy of the LHC. He was ready for a change. "I've had a lot of experience building hardware," he said, "and as experiments get more complex, the amount of your life that you have to dedicate to achieving some particular goal is growing larger and larger. I felt that at CERN I had just spent several years of my life building data-acquisition electronics, ensuring that an upgraded experiment could record its data on tape. I wanted to get away from that for a while. We all get very specialized in these complex endeavors, but I don't like being specialized and I try to resist it wherever possible."

The UA2 experience was not the only treasure that Kevin brought back to the USA. At CERN, he met and married fellow experimentalist Sandra Ciocio, herself a planetarian who followed her Rome doctorate with postdoctoral experience at Boston University before moving to CERN.

So it was that years before the first protons would speed around the Large Hadron Collider, Kevin sat in Berkeley before the nineteen-inch color display of his Silicon Graphics Personal Iris workstation, calculating the odds for the outcome of a head-on collision between two protons. A three-dimensional plastic topographic map of Switzerland, at 1:500,000 scale, a souvenir of his Geneva days, relieved the institutional-beige wall behind the monitor. To the left of the color monitor was Kevin's long-term memory. Four sturdy-looking metal boxes of different design, each the size of a couple of cereal boxes, whir and groan, pilot lights winking on and off. Rectilinear green numbers dart across the black face of an eight-millimeter videocassette deck. Two other boxes house hard discs; a compact-disc read-only-memory device sits atop the stack. Cartons of videocassettes are scattered about the room.

The CD player is used occasionally to install new software. At other times, as the headphones dangling from the front panel prompt the visitor to discover, Kevin plays 10,000 Maniacs to block out interruptions and the noise of the computer's fan. "I enjoy trying to link together the physics and the detector," he says. "There are so many experimentalists these days whose goal is just to build something and have it function at some level. It almost wouldn't matter whether it had a real purpose in writing a physics paper or not. For me, doing these simulations is a way for an experimentalist to think directly about physics. It's the only way to reach out and touch the laws of nature."

In September 1993, the United States Congress abandoned construction of the Superconducting Super Collider. By that time, much of the physics program had been laid out through Monte Carlo efforts like Kevin's. When members of the SSC collaborations joined experiments at the LHC, many of their insights flowed directly into studies made by their European counterparts. The Berkeley contingent merged into the ATLAS Collaboration, an international team that would eventually number more than 3,000.

At the LHC, the chance is one in a million that a *W* boson comes out of a collision, less than one in a billion that a Higgs boson emerges. With its single spinner, *All-Star Baseball* doesn't tell the full story of the bat striking the ball and the ball's subsequent passage through the infield, only the summary of whether it is caught or falls in for a hit. Kevin's electronic spinner can choose a million random numbers a second. His program uses these numbers to simulate the collision, reckon by probabilities which hundred or two hundred particles will emerge, and follow each one as it traverses the computer's image of the detector.

Some particles trickle out of the collision like bunts, curling up in the magnetic field, never making it out of the tracker, the detector's infield. Others reach into the shallow parts of the outfield, the electromagnetic calorimeter. The longer shots land deep in the hadronic calorimeter. High-energy muons make it further, ground-rule doubles that finally stop in the last meters of steel, and home runs, knocked out of the detector ballpark. Neutrinos fly out of the collision at the speed of light, not even seen by the team in the field.

The program records the result and spins again. It runs day and night, previewing the physics to be done and a detector's response. Once the detector comes into being, simulations will help its builders understand its performance, learn to live with its quirks, and deal with the unexpected. Monte Carlo is the point of contact between the ideas of particle physics and the realities of detector technology.

Designing a detector for an experiment in particle physics is a lot like assembling a sports team. Detector components must accomplish many distinct tasks and respond to many disparate situations, like ballplayers at different positions. Funding agencies regularly remind experimental collaborations that they cannot afford an all-star component at every position. Moreover, even all-stars are human, even the finest detector element is fallible. Just as once in a while a first baseman lets a batted ball skip through his legs, a tracker will occasionally fail to record a passing particle. The fans' favorites are not always the best players for a team, and what the group from Euphoria State University wants to build may not be what a detector really needs. The task of detector simulation is to understand how diverse components perform together, how well one subsystem complements another, how much redundancy is enough. In baseball terms, it is a tool to help the collaboration field the best team their money can buy.

Seated at his terminal, Kevin seems serene. His gestures and his speech are so composed that it is easy to imagine that he never even wastes a thought, let alone a motion. Kevin wears his passions on the inside, but his curiosity—his need to know—is a beacon. "Understanding the next piece of the standard model, electroweak symmetry breaking, that's what drives me to want to build one of these detectors. I find it an intensely interesting question what the basic particles and forces really are and what is the mysterious thing that gives mass to the particles."

"When I first heard of Monte Carlo," Kevin says, "it seemed so natural to me that I used it everywhere. I've probably written thousands of Monte Carlo programs to ask little questions, often statistical questions. It's very much an experimentalist's kind of tool, probably because we tend to be people who think a little bit less analytically and more in terms of how events are governed by a certain set of rules. We don't necessarily want to deduce the consequences of those rules by pure thought, but we're content to build them into a machine and watch what comes out. Monte Carlo is like building an experiment. You construct a certain set of things that you hope will behave in a certain way. Then you turn it loose and see what happens. The difficulty is that it takes almost as long to build a good simulation as it takes to build a detector. And then you still have to prove that the simulation is right.

"Monte Carlo is a naïve person's approach in a sense. Anyone can write a Monte Carlo which simulates step-by-step what happens when you put a particle somewhere in a detector and it starts going out. You track it through the magnetic field and make it bend a little bit. It's just an attempt to model step-by-step what's going on from simple principles. If you put in all the atomic and nuclear physics we know, all 7,000 tons of detector, and traced everything that happened there, you'd have to get the right answer, unless there were coding bugs. To do that kind of simulation doesn't require any great genius, only persistence, but we can't do those kinds of simulations, because there just isn't a big enough computer. The trick is always to make the right approximation."

Kevin's principal collaborator on the SSC and LHC simulations was Ian Hinchliffe, a Berkeley theorist who wrote the bible of supercollider physics with Estia Eichten, Ken Lane, and Chris Quigg. Ian is an Oxford-educated Yorkshire lad—okay, a terrier—who is as frenetic as Kevin is serene, as ruffled as Kevin is tidy. When readers started referring to "Supercollider Physics" by the authors' initials as "E-H-L-Q," Ken Lane protested, "It's pronounced *elk*. The *H* is silent!" The concept of a Silent Hinchliffe was so improbable and delicious that the shorthand EHLQ, pronounced elk, stuck instantly.

One of Ian's great joys is finding out new things by interpreting experiments, a style of doing science he learned from his first physics teacher. Mr. Chester was near retirement when he encountered the eleven-year-old Master Hinchliffe, but he was still fresh with enthusiasm for physics as an experimental science. He performed demonstrations in front of the class, then asked questions that stirred up his pupils to argue about what the experiment meant.

"From that," Ian relates, "you got a flavor of how things worked. It was not like a history lesson, it was *alive!*"

Ian became a theorist because "I was always a bit klutzy, I think, and I was always numerate. When I was an undergraduate, I didn't know a big collaboration could exist. I thought you had to go into the lab alone in the morning and breathe on the apparatus just right to make it work. Maybe if I had been a CERN summer student, I might have seen how I could have been an experimenter." Ian built a bare-bones software framework, PAPAGENO, that turned theorists' descriptions of physics processes into lists of arrows that represent the flight of particles out of a collision. Like Bob making a new Ted Williams card for *All-Star Baseball*, Ian made new rules for the fantasy detector game that he and Kevin would play, in parallel with many other teams around the globe.

Kevin discusses the division of labor: "There's quite an industry of collaboration between theorists and people like me. It can be only a matter of weeks between a theoretical preprint appearing in Italy and a piece of code sitting in my workstation generating events. Then I take them and try to understand how might I, a naïve physicist, do an analysis looking for some piece of physics in the future, and how much would I be limited by the detector performance when I did it."

Even in those early days, Ian, the theorist, had strong views about how the detector should look, while Kevin, the experimentalist, also had strong views about what physics goals are important. How can you tell that Ian is a theorist and Kevin is an experimentalist? For one thing, they have very distinct aesthetics. Experimentalists get pleasure from the process, they get excited about how far they can push the technology. Theorists look at detectors in very much more abstract terms, as instruments that see the Feynman diagrams that theorists are thinking of all the time.

Kevin says, "An experimentalist who's spent years and years painstakingly building and testing detectors has a different sense of them as a living accumulation of blood, sweat, and tears. Ian does have some sense of what's real and what's not real, so his comments aren't as off the wall as they might be. But somehow you can just tell that he's never built anything."

Part of the difference in experience is that Kevin knows that *things* don't always work. "A process of education takes place as you build one of these very complicated systems. I started off more interested in theory, but there's something I find very fascinating about the collaborative aspect of building a large detector—a whole lot of people doing a whole lot of different kinds of things, and in the end, this amazing object coming out of it that can ask physics questions. There's a lot that goes on in building something like that. It's not so much what can be built in an engineering sense—that what Ian is thinking of would fall over, whereas what I'm thinking about would not. It's understanding what can be constructed in the sense that this group of experimental physicists could actually create such a thing and expect to have it work when they're done—and somehow debug it."

Most of the work that Ian was then doing was to try to understand how ATLAS will respond to specific kinds of physics. "I'm always driven by the worry," he frets, "that somebody might be making a decision somewhere which is going to compromise the physics of this detector, and that they're making that decision for a hardware reason, maybe, because they're interested in some particular technology, or worse, for some geopolitical reason, or just because it's cheaper. How do you balance the books and the physics at the same time? You have to ask the right question: What's the issue? What matters? Maybe it doesn't make any difference—then you choose the solution more people like, or the cheapest one."

"During the design and optimization, things are actually quite fluid," says Kevin. "Once you're really deep in the guts of building something, reality intrudes, there is no way to change anything very quickly. Every time you think about changing something, you also think about the months of effort that are required to accomplish it. [At the beginning], it's a time when there is little resistance to change. For an experimentalist, that's kind of exciting. Theorists work in that world all the time. You start off as an experimental graduate student spending interminable hours building things, and you become very conservative, I guess. When you don't have to be quite so conservative, it's fun."

If the simulation questions the wisdom of a certain detector component, "There is a rational decision-making process—and most of the time it fails," Kevin says with the flicker of a smile. "We have tried hard in this collaboration to make decisions on technical grounds. It's just that there are stronger arguments that play roles as well, a lot of political realities that can take precedence. I don't have much doubt that if you came up with an overwhelming technical objection, people would listen to it. But there is definitely an equilibrium between technical correctness and institutional pressures that has to be reached in some way."

Today, highly evolved Monte Carlo physics simulations lie behind every detector design, every measurement, every search and discovery. The evocatively named PYTHIA renders homage to the high priestess who served as the oracle at the Temple of Apollo at Delphi, while SHERPA (Simulation of High-Energy Reactions of PArticles) and HERWIG (Hadron Emission Reactions With Interfering Gluons) settle for more pedestrian acronyms. Such programs, built through years of sustained effort, aim to give the best achievable representation of established physics and background reactions and of possible new phenomena. Their authors are community treasures!

During the analysis and evaluation of collision data, experimenters use simulations of both physics and detector performance to conduct "pseudoexperiments" in order to assess the probability and significance of fluctuations and outliers. How could the detector and analysis protocol mislead? How could ordinary processes mimic something astonishing?

Financial planners have adopted their own version of pseudoexperiments, conducting hundreds of simulations with a range of plausible assumptions over time to estimate the robustness of a retirement portfolio. If you should be told that 95 percent of simulations leave you solvent at the age of ninety-five, you should not only ask how comfortable you are with the one-in-twenty risk of running short of cash, but also what the simulations might have missed, what corners they might have cut. Quantitative analysts Emanuel Derman and Paul Wilmott, both physical scientists before they became celebrated quants, put forward a Modelers' Hippocratic Oath in the aftermath of the 2008 financial crisis. Several elements speak directly to the art of particle physics simulations:

> I will remember that I didn't make the world, and it doesn't satisfy my equations.
>
> I will never sacrifice reality for elegance without explaining why I have done so.
>
> Nor will I give the people who use my model false comfort about its accuracy.
>
> Instead, I will make explicit its assumptions and oversights.

Adherence to these precepts can save both fortunes and scientific reputations!

There are questions about detector performance that even the most complete simulation cannot answer reliably. "Electron identification—distinguishing real electrons from charged pions—is excruciatingly difficult to simulate," says Kevin. "In principle, we know all the details of how particles interact electromagnetically with the material in the detector. Nevertheless, if with a certain set of filters we can throw away 99.9 percent of the pions that come into our calorimeter, we don't really trust our simulations to tell us what that remaining tenth of a percent looks like. They're very unusual events that involve weird combinations of things that we can't predict very well. Eventually, we have to find a way to *measure* the contamination of pions in the sample we call electrons."

The slogan "Garbage in, garbage out" is a reminder that a simulation can be corrupted by faulty input—a wrong probability for the Higgs boson to decay or for the Immortal Babe to strike out. Physicists, quants, and fantasy baseball players need also to remember that simulations are limited by the information put in: incomplete truth in, partial story out. But the allure of human intuition is potent: Long after the definitive baseball simulation shows that the 1955 Dodgers wouldn't have had a chance against the 1927 Yankees, Bob will believe that, on the diamond, the Boys of Summer would have found a way to win.

17

VALIDATION

As June of 2012 turned to July, particle physicists around the world were in a collective state of eager anticipation. At the opening session of the International Conference on High Energy Physics, scheduled for the Fourth of July in Melbourne, Australia, two experimental groups working at CERN's Large Hadron Collider were to present updates on the search for the Higgs boson. All of us fervently wished that it might be the moment at which the Standard Model of particle physics—quarks, leptons, quantum chromodynamics, and the electroweak theory—could be called complete—or else what was written in our textbooks might be swept aside.

And it was not only particle physicists who were paying attention. A month before, Chris had set out from Hoek van Holland on the North Sea on the first leg of a 2,289-kilometer trek that leads to the French Riviera in Nice. Over the next three weeks on the E2 path, it seemed that everyone who came to know his off-the-trail identity was eager to learn how the experiments were going and when CERN would find the boson.

How had we come to this threshold? The first step toward validating the electroweak theory came with the discovery in 1973 of the new weak interaction that the theory predicted, the neutral-current interaction mediated by the Z boson. A single bubble-chamber picture captured a muon neutrino scattering from an electron, a reaction forbidden by the established theory built on Enrico Fermi's ideas. Systematic study of neutrino interactions indicated that the general features of the novel force agreed with the new electroweak theory. By 1978, electron-scattering studies at the Stanford Linear Accelerator Center confirmed that the neutral-current interaction violated parity—distinguished left from right—as the theory specified. Next came the proton-antiproton colliders at CERN and Fermilab that enabled the discovery and close examination of the predicted force carriers, the W and Z. Through the 1980s, neutrino-scattering experiments, electron-positron collisions at energies up to 32 GeV per beam, and other studies supported the one-act version of the electroweak theory as a reliable approximation to reality. With a nod to Feynman's graphical language, we refer to the one-act description as "tree-level," because the diagrams contain only branches, but no loops representing quantum corrections.

In 1981, even before the W and Z bosons had been discovered in proton-antiproton collisions, the CERN Council approved construction of the Large Electron-Positron collider ring known as LEP. Large was no exaggeration; the ring is twenty-seven kilometers around, more than four times the circumference of the proton synchrotrons that became proton-antiproton colliders at CERN and Fermilab. It was Europe's largest civil engineering project prior to the Channel Tunnel. Four LEP experiments would eventually study some seventeen million Z bosons, produced by tuning the combined energies of the electron and positron beams to match the Z-boson mass.

In the United States, the Stanford Linear Collider, or SLC, was approved for construction in 1983. It was a novel single-pass collider, in which bunches of electrons and positrons had but one encounter, rather than circulating in rings that would give them many chances to interact. Because the particles do not run around a ring in which they can meet the same accelerating stations many times, much of the length of a linear collider must be devoted to accelerating cavities that the particles pass through only once. That is a disadvantage that is compensated by the fact that electrons bent in a circle lose energy in the form of synchrotron radiation—they radiate when accelerated (bent) while electrons moving in a straight line do not. Linear colliders offer a path to higher electron energies that storage rings cannot match. The SLC was intended both as a physics device and as a demonstration that linear colliders are feasible and advantageous for high energies. Although the SLC yielded only 600,000 Z bosons, the straight-line design made it possible to control the alignment of the electrons' spins in a way that enabled complementary and competitive measurements of the Z properties.

Almost as soon as the Z factories began to operate in the summer of 1989, they refined our knowledge of the neutral gauge boson's mass. Today, thanks to the LEP experiments, the Z-boson mass is determined to be 91,187.6 MeV, with a precision of 2.1 MeV, which is to say less than one part in 43,000. To reach this precision, CERN scientists and engineers had to account for the tidal influence of the sun and moon as well as seasonal deformations of the LEP ring due to the changing water level in Lake Geneva and the amount of rainfall in the vicinity. They also had to track down a puzzling electric current induced by the passage of the French high-speed train, the TGV, on the Geneva-Lyon line.

Very quickly, by measuring the Z lifetime—equivalently, by determining the width of the Z resonance—LEP showed that decays into neutrino-antineutrino pairs corresponding to the electron, muon, and tau families fully account for the "invisible," or undetectable decays. No more flavors of neutrinos—or other feebly interacting particles—are indicated. Today, the number of neutrino species inferred from LEP experiments agrees with the expected value of three within a half-percent uncertainty.

When we told the parable of the picophysicist working in the interstices of a lump of magnetized iron in chapter thirteen, "An Electroweak Theory," we saw that our miniature counterpart would have to be resourceful to establish that the underlying laws of nature did not single out a preferred direction in space. According to the electroweak theory, we too live in a world of bias, and must be resourceful to find evidence for the symmetries of nature's laws. For the picophysicist, the symmetry is rotational invariance in three dimensions. For us, the quarry is the gauge symmetry hypothesized by Glashow, Weinberg, and Salam. That symmetry is hidden at a distance scale about a billionth of a billionth of the human scale, in the nanonanoworld or, to use the standard prefix, the *atto*world. So we must think like attophysicists or, to give ourselves space to work, *zepto*physicists a thousand times smaller, to look for signs that the gauge symmetry is exact. That means that we must conduct experiments on the electroweak energy scale.

An essential part of the evidence we seek is found in one of those intricate interplays that take place in gauge-theory calculations, extravagant behavior in one element that is tamed by extravagant behavior in another. The case of interest is the creation of a pair of gauge bosons, W^+ and W^-, in collisions of electrons and positrons, which the LEP experiments could study when the collider energy was raised above the Z-boson region around 91 GeV up to 160 GeV, and then in steps up to 209 GeV. At the one-act level of the electroweak theory, three mechanisms—three Feynman diagrams—drive the reaction. The electron and positron can exchange a neutrino, each emitting a W boson; or they can combine to form either a virtual photon or a virtual Z boson, which then disintegrates into W^+ and W^-. Taken on its own, any one of these would give rise to a reaction rate, or cross section, that rises without bound at increasing energies: the probability for interaction would exceed 100 percent. In our calculations, the bad behaviors of the three pieces compensate for each other so the resulting prediction behaves benignly at high energies. When the LEP experiments carried out the measurements, this is exactly what they found—perfect agreement with the well-behaved predictions of the electroweak theory. The gauge symmetry based on weak isospin and weak hypercharge may be a secret symmetry, but it is there!

As the electroweak theory passed more and more tests, with experiment confirming the predicted interactions of the new, neutral Z boson with each of the known quarks and leptons, experimental measurements became so refined that the one-act, tree-diagram predictions were too crude to match the experimental precision. Two-act calculations, involving Feynman diagrams that contain loops representing quantum corrections, became the order of the day. Theorists systematically included contributions from all the known particles.

The lepton constituents included the long-known electron and muon and their associated neutrinos, plus a newcomer called tau and the tau neutrino. The heavy charged lepton tau (τ) was discovered in 1975 by Martin Perl and collaborators in the same SLAC experiment that discovered the J/ψ particle and charmed mesons. It weighs 1,777 MeV, nearly 3,500 times the electron mass, and lives on average less than a third of a picosecond.

Evidence for a third neutrino, associated with tau, was circumstantial but hard to gainsay. In the summer of 2000, twenty-five years after the discovery of tau, the DONUT (Direct Observation of NU-Tau) experiment at Fermilab showed explicit evidence that tau neutrinos interacted with matter to produce tau leptons. At the neutrino energies of this experiment, the tau typically decays within two millimeters of its creation point into a single charged off-spring and one or more neutrinos. To locate and resolve those short tracks, the experimenters turned to a high-resolution detector from the infancy of our subject, photographic emulsions. The emulsions were augmented by a fiber-optic tracker that provided medium-resolution tracking and a time stamp for each event, backed up by components that could identify electrons and muons. Finding a handful of these short-track events was a stunning technical achievement that confirmed the tau's partner as a third neutrino species.

In the list of quarks, there was a missing entry, a hole in our table of the fundamental constituents. The tally consisted of the original up, down, and strange, the charm quark of November Revolution fame, and the b quark, called bottom or beauty, discovered in the b-quark–anti-b-quark Upsilon states at Fermilab in 1977 by a team led by Leon Lederman.

Mesons made of a b quark and a light antiquark were first observed at the Cornell Electron Storage Ring in 1983. If, as Glashow, Iliopoulos, and Maiani reasoned, quarks and leptons must appear in matched pairs, then we should expect a charge +2/3 partner for the charge −1/3 b quark. Taken together, information on the b quark from many laboratories proved that it was the lower member of a weak-isospin pair, so that a top quark really did have to exist. By 1989, when SLC and LEP began to take data, experiments at the SPS Collider and the Tevatron could say that, if top existed, its mass had to exceed 50 GeV. There is no established theory for the values of the quark and lepton masses. A few theorists had asserted, with no discernable justification, that since the bottom-quark mass was about three times the charm-quark mass, top should weigh in at about 15 GeV. They were disappointed.

If the top quark had to be there, then it had to be included in the quantum corrections to all the quantities that could be measured with a high degree of precision—about two dozen, in all. But what value to choose for the top mass? A very productive approach was to treat the top mass as an adjustable parameter in the calculations and to ask for each observable what value gave the closest agreement between prediction and experiment; in other words, to use the quantum corrections as an indirect way to probe the top-quark mass. If the inferred values were all over the map, that would signal an inconsistency between the electroweak theory and experiment. If they all lined up, then there would be a sharp test of the electroweak theory once the top mass was measured directly. In the event, the various inferred values were highly consistent, though not yet highly precise. By 1994 they pointed to a top mass between 150 and 200 GeV, far greater than that of the reigning heavyweight, the 4.5-GeV bottom quark.

Two decisive experiments were carried out at Fermilab's Tevatron Collider, in which a beam of 900-GeV protons collided with a beam of 900-GeV antiprotons. Creating top quarks in sufficient numbers to claim discovery demanded exceptional performance from the Tevatron, because less than one interaction in ten billion resulted in a top–anti-top pair.

Observing traces of the disintegration of top into a b quark and a W boson required enormously sophisticated detectors and extraordinary attention to experimental detail. Both the b quark and the W boson are themselves unstable, each having many multibody decay modes. The b quark's lifetime is about 1.5 picoseconds, long enough for the b to join with a light antiquark to form a B meson, which can travel about a millimeter from where it is created. In the discovery runs, the CDF (Collider Detector at Fermilab) Collaboration pioneered the use of a silicon microvertex detector in the collider environment, which enabled the identification of *B* mesons by their decay vertices displaced from the collision point. Both CDF and their friendly rivals in the DZero Collaboration could recognize *B* mesons through their decays into electrons or muons. The *W* boson lives for less than a trillionth of a trillionth of a second—a yoctosecond, in the official nomenclature. Two-thirds of the time, it decays into a quark-antiquark pair, which manifests itself as two narrow jets of hadrons. The remaining third of *W* decays are into lepton pairs—a conspicuous high-energy charged lepton accompanied by an invisible neutrino.

From 1989 on, the Tevatron experiments raised their lower bound on the top-quark mass by about 15 GeV per year. By 1995, both collaborations presented persuasive evidence for a top quark with a mass near 175 GeV, consistent with the value inferred from quantum corrections to a whole suite of electroweak observables. At that mass—more than 300,000 times the electron's mass—the top quark lives on average half a yoctosecond. The top quark is so much more massive than the other quarks and leptons that it is natural to ask whether it is the outlier, or the only fermion with a normal mass. The latter possibility is suggested by the fact that, if the Higgs field is indeed responsible for the fermion masses, its coupling strength to top is very nearly one, clearly a natural choice, while its couplings to the other quarks and leptons are far smaller. We don't yet know the answer to this riddle. Four decades of top-quark studies at the Tevatron and at the Large Hadron Collider confirm that top behaves just as the electroweak theory says it should.

Experiments now fix its mass at just under 173 GeV, with an uncertainty of less than a fifth of a percent. Remarkably, a particle that has no size and no internal structure that we can resolve weighs nearly as much as a highly composite gold nucleus.

The first modern searches for the Higgs boson—those that entered what we now know was fertile terrain—were carried out by the four LEP experiments between 1995 and 2000. Theorists, unconstrained by the need for apparatus, had been on the case for two decades. It is, after all, our job to take ideas seriously and to see where they lead and how to test them. In a prescient late-1975 article titled "A Phenomenological Profile of the Higgs Boson," John Ellis, Mary K. Gaillard, and Dimitri Nanopoulos of the CERN Theory Division discussed the production, decay, and observability of the spin-zero Higgs boson expected in the Weinberg-Salam model. A few properties of the hypothetical boson had been worked out in the decade following Weinberg's paper, but theirs was the first systematic analysis of how to conduct a search. The authors stressed that the Higgs boson might not exist; the electroweak symmetry might be hidden in other ways. The confirmation or exclusion of the Higgs boson would be a fork in the road toward a credible theory of the weak and electromagnetic interactions. No doubt influenced by experimental capabilities of the moment, they concentrated their most detailed considerations on Higgs-boson masses of 10 GeV or less. Their paper concluded with "an apology and a caution. We apologize to experimentalists for having no idea what is the mass of the Higgs boson . . . and for not being sure of its couplings to other particles, except that they are probably all very small. For these reasons we do not want to encourage big experimental searches for the Higgs boson, but we do feel that people performing experiments vulnerable to the Higgs boson should know how it may turn up." That is not exactly a clarion call, but it is notable that they compiled and extended what could be said about the hypothetical particle. Apologies and cautions notwithstanding, they put the Higgs search prominently on the experimental agenda.

Although he had digested "A phenomenological profile . . ." Chris found his way to the Higgs boson by a crooked path. Toward the end of 1976, he had completed his guide to finding the W and Z bosons, which helped inform the design of detectors for CERN's proton-antiproton collider. After sending his work off for publication, he must have seemed a little bit at sea. His colleague Ben Lee came to him and said, "You know, you should think about weak interactions at very high energies. Every theorist has to do it. It's like the chicken pox. It's better to get it over with while you're young."

Chris recalls: "I read the existing literature and discussed with my Fermilab colleague Hank Thacker how we might approach what seemed to be a very amorphous problem. Without a clear plan at first, we started calculating

Feynman diagrams, beginning with the simplest ones from the Fermi theory and moving on to Weinberg-Salam. When we began, it was as much entertainment as serious labor. Ben soon noticed how much we were enjoying ourselves, and joined in. If your only experience with a blackboard is being handed a piece of chalk and told to work a problem in front of your math class, it may not be your idea of a good time. But working out ideas and equations in the company of a colleague or two is one of the great social pleasures of a life in theoretical physics. Many of the walls in the Fermilab Theory Group are lined with floor-to-ceiling panels of sheet metal coated with chalkboard paint. The Conference Room in which we worked holds seven of these four-foot-wide panels, so we could write equations for a couple of hours before pausing to copy the work into our notebooks. Then we would go off on our own and look for improvements or new directions.

"At some point, we decided to put the prospect of immediate experimental consequences aside; we dropped the leptons and quarks and asked what would happen if we could make two W bosons—not yet discovered—collide at high energies. The parts of the gauge bosons that arise from spontaneous symmetry breaking mattered most for our thought experiment at high energies. Gauge symmetry again enforces a marvelous interplay among different Feynman diagrams, so any unbounded growth with energy that one diagram might imply is compensated by another. We learned that the Higgs boson would play an essential role in making WW scattering behave properly at high energies.

"A day came when we felt that the answer was on our blackboard, but the question escaped us. Providentially, our colleague Bill Bardeen walked into the conference room brandishing a preprint from Tini Veltman in Utrecht titled 'Second Threshold in Weak Interactions,' suggesting new physics associated with the Higgs system. There was our question! We turned to the blackboard panel on which the high-energy limit of our thought experiment was a constant value—benign behavior—set by the Higgs-boson mass. We could infer a constraint on the mass by requiring that the probability of an interaction not exceed 100 percent. That critical value is very nearly 1 TeV. It is a tipping point: either something resembling the Higgs boson would be found with a mass below 1 TeV, or the weak interaction would become strong, displaying such new phenomena as WW resonances and multiple productions of W bosons.

"On that day in 1977, I began to feel like a child with an imaginary friend. My imaginary friend, called the *Higgs field*, accompanies me everywhere and is responsible for many things that happen in the world. Naturally I've believed that my imaginary friend is real. But because I am a physicist, I have wanted to find an experimental test."

The observation that something new must happen on the 1-TeV energy scale crystallized a suspicion that if real experiments, not just thought experiments, could take us there, we could learn what made the electroweak symmetry a secret symmetry. Reaching that promised land would take several steps. Like the picophysicist, we are too small to heat up the universe and restore the electroweak symmetry. We can hope to disturb the universe enough in a small region to excite the Higgs field and create a Higgs boson.

The Ellis-Gaillard-Nanopoulos paper spurred many experimenters to look wherever they could for signs of a Higgs boson. Nevertheless, because most of anticipated signals were minute and others beset with theoretical uncertainties, essentially the whole range of Higgs boson masses, from very small values up to the highest masses accessible, was still to be explored when the electron-positron colliders at SLAC and CERN came on the scene.

LEP's much higher event rate made CERN the place for Higgs hunters to be. In the Z-factory phase, the target was a Z boson that decayed by emitting a Higgs boson and leaving behind a shadow Z with less mass than the real thing observed by its disintegration into two jets of hadrons, a pair of electrons or muons, or even "nothing"—a neutrino-antineutrino pair. The most telling signature turned out to be the Higgs boson decaying into a b-quark–anti-b-quark pair (the dominant decay mode) recoiling against nothing. At the end of the first LEP phase, the four experiments combined their analyses to conclude that the mass of a textbook Higgs boson would have to exceed 65.6 GeV.

When LEP moved above the energy of the Z boson, the experimental signature became a Higgs boson produced along with a real Z boson. Starting in

1995, each increase in energy opened new terrain. There were also new backgrounds to combat: the reaction electron + positron yields two Z bosons and produces signatures that mimic the Higgs + Z final state. Until the year 2000, none of the experiments had produced a significant excess of events compared with the background expectations; the absence of a signal implacably raised the lower limit on the Higgs-boson mass in steps up to 108.6 GeV.

Once theorists could put the observed top mass into their calculations, it was tempting to look for signs of the more subtle quantum corrections that involved the Higgs-boson contributions. As for top before its discovery, comparing calculations with measurements could indicate what the Higgs-boson should be. By 2000, the constraints favored a Higgs-boson mass below 240 GeV, if guidance from the electroweak theory could be trusted. That qualification is essential; we did not know that the Higgs mechanism was the right answer to electroweak symmetry breaking, so experiments had to attend to other possibilities, including some in which the weak interactions became strong at high energies.

That summer, with LEP running near its top energy, brought a flurry of excitement. The ALEPH experiment—one of four operating at LEP—had recorded one suggestive Higgs-boson candidate at a mass of about 114 GeV. By early September, when LEP's retirement from service was scheduled, ALEPH had found two more tantalizing events. Was this how the evidence for a discovery might start to accumulate? CERN management, eager to get on with installing components for the Large Hadron Collider in the LEP tunnel, granted a brief extension of LEP running until early November. The stay-of-execution increased the data sample at the highest energy by 70 percent. When the experiments presented their final analyses, ALEPH showed a three-standard-deviation excess over expected background near 115 GeV, but data from the DELPHI, L3, and OPAL experiments reduced the overall significance. The chance that the observation was a statistical fluke was calculated at 9 percent. We wouldn't board a flight that had a 9 percent probability of not arriving safely, and we certainly wouldn't board a hundred flights at the same odds, because our chance of surviving would be only one in 12,500.

Nor would we hang our professional reputations on a data plot that had a 9 percent probability of being misleading if we were in the business of making many plots. LEP's final contribution to the Higgs search was a new lower mass limit of 114.4 GeV. With that, the LEP collider and experiments were dismantled to make way for the LHC leaving the field—for the moment—to the Tevatron Collider at Fermilab.

Teasing a signal out of the high background rate at LEP was not easy; the challenge is even greater at the Tevatron, where at most one event in twenty billion would produce a Higgs boson. Members of the CDF and DZero experiments learned to focus on events they could select efficiently and separate reliably from backgrounds. At masses around 120 GeV, the most favorable case was a Higgs boson produced in association with a W or Z, followed by the Higgs decaying into a bottom–anti-bottom quark pair and the W decaying into an electron or muon plus missing energy—a neutrino—or the Z decaying into a pair of electrons or muons. For larger masses, around 160 GeV, the most sensitive reaction involved two gluons combining through an intermediate loop of top quarks to form a Higgs boson that decayed to a pair of W bosons, each of which decayed into an electron or muon and a neutrino. Many innovations were required to reach a meaningful sensitivity, among them the efficient identification and measurement of B mesons and the application of machine-learning techniques to improve the discrimination between signal and background.

The Tevatron Collider ceased operation at the end of September 2011, having been surpassed in both energy and event rate by the Large Hadron Collider, but energetic data analysis continued. By the summer of 2012, the Tevatron experiments combined to exclude Higgs masses between 149 and 182 GeV. There was a positive indication, too: the sample containing two b-quark tags showed an excess in the mass range between 115 and 140 GeV. Had there been a narrow peak, experimental resolution would have blurred it. There was about one statistical chance in 550 that the Tevatron excess represented a fluctuation, but it did have the earmarks of the associated production of an electroweak gauge boson plus a Higgs boson that decayed into a pair of b quarks. We estimated that with two or three times the data sample, the Tevatron events might have grown to a discovery.

But the clock had run out!

The Large Hadron Collider is a fantastic device. Counter-rotating bunches of protons cross more than thirty million times a second, leading to approximately a billion proton-proton collisions per second, each creating perhaps a hundred particles. It has operated at beam energies as high as 6.8 GeV, just short of the 7-GeV design specification. And yet the machine was regarded as the easy part of the LHC physics program. When the project was taking shape, many elders of the experimental world opined that we might build the accelerator, but we had no idea how to build detectors that could exploit it. They were not wrong to be daunted, but they underestimated the creativity and dedication of the physicists, engineers, and others who accepted the challenge, and they failed to anticipate the technological advances that made new solutions imaginable.

The Higgs-boson search-and-discovery at the LHC was carried out using two general purpose detectors, ATLAS and CMS. We can liken these detectors to cameras capable of recording forty million hundred-megapixel images every second. They are truly wonders to behold, conceived and executed by 3,000-member teams over fifteen years. Each is installed in a vast cavern, about one hundred meters below the surface. ATLAS is adjacent to the main CERN campus in Meyrin, Switzerland; CMS lies 8.5 kilometers across the LHC ring in Cessy, France.

CMS is named for its central analysis magnet, which has the form of a cylindrical shell thirteen meters long and six meters in diameter, with its axis along the beam direction. In common with ATLAS, it incorporates the traditional elements of tracking, electromagnetic and hadronic calorimeters, and muon detectors. A notable innovation is the tracker made entirely of silicon detectors—pixels nearest the beam, strips further out, that report nearly a hundred million signals of the activity they record. The active area of the tracker would cover a singles tennis court! Overall, the CMS detector, which occupies a cavern six stories high, is 21.6 meters long and 14.6 meters in diameter. Its total weight is 14,000 tonnes—two Eiffel Towers.

When circulating beams are imminent in the accelerator, everyone must vacate the cavern that houses the detector. Bright fluorescent work-lights are switched off, and only the red and green LEDs on the detector—some steady, some twinkling—illuminate the cavern. With a roar, an air conditioning system forcefully exchanges the air inside the underground areas. Sounds at different pitches emanate from the cooling pumps that push chilled water or compressed carbon dioxide through a network of pipes within the detector. Collisions of the beams, which take place in the ultrahigh vacuum of the LHC, make no sound at all.

The ATLAS detector occupies a considerably greater volume than CMS, but weighs only half as much. It is forty-four meters long and twenty-five meters in diameter, assembled within a ten-story-high vault. The iconic features are the twenty-five-meter-diameter muon wheels at opposite ends of the detector and the eight superconducting coils, about twenty-five meters long and twenty meters in diameter, that create a magnetic field to bend charged-particle tracks.

The Tevatron collaborations, which grew to about 500 members each, had become thoroughly cosmopolitan: CDF researchers came from about sixty institutions in fifteen countries, while the DZero members represented about ninety institutions in eighteen countries. The LHC teams took international participation to a new level. During the search for the Higgs boson, ATLAS and CMS both had representation from approximately two hundred institutions and forty nations. The ATLAS and CMS experiments, which must be leading candidates for the most technologically challenging scientific instruments ever built, achieved their design goals after the first few months of data taking. That these amazing detectors could be integrated from pieces designed and manufactured all over the world and still work nearly flawlessly is astonishing. That individuals from so many national origins and scientific traditions could make common cause to accomplish the nearly impossible is inspiring.

Forward-looking scientists in Europe and North America conceived CERN after the Second World War to promote unity in postwar Europe through an open world-class laboratory for physics research that would arrest the brain drain to America. Long an icon of European cooperation, it now reaches to all the inhabited continents. For its contributions to international understanding, CERN is worthy of the Nobel Peace Prize, intended to recognize those who have "done the most or the best work for fraternity between nations, etc."

To enable the widely dispersed collaborations to function productively and to help collaborators far from Geneva to feel connected to work at the laboratory, CERN and its partners developed the Worldwide LHC Computing Grid, a global infrastructure that has grown to 170 sites in 42 countries. The grid's resources for data storage and analysis are integrated into a single system available seamlessly to all scientists involved in the LHC. Users all over the world have access to the experimental data shortly after they are recorded.

During power tests of the chains of LHC magnets on September 19, 2008, a fault occurred in the electrical connection between a dipole (bending) magnet and a quadrupole (focusing) magnet. An electrical arc developed, breaching the cryogenic lines and releasing a powerful helium geyser into the tunnel. Some magnets were pushed off their mounts, leaving a scene reminiscent of a train derailment. Many systems, including those designed to protect magnets in the event of quenches—departures from the superconducting state—did work as planned, and no one was injured. Nevertheless, "the Incident," as it came to be known, was a graphic reminder of how many elements had to function perfectly, both individually and in concert, in the largest collider ever undertaken. Several dozen magnets had to be replaced, and many electrical and mechanical improvements were implemented to prevent a recurrence.

The Large Hadron Collider era began with a brief shakedown run in 2009 with colliding beams of 450 GeV, the energy at which protons are injected from the Super Proton Synchrotron. We saw the first results at a birthday symposium for Chris in December when Fabiola Gianotti, leader of the ATLAS experiment, began her talk by saying that for the first time she would show no simulations, only real data. Even at a fraction of the planned collision rate and beam energy, it was a thrilling display of the potential for discovery of the collider and detector and an unforgettable birthday present!

The physics campaign launched in April 2010 with beams of 3.5 TeV, surpassing the nearly 1 TeV energy of the proton and antiproton beams in Fermilab's Tevatron. The first task for ATLAS and CMS, as they established the performance of the detectors, was to "rediscover" a half-century of particle physics history. This they accomplished in less than one year, reproducing many landmark discoveries.

At an end-of-year "LHC Jamboree" in December 2011, the two experiments presented the first "tantalizing hints" of a new particle—an excess of unusual events in the mass range between 120 and 130 GeV. A press release from the laboratory was circumspect: The new work was "sufficient to make significant progress in the search for the Higgs boson, but not enough to make any conclusive statement on the existence or nonexistence of the elusive Higgs." The leaders of the experiments scrupulously avoided premature claims, but expressed confidence that refined analyses and additional data would resolve the question during 2012. To boost the rate at which Higgs bosons might be produced, the CERN accelerator team raised the LHC energy to 4 TeV per beam in January.

And so we come to July 2012. The Higgs update would be presented at CERN in Geneva, which many took as a hint that a discovery would be announced. Bob was in Paris, preparing to watch the Fourth of July live stream of the presentations with his summertime colleagues in the nuclear and high-energy physics laboratory on the Jussieu campus near the left bank of the Seine. He was studying top-quark properties with members of the lab's ATLAS group. On July 1, Chris was about to depart from the Chicago suburbs for Melbourne, where he would watch the seminars broadcast from CERN and meet the principals when they arrived from Geneva. Shortly after noon, sirens sounded a storm warning; within minutes, a loud crash shook the house. Sprinting to his study to inspect the source of the sound of falling water, he saw the severed limb of a silver maple tree protruding from the ceiling, pointing ominously at the center of his desk. "I took this as a sign that the gods were jealous because we were closing in on the secrets of the universe," he says. Melbourne would have to wait. He would witness the "Latest update in the search for the Higgs boson" starting at two A.M. on the Fourth in the company of his Fermilab colleagues, including many members of the CMS Collaboration, more than normally sleep-deprived because of the push to produce results in time for the event.

Every one of the 400 seats in CERN's Main Auditorium was filled. With the exception of a few witnesses live-blogging the event, no one—for once—had a laptop open to check email or peek at computer output. The audience was positively rapt. The same atmosphere prevailed at Fermilab, in Paris, and doubtless everywhere else that physicists had gathered to hear the news. No one wanted to miss a syllable, or to blink when a crucial piece of evidence appeared.

Although the mass of the Higgs boson is not predicted by the electroweak theory, all of its other properties are predicted once the mass is specified. If the excess around 125 GeV were real, it was a gift from nature: many different decay modes might be studied, leading over time to a very complete profile of the new particle. In the discovery seminar, the leaders of the CMS and ATLAS experiments emphasized their observations of two different candidates for Higgs-boson decays. A Higgs boson decaying into two pairs of leptons (either electrons or muons) through its coupling to two Z bosons would directly implicate the new particle's role in electroweak symmetry breaking and give a good fix on its mass. A decay into two photons would require a quantum loop of top quarks or W bosons, and so would begin to explore the anatomy of the theory; it too had the potential to determine the mass precisely.

Along with less significant supporting evidence from other decay products, both collaborations presented strong signals in these two favorable modes. Taken together, the four-lepton and two-photon signals surpassed the five-standard-deviation threshold for discovery independently in ATLAS and CMS. The observation of two distinct decay modes in two complementary detectors gave us confidence even beyond the statistical significance of each. Along the way, both experiments had confirmed the reliability of standard-model simulations of backgrounds and of benchmark signals.

An hour and forty-two minutes into the proceedings, Rolf-Dieter Heuer, CERN's director general, pronounced, "As a layman, I would now say, 'I think we have it.'" (He was no layman!) With that, jubilation erupted, and a prolonged standing ovation ensued. Four of the theorists whose insights into hidden symmetries nearly a half-century before had set in motion the efforts that led to this triumph—François Englert, Gerald Guralnik, Richard Hagen, and Peter Higgs—were present to savor the moment. So too were past leaders of CERN and many (but only a small fraction) of those who had realized the collider, the experiments, and the theory that we call the standard model of particle physics.

Bob remembers the experience this way: "I was in Paris, at LPNHE, watching on the screen in the main lounge, but feeling as if I were there myself at CERN. We certainly thought we knew that the announcement was coming. I wonder if a proper comparison would be with the NASA team watching the moon landing. I remember being very impressed with the quality of the presentations and amused that this was being widely watched by many people who wouldn't understand a word. I really think that this was an extraordinary achievement. First, the theoretical achievement was itself remarkable. That our basic principles could lead us to predict the existence of a particle totally unlike all those we had found previously is amazing. It is one thing to predict a fourth quark when you have three, or a fifth and sixth when you have three charged leptons, but to predict a neutral scalar fundamental particle when none had been seen is quite different. On the experimental side, the accomplishment was perhaps even more remarkable. They were working at event rates that had not so long ago been viewed as unmanageable. To find the two-photon signal over the enormous background really shows something special about particle physics. The four-lepton observations were impressive, beautiful, but somehow much more straightforward because the backgrounds were easier to deal with."

Chris recalls watching from Fermilab: "The presentations were crisp, the evidence was compelling, and I was moved beyond words that mere mortals had come together to do something truly extraordinary, that we knew a bit more about how the world works. As I had hoped all along, my imaginary friend is real! Science strives for objectivity, but people matter. The ATLAS speaker, Fabiola Gianotti, is one of my closest friends in physics; the CMS speaker, Joe Incandela, learned about the Higgs boson in my course at the University of Chicago. I knew that there would be much more to come, for physics and for each of us, but that was a moment of sweetness and solidarity. I would see them soon in Melbourne, where we were sure to discuss what would come next, for both theory and experiment. My CMS colleagues invited the Fermilab audience to celebrate the discovery with cake and sparkling grape juice. In fact, we required no chemical stimulation to savor the occasion."

Rolf Heuer's "It" was certainly a new neutral boson that shared characteristics of the textbook Higgs boson. Years of effort would be devoted to testing how closely observations meet expectations. The evidence continues to accumulate just as it would if the new particle were indeed the standard-model Higgs boson.

Let's take a moment to recall why discovering the agent of electroweak symmetry breaking is so important. The diversity of our world, relief from a tedious sameness, is due in no small measure to the fact that secret symmetries—more perfect symmetries than the ones we see directly—are hidden by circumstance from direct observation.

A world of perfect symmetry is a surpassingly egalitarian world. Matter particles and force carriers all dash about at the speed of light, exchanging information in brief encounters, never stopping to form lasting associations. Liaisons are here today, gone today. There are no atoms, no intricate structures, and no physicists. All particles are siblings and all forces are one. It is a world of complete disorder—not to say anarchy, for symmetry rules with so heavy a hand that it imposes an unrelenting sameness, a stability in mutability. Everything is interchangeable. It is a perfectly boring world.

The symmetry that links the weak force and electromagnetism is just such a secret symmetry. In our world, these two forces behave very differently. Launch a radio wave or a beam of light, and it will propagate forever, or until it encounters an obstacle. We say that its range is infinite. In contrast, the weak interaction's sphere of influence reaches only about a millionth of a billionth of a meter. And yet our experiments show that the symmetry relating the two forces really is there.

Now, if you want to build an interesting world like ours, it is a good thing that the symmetry is hidden. If it were not, then electrons would have no mass at all, and they wouldn't be held in any finite orbits around whatever nuclei might exist. That would mean no chemical bonding, no solids or liquids, and no template for life! Little wonder that learning how the *electroweak symmetry* is hidden has been a preoccupation of particle physics for decades.

Eleven years after the discovery, ATLAS and CMS have compiled an extensive dossier for the new particle, which we feel secure in calling the Higgs boson. The spin of the boson is zero and characteristics of its decays align with theory. There is no evidence for other Higgs-like particles, whether electrically charged or neutral. The Higgs boson's mass is known precisely as 125.25 GeV with an uncertainty of 0.17 GeV. That determination leaves no known unknowns in the two-act (or one-loop) quantum corrections to many observables. The agreement between calculation and measurements at a level of one part per thousand means that the electroweak theory is far more than a promising first approximation.

The Higgs has been observed to decay into pairs of *W* bosons (with 22 percent probability according to theory), *b*-quark pairs (55 percent), tau-lepton pairs (6.3 percent), in addition to the discovery modes two photons (0.2 percent) and pairs of *Z* bosons (3 percent). Each of these has been established in four distinct production modes, the most important of which is two gluons (one from each colliding proton) creating a loop of top quarks that emits a Higgs boson. The other mechanisms are the production of a Higgs boson together with a *W* or *Z*, which was favored for the searches at the Tevatron and LEP, two gluons producing a pair of top quarks from which a Higgs boson is radiated, and a pair of weak bosons radiated from quarks in the protons that combine to form a Higgs boson.

The importance of the last reaction for hadron colliders was first recognized by Bob and Sally Dawson, now at Brookhaven National Laboratory, in

1984. This "vector-boson fusion" mechanism will enable us to realize in the laboratory the WW-scattering thought experiment imagined by Chris and his colleagues. All the production rates observed so far agree with standard-model expectations. The LHC experiments have also seen indications of decay into muon pairs (0.02 percent) and other rare modes. The electroweak theory predicts that the couplings of the Higgs boson to other particles are proportional to particle mass; this has been verified, within small uncertainties, for top, Z, W, b, and tau, and is indicated for the muon. The link between coupling strength and mass sets the Higgs boson apart from everything that came before. Higgs boson aside, equal treatment of electron, muon, and tau—lepton universality—has been an established pattern in experiments for decades.

This body of evidence points to the conclusion that the Higgs field envisaged in the electroweak theory does hide the electroweak symmetry, distinguishing electromagnetism from the weak interactions, and that, in the process, it gives masses to the electroweak bosons W and Z. Interactions with the Higgs field also give masses to at least the heaviest quarks (top and bottom) and the heaviest lepton (tau). We do not know whether other agents might make minor contributions to these tasks—our experiments will continue to search—but the 125-GeV Higgs boson certainly plays the dominant role. Still to be explored is whether the Higgs boson keeps the electroweak theory from misbehaving at high energies.

The discovery of the Higgs boson is both a climax and a new beginning. To turn the Higgs boson from a sensation into an incisive tool we must explore some great questions. First, what is the origin of electroweak symmetry breaking? Second, what determines the values of the quark and charged-lepton masses? Third, what determines the mass of the Higgs boson itself? And fourth, in what other realms of physics might the paradigm of the "Higgs mechanism" apply?

18

TRADING PLACES

In Robert Louis Stevenson's 1886 gothic novella, *Strange Case of Dr. Jekyll and Mr. Hyde*, the upstanding Dr. Henry Jekyll can transform himself into the uncaring, self-indulgent Mr. Edward Hyde by swallowing a special potion. By this supernatural device, Stevenson created a compelling story, playing on our sense that—on occasion—radical transformations of individuals do occur. A person we think we know can be revealed as a very different person with quite opposite character. Becoming Hyde, the good doctor liberated himself to express his basest urges without feeling guilt.

That opposites must exist is one of the most striking consequences of quantum theory and special relativity. From principles that seem unassailable, we can prove that for every species of particle there must be an *antiparticle* with identical mass and spin, but opposite electric charge. The first inkling that particles must have antiparticle counterparts came with Dirac's prediction of an antielectron (now called positron) in 1931. Anderson's capture of the first specimen in 1932, followed by Blackett & Occhialini's observation of electron-positron pair creation, would seem to have proved the case as we saw in chapter 4, "Heavens." But Dirac's original argument seemed very specific to electrons, and it was not obvious to all that it should apply to particles like the proton, which Dirac's equation does not describe completely. Moreover, our world is built only of protons, not equal numbers of protons and antiprotons—this is still a leading conundrum today—and no antiprotons had been found in cosmic-ray studies, so it was respectable to have doubts about antiparticles in general.

The energy of the 6-GeV proton synchrotron in Berkeley, the Bevatron, was chosen so that collisions with a stationary target could produce a proton-antiproton pair, provided that the antiproton did exist. A 1955 experiment led by Emilio Segrè and Owen Chamberlain found thirty-nine examples of negatively charged particles with the proton's mass. The confirmation of the antiproton—which would annihilate into a flash of energy on encountering a proton—elicited both apprehension and mirth. On October 19, the *Berkeley Daily Gazette*, headlined its account of the announcement, GRIM NEW FIND AT UC. Harold P. Furth, a pioneer of the American quest to harness fusion energy, marked the discovery with a poem published in *The New Yorker* of November 10, 1956. In "Perils of Modern Living," he fantasized that Dr. Edward Teller, head of the hydrogen bomb program, might someday meet a twin made of antimatter: "Their right hands/Clasped, and the rest was gamma rays." Chamberlain and Segrè's work was seen as establishing in general the correspondence between particles and antiparticles. The worldwide success of Dan Brown's mystery-thriller, *Angels and Demons*, testifies to the enduring hold that antimatter exerts on the popular imagination.

The higher energy of Brookhaven National Laboratory's Alternating Gradient Synchrotron brought the discovery of the antideuteron, an anti-nucleus composed of one antiproton and one antineutron. The heaviest antinucleus yet observed, the anti-alpha particle—two antiprotons bound to two antineutrons—was first detected in high-energy gold-gold collisions at Brookhaven in 2011. Antihydrogen atoms made of an antiproton and a positron are under active scrutiny at CERN in Geneva, to investigate how precisely their properties line up with those of hydrogen. This is a key test of our understanding of the particle-antiparticle relationship, and thus of the seemingly unassailable principles that underlie that understanding.

Where Stevenson's "shilling shocker" calls for us to suspend disbelief is not the opposing personalities of Jekyll and Hyde, but the apparent transformation of one into the other through the medium of Dr. Jekyll's potion. How this actually worked is, of course, not explained, for the simple reason that it cannot happen in real life. In the quantum world, however, surprising things occur that would in the macroscopic world require magic—or at least an author's fantastical imagination. In some circumstances, a particle that turns into its opposite is not just possible, but unavoidable.

Each particle collision observed at an accelerator laboratory is a mystery story in miniature. We are presented with the debris and the forensic challenge of figuring out what happened, after the fact. Some of the particle identifications are easy to make. If the detector sits in a magnetic field, the outgoing particles with a positive charge will curve one way and those with a negative charge the other way. A charged particle that passed through the detector seemingly unimpeded must have been a muon. Neutral particles leave no tracks and so are more difficult to identify. A photon only shows up when it encounters matter and makes a shower of electrons, positrons, and more photons.

In some instances, the nature of a "living" particle can be discerned as it makes its way through the detector before decaying. Are we looking at a positive pion or a proton? Typically, we can measure the momentum of the particle, but no more. If we knew the velocity as well, we could distinguish one from the other because combining the momentum and velocity would tell us the mass; the masses of the pion and proton are very different.

Some detectors have components for measuring velocities, at least if they are not too near the speed of light. For many particles, their true identities can be found only through postmortem examination. The particle pathologist examines the remains following the particle's decay. Often not all of the remains will have been collected. Moreover, many particles decay in a dozen or even a hundred different ways. In the absence of complete information, what counts are telltale clues to help sift through a long list of possibilities.

Among the commonplace collision products are the neutral kaons and it is their character that is worthy of a tale by Robert Louis Stevenson. The kaons earned their designation as *strange particles* by clinging to life, persisting for about a nanosecond. That billionth of a second seems pretty brief to us, for whom the proverbial blink of an eye is about one-third of a second, but it was at least a trillion times longer than would have been expected. If Harry Moseley ordered the elements, it was Murray Gell-Mann who, more than anyone else, ordered the elementary particles. He named the strange particles, attributed to them the property called strangeness, and classified two pairs of strange mesons: the charged and neutral kaons K^+ and K^0, with strangeness plus-one, and their antiparticles K^0-bar and K^- with strangeness minus-one.

It occurred to Gell-Mann and his collaborator Abraham (Bram) Pais that K^0 and K^0-bar would inevitably play Jekyll and Hyde, for either could shed its strangeness and turn into a pair of pions that might recombine to form either a K^0 or K^0-bar. The speed at which they trade places puts the Jekyll-Hyde personas to shame; they make the switch about five billion times per second, stopping only when, in one guise or the other, the neutral kaon decays.

In *Strange Case* a single person is both Dr. Jekyll and Mr. Hyde, but the K^0 and K^0-bar are two distinct entities. Gell-Mann and Pais explained that they would actually mix to form two properly behaving particles, each obeying an exponential decay law. One mixture would be, so to speak, Dr. Jekyll plus Mr. Hyde, while the other would be Dr. Jekyll minus Mr. Hyde. Would anyone name a novella *Dr. Jekyll Minus Mr. Hyde*? Literary license extends only so far. But in quantum mechanics two things can be combined in any proportion, plus or minus (and even with complex-number ratios).

If Jekyll and Hyde were to exchange roles, Dr. Jekyll plus Mr. Hyde would remain Dr. Jekyll plus Mr. Hyde, but Dr. Jekyll minus Mr. Hyde would instead become Mr. Hyde minus Dr. Jekyll, which is just the opposite. Gell-Mann and Pais showed that the symmetric plus combination could decay to two pions, but the antisymmetric minus combination could not. If Dr. Jekyll tried to decay to two pions, the minus contribution from Mr. Hyde would just cancel that tendency. The minus combination, lacking the chance to decay to two pions, would live longer than the plus combination.

The shorter-lived combination is simply called K-short and the longer-lived, K-long. Send a beam of negative pions with sufficient energy into a block of material and some collisions of a pion and a proton will make a Lambda (Λ) particle—a strange cousin of the proton—plus a neutral kaon, K^0. Let's identify the K^0 as Dr. Jekyll, which means K-short plus K-long. (If we add K-short to K-long, the Mr. Hyde pieces just cancel out.) The only part of a K^0 that can live a long time is the K-long. This tells us that any K^0 that is observed a long way from where it is produced should not decay into two pions. Once we accept the idea that K^0 and K^0-bar are mixtures of K-short and K-long, this is straightforward enough, even if unfamiliar in everyday life.

Now for a little weirdness: put up a detector near the target in which the K^0 was created and you will observe decays characteristic of the K^0. Some will include a positron along with other particles. The positron is a tell that what decayed was indeed a K^0. Move your detector further away, and you will sometimes see electrons emitted instead of positrons—a marker for K^0-bar. We observe only K^0 at production, with K^0-bar materializing downstream; the Dr. Jekyll neutral kaon has turned into its opposite, the Mr. Hyde *anti*-neutral kaon. No potion, no magic, just quantum physics played out on a macroscopic scale.

We say that the K^0 has oscillated into K^0-bar, because the composition of the neutral kaon beam changes continuously—the share of K^0-bar rising and falling as the square of a sine curve—as we move the detector downstream of the production point. The K-short and K-long have different masses, and so their components do not develop in lockstep. We can say that the internal clocks of K-short and K-long advance at different rates; the different measures of time give rise to the evolving mixture of K^0 and K^0-bar.

Only neutral particles can play at Jekyll and Hyde. A proton with positive charge certainly cannot become an antiproton with negative charge, because electrical charge is conserved. Only special neutral particles can trade places so that we notice. The neutral pion is its own antiparticle, and so is the photon. It would be a boring story if the character took a potion and remained unchanged in every particular!

In a 1957 paper, Bruno Pontecorvo asked whether "there exist other 'mixed' neutral particles (not necessarily 'elementary') besides the K^0 meson, which differ from their antiparticles and for which the particle-antiparticle transitions are not strictly forbidden." Following Gell-Mann and Pais's reasoning, he argued that muonium, an exotic atom made of a positive muon and an electron, might oscillate into antimuonium, made of a negative muon and a positron. According to what we know now, this should not happen, because there is no counterpart to the virtual two-pion state that links K^0 and K^0-bar. Particles discovered in the 1970s do qualify. The meson made from a charm quark, c, and an anti-up quark is called D^0 and it can trade places with the D^0-bar. Likewise, the B^0-bar meson, made of a b quark and an anti-down quark, can trade places with its antiparticle, the B^0. The neutral Ds and Bs are truly stars at the game, switching particle-antiparticle roles about a quadrillion times per second.

Remarkably, Pontecorvo was himself something of a dual personality à la Jekyll and Hyde, though he made the switch just once, in 1950. He began research as part of Enrico Fermi's group in Rome while just twenty-one years of age. Thereafter he joined the Paris group headed by Frédéric Joliot and Marie Curie's daughter Irène, the team that first showed how radioactive nuclei could be created in the laboratory. While in the Fermi and Joliot-Curie laboratories, Pontecorvo made fundamental contributions to nuclear physics in its early days.

Having grown up in a prosperous Italian Jewish family, Pontecorvo fled Paris in 1940 ahead of the advancing Nazis, following the departure of his wife and young child. With difficulty they made their way from France to Spain and then to Portugal. Their emigration to the United States was facilitated by an Italian colleague, Emilio Segrè, then in Berkeley. Segrè succeeded in finding Pontecorvo a job, but it wasn't in Berkeley and it wasn't in nuclear physics research. Instead, he was hired to prospect for oil, working out of Tulsa, Oklahoma. This was not as outlandish as it might sound, because radioactivity was already in use to identify likely oil reserves.

Pontecorvo found his chance to return to nuclear research in 1943 when the British, as part of their effort to develop an atomic bomb, set up a group in Montreal, later moving to Chalk River in the Canadian province of Ontario, to pursue the development of a nuclear reactor following Enrico Fermi's success in Chicago. To make a nuclear reactor, you need to slow down the neutrons generated in nuclear fission, increasing their ability to make still more neutrons and initiate a chain reaction. Fermi slowed down ("moderated") the neutrons with graphite, a form of pure carbon. Taking a suggestion from the Joliot-Curie team, the Canadians chose to use heavy water, in which oxygen combines with deuterium, a heavy form of hydrogen, as a moderator. Because heavy water was identified as a strategic substance for advancing nuclear research, the French liberated the entire world's supply—a precious 185 kilograms—from the Norsk hydroelectric plant at Rjukan to keep it out of the hands of the invading Nazis. After Germans commandeered Rjukan, Norwegian partisans destroyed the plant and its store of heavy water, a story retold in the Kirk Douglas film, *The Heroes of Telemark*. The French heavy water made its way from Paris to the United Kingdom, and eventually to Chalk River.

While in Canada Pontecorvo elaborated a method for detecting neutrinos, a feat that had seemed well-nigh impossible ever since Pauli postulated the feebly interacting particle in 1930. Taking cues from Chalk River colleagues, he noted that a neutrino hitting a chlorine nucleus could transmute it into a radioactive argon nucleus plus an electron. All you had to do was expose a big vat filled with a chlorine-rich dry-cleaning fluid to an intense beam of neutrinos and then extract the individual atoms of argon made in the reaction. Initially, Pontecorvo imagined that this could be done with the neutrinos produced copiously by a nuclear reactor, a much-enlarged version of the one under construction at Chalk River.

In 1956, a team led by Clyde Cowan and Fred Reines observed *antineutrino* interactions at a powerful heavy-water-moderated reactor in the Savannah River Plant in South Carolina, a site that produced plutonium and other materials for nuclear weapons. Using 200 liters of water as a target, they were able to detect both the positron and neutron produced in the collision of an antineutrino with a proton—a reaction called inverse beta decay. Also at Savannah River, Ray Davis executed the chlorine experiment and convincingly found no events, with a sensitivity sufficient to establish that the neutrino and antineutrino are distinct particles. Reactors produce antineutrinos, not the neutrinos required to convert chlorine to argon. Davis devoted the rest of his career to the search for neutrino-chlorine interactions, with a revolutionary outcome. Pontecorvo, for his part, would have much more to say about neutrinos.

Outside the laboratory, Bruno Pontecorvo was quite the man-about-town. Our mentor, Dave Jackson, recalled him as "someone I had known from afar at Chalk River in 1947. Handsome, flirtatious, very Italian, he was the heartthrob of all the single women in Deep River [the company town]. He and his beautiful, vivacious Swedish wife Marianne were prominent on the tennis courts."

Pontecorvo remained in Canada until 1949 when, turning down myriad faculty offers from American universities, he chose to move to the British Atomic Energy Research Establishment at Harwell near Oxford, directed by John Cockcroft, who had led Chalk River. His tenure there was short, a prelude to a transformation. In September 1950, at the end of a summer vacation in Italy, Bruno Pontecorvo and his family disappeared. Unknown to his colleagues in the West, Dr. Pontecorvo had changed identities and become Bruno Maksimovich, "The Professor." He and his family would spend the next decades in the Soviet Union—in Moscow and at the Joint Institute for Nuclear Research in Dubna (the Soviet bloc's answer to CERN). Five years passed before his new life was revealed to the world.

In the introduction to *Half-Life*, his biography of Pontecorvo, Frank Close observes that "His personality was . . . divided into two complementary halves. On the one hand there was Bruno Pontecorvo, the extroverted, highly visible, brilliant scientist, and on the other was . . . Bruno Maksimovich, the enigmatic, shadowy figure who was secretly committed to the communist dream." Whether he had ever been a Soviet spy is unresolved.

Though forbidden to travel to the West, Pontecorvo was free to continue his research in the Workers' Paradise; he focused on weak interactions and especially on the enigmatic neutrino. He speculated in his 1957 paper that neutrinos and antineutrinos might trade places just as K^0 and K^0-bar do, but neutrino-antineutrino oscillations have never been observed. Thirty years would pass before his conjecture that particles other than the neutral kaons could trade places bore fruit in the case of neutral B mesons. A different Pontecorvo insight, also concerning neutrinos, came first.

In the late 1950s, physicists began to suspect that there might be two kinds of neutrinos. Nuclear beta decay, which stimulated Pauli's invention of the neutrino, typically yields an electron and an antineutrino. A charged pion decays almost exclusively to a muon and a neutrino. Are the neutrinos associated with the muon and electron the same?

Pontecorvo in 1959 and independently Mel Schwartz at Columbia University in 1960 proposed an experiment to find out: direct a beam of neutrinos from pion decay onto a massive target and see whether the few neutrinos that interact turn into muons alone or into both muons and electrons. If there were just one kind of neutrino species, a neutrino beam should make both muons and electrons. When a team led by Leon Lederman, Schwartz, and Jack Steinberger performed the experiment in 1962 at Brookhaven National Laboratory's new Alternating Gradient Synchrotron, they observed only muons, showing that the neutrinos that accompany muons are different from those that accompany electrons.

The "two-neutrino experiment" suggested family relationships: the electron paired with its electron neutrino, the muon with its muon neutrino. At once, Ziro Maki, Masami Nakagawa, and Shoichi Sakata, working in Nagoya, Japan, asked whether electron neutrinos and muon neutrinos might mix with each other. By 1967, Pontecorvo was back on the case. He suggested that the electron neutrino and muon neutrino might oscillate as K^0 and K^0-bar do: start with one kind and it could turn into the other. He gave examples of oscillation searches at reactors and accelerator laboratories. Looking back to a Chalk River colleague's observation that it should be possible to detect neutrinos created at the center of the sun, he speculated that solar neutrinos might yield evidence for oscillations.

The sun's heating is the result of making helium out of hydrogen, or more accurately making helium nuclei from protons. This isn't easy to do. If you've ever tried to start a fire by rubbing sticks together, you know what it's like. You need to keep trying until it finally catches. At the center of the sun, where the temperature is fifteen million kelvins, protons keep smashing into each other, over and over again, and just very occasionally—once in nine billion years, on average—a tiny thermonuclear flame begins: the two protons fuse into a deuteron plus a positron, an electron neutrino, and released energy. The reaction rate is so slow because it involves the weak force. Further reactions lead to helium. To prove that fusion powers the sun, physicists would have to detect the solar neutrinos.

Nuclear astrophysicists can figure out how many neutrinos are made this way because we know how much energy the sun produces, and the neutrinos take away a known share. The answer is that there are really a lot of neutrinos coming from the sun. About a hundred trillion of them pass through your body every second. Fortunately, all these neutrinos don't do anything to you; only one is likely to interact in your body during a lifetime. Unfortunately, they don't carry enough energy to convert chlorine into argon.

Other, rarer processes in the sun do produce neutrinos with more than enough energy for the transmutation of chlorine. The rates for these more complicated processes are harder to calculate, making predictions for an experiment less certain.

Ray Davis, who had pioneered the chlorine technique at Savannah River, led a small team with a big tank of cleaning fluid—100,000 gallons (378,000 liters) of it—shielded from cosmic rays nearly a mile underground at the Homestake gold mine in Lead (rhymes with bead), South Dakota. In 1968 Davis reported that he had failed to find the expected signal. Pontecorvo and Vladimir Gribov took the reduced rate as a possible sign that electron neutrinos were oscillating into muon neutrinos, which Davis would not detect. Experiment and prediction improved over time, and in the end the measured rate was less than half what was expected. The experimental method was remote from the traditional techniques of particle physics, so it was hard to know how to judge the reliability of the Homestake result.

In 1976, Davis and John Bahcall, who led the efforts to calculate the sun's rate of neutrino production, summarized the situation: "The attitude of many physicists toward the present discrepancy is that astronomers never really understand astronomical systems as well as they think they do, and the failure of the standard theory in this simple case just proves that physicists are correct in being skeptical of the astronomers' claims. Many astronomers believe, on the other hand, that the present conflict between theory and observation is so large and elementary that it must be due to an error in the basic physics, not in our astrophysical understanding of stellar evolution."

Excuses were proposed and fingers pointed. Maybe the experimenters were not extracting the argon atoms as efficiently as they thought. Maybe the sun's center had cooled; this wouldn't affect the surface temperature for a long time and earthlings wouldn't yet have gotten the news. Davis and Bahcall gave only passing notice to Pontecorvo's proposal of neutrino oscillations. This perhaps revealed the prevailing prejudice that weak-interaction theory has a more elegant structure if neutrinos have zero mass. Neutrinos are so light that physicists have even now not succeeded in measuring their masses. That fact and other evidence about their properties led some workers, starting in the late 1950s, to set their masses to zero by decree. A massless particle—the photon, for example—moves always at the speed of light; its internal clock never ticks, and without clocks ticking at different rates, there can be no oscillation phenomena.

Using the chlorine-conversion technique, Davis was not able to detect events in real time; he harvested the argon atoms every few weeks. That meant that he couldn't prove that the neutrinos came from the sun. In 1986, a new experiment was commissioned under about a kilometer of rock overburden in the Mozumi mine near Kamioka, in a mountainous region near the west coast of Honshu, the largest of the Japanese islands. Kamiokande-II was a tank, holding more than 2,000 liters of water, that recorded the arrival time, direction, and energy of the tiny fraction of solar neutrinos that scattered on electrons in the water. The detector exploited a phenomenon discovered in 1934 by the Soviet physicist, Pavel Cherenkov. In a medium like water, the speed of light is lower than in a vacuum and so energetic particles can travel faster than light in the medium—though not, of course, faster than light in a vacuum. When a charged particle passes through water faster than the reduced speed of light, an electromagnetic shock front creates a cone-shaped burst of light—sharply defined for a muon, fuzzy for an electron. The Kamiokande-II tank was lined with twenty-inch photomultipliers spaced a meter apart; they recorded the arrival time, location, and amount of light, from which the experimenters could reconstruct the cone and the electron's direction, and so determine where the neutrino had come from.

By 1988, the Kamiokande-II team had accumulated enough events to indicate that the neutrinos they detected came preferentially from the sun. Within a decade, using an advanced detector called Super-Kamiokande, the team produced a map of the sky seen not in light, but in neutrinos, showing the sun as the brightest object. Think of the achievement: from their observatory a kilometer underground, human beings have seen the sun—even when it is shining on the far side of the Earth at night—by observing neutrinos, which our human senses do not register at all. Super-Kamiokande is impressive in scope: 50,000 tons of ultrapure water contained in a tank thirty-nine meters in diameter and forty-one meters high, viewed by approximately 13,000 photomultipliers. An order-of-magnitude larger successor, Hyper-Kamiokande, is in the works.

In 1988, the Kamioka group found other peculiar neutrino behavior. In this case, the neutrino source was cosmic-ray interactions, which had been relegated to a supporting role in particle physics since the Cosmotron accelerator turned on in 1953. When cosmic rays that approach Earth from all directions—most of them very-high-energy protons—collide with nuclei in the atmosphere, electrically charged pions are produced in great numbers.

Each of those decays into a muon and a neutrino and the muon subsequently decays into an electron (or positron) plus one muon neutrino and one electron neutrino. (Here we don't distinguish neutrinos and antineutrinos; neither does the water-Cherenkov detector.) If you installed a neutrino counter somewhere on Earth, you might then expect to measure two muon neutrinos for every electron neutrino. The neutrinos would come at you from every direction, not just from the sky above, but from the sky below as well. Neutrinos produced on the opposite side of the Earth feel little resistance zipping through 12,700 kilometers of matter. Kamiokande-II counted about one muon neutrino for every electron neutrino. The electron neutrinos arrived in the expected numbers, but there was a dearth of muon neutrinos.

The enhanced sensitivity of Super-Kamiokande showed in 1998 that neutrinos from above followed the expected two muon neutrinos to one electron neutrino pattern, but that too few muon neutrinos arrived from below. How were these upward-going muon neutrinos different from the downward-going ones? They had traveled farther. The neutrinos were created in the atmosphere and the downward-going neutrinos didn't have far to go to reach the detector, but the upward-going ones traveled an extra distance, from the other side of the globe. It was tempting to think that the muon neutrinos had "oscillated away." But they couldn't have oscillated into electron neutrinos, because no extra upward-going electrons were observed. Super-K was able to show in 2000 that the missing muon neutrinos had turned into the third species, unknown in Pontecorvo's time: tau neutrinos.

These are the neutral partners of the tau lepton, the very heavy big brother of the electron and muon. Further analysis in 2004 established that the phenomenon wasn't a one-time identity swap, but a true oscillation back and forth. It is a quantum phenomenon not on an atomic distance scale, but on the truly macroscopic scale, up to an Earth diameter. This unambiguous demonstration of neutrino oscillations, with its implications that neutrinos have mass, cast the solar neutrino puzzle in a new light.

The discovery of atmospheric neutrino oscillations motivated long-baseline experiments using accelerator beams to study the phenomenon under controlled conditions using nearly pure beams of muon neutrinos or antineutrinos, rather than the mixture generated by cosmic-ray interactions in the atmosphere. Experiments in the United States, Japan, and Europe have directed neutrino beams up to about 750 kilometers through the Earth to distant detectors. Together with long-baseline experiments at reactors, they have greatly refined our knowledge of oscillations, especially between muon neutrinos and tau neutrinos.

And what of the solar neutrino puzzle? By 1995, Davis and company had recorded 2,200 solar-neutrino interactions—a strong signal, but only one-third of the calculated rate. Two other experiments chose gallium as the target in place of chlorine because it is sensitive to the bulk of the neutrinos, the low-energy ones that arise in the initial step that binds two protons into a deuteron. Again, the yield failed to meet expectations, but not by as big a factor. By 2001, Super-Kamiokande had observed more than 18,000 examples of neutrino scatters from electrons and inferred a flux of neutrinos from the sun only 45 percent of the predictions. Still, no one had discovered what became of the missing electron neutrinos. Could John Bahcall's solar-neutrino calculations be trusted? The arc of history would place heavy water from Chalk River at center stage.

Scientists constructed the Sudbury Neutrino Observatory (SNO) two kilometers deep in the Creighton nickel mine in northern Ontario. The detector is a transparent acrylic vessel twelve meters in diameter that held a thousand liters of heavy water with a value of $300 million Canadian. Far smaller than Super-Kamiokande, SNO had the advantage that it could observe two separate reactions that broke up the deuterium nuclei in the heavy water.

One, dissociation by the charged-current (W-boson) interaction, is sensitive only to the electron neutrinos arriving from the sun and the other, dissociation by the neutral-current (Z-boson) interaction, is sensitive to all three neutrino species. Comparing the rates for the two reactions proved that electron neutrinos produced at the center of the sun were, as suspected, arriving at Earth as a mixture of electron, muon, and tau neutrinos. Now all the pieces fit together: the solar (electron) neutrino deficit is real, neutrino production in the sun is understood, and electron neutrinos mix with the others.

The neutrino-oscillation story is not the simple (if supernatural) tale of Jekyll and Hyde but rather more like the Marx Brothers routine in *Duck Soup*, in which Chico and Harpo keep switching hats with a hapless and sometimes hatless lemonade-stand dealer. Every time the lemonade man reaches for his hat, Harpo switches one for another. It is as if the three characters are mixtures of neutrinos with well-defined masses, like the K-short and K-long, and the switching hats indicate the ever-changing manifestations as electron neutrino, muon neutrino, and tau neutrino.

Madcap analogies aside, the neutrino-oscillation story is wonderfully rich and challenging. The low-energy neutrinos produced in the center of the sun, we now know, emerge as electron neutrinos and oscillate on their way to Earth. The more energetic neutrinos detected in the chlorine and water experiments are changed by interactions within the dense environment of the sun into a definite-mass mixture that we call neutrino-2, propagate to Earth without further change, and reveal the mixture when they arrive in our detectors.

How small are the small neutrino masses? Oscillation rates tell us differences of the squares of the neutrino masses, not the masses themselves. Combining many different experiments using solar and atmospheric neutrinos, neutrinos from nuclear reactors and accelerator beams, the differences showed that two masses are very nearly the same and somewhat separated from the third. Which was lighter, the two similar ones or the separate one? We still don't know. The neutrinos are, in any event, far lighter than the other fundamental particles. It is possible that the heaviest of the three masses is about fifty thousandths of an electron volt, one hundred million times smaller than the electron's mass.

The standard model of particle physics incorporates the idealization that neutrinos have no mass. In the discovery of neutrino oscillations, we have not just a hint, but confirmed evidence for a new phenomenon. Where will experimental evidence lead next?

19

NUMINOUS FIRE

Long ago, when belief in magic was widespread, our ancestors thought that miracles wrought by the gods punctuated daily life. The conviction that natural phenomena resulted from the cavalier acts of fickle deities was in time undermined by attentive observations.

Five thousand years ago the ancients had charted the courses of the constellations through the heavens and had begun to devise accurate calendars. The daily return of the sun, the turning wheel of the seasons, and the regular motions of the stars and planets all point to nature's reliability. Experience instructs us to conceive of the universe as an orderly, rather than a capricious, place—a place we can hope to understand. Nature's laws are trustworthy: within our collective experience, they hold at all times and in all places. The force of gravity that governs the motions of the stars and planets is the same force that makes objects fall to Earth. This universality of nature's laws means that in terrestrial laboratories we can learn the physical laws that govern the structure of the universe, and that what we learn by studying the heavens can be applied here on Earth.

In what has been hailed as the most brilliant PhD thesis ever written in astronomy, Cecilia Payne wrote in 1925 that "[t]he application of physics to the domain of astronomy . . . seems to possess almost unbounded possibilities. In the stars we examine matter in quantities and under conditions unattainable in the laboratory." Armed with a better understanding of atomic spectra than most of her astronomical contemporaries, the twenty-four-year-old Englishwoman deduced the relative abundance of the elements by analyzing the spectra of stellar atmospheres.

Since stargazing gave way to organized observations of the heavens, astronomy has taught us much about the history, composition, and structure of the universe—the domain of cosmology. At the beginning of the 20th century, an era of great telescopes enabled one of the founding discoveries of modern cosmology—the recognition that galaxies are separate "island universes" composed of billions of stars, located at great distances from our own Milky Way galaxy. The decisive evidence was presented by Edwin Hubble of the Mount Wilson Observatory as 1924 turned to 1925. Then, in 1929, by examining the spectra of light arriving from the galaxies, Hubble found evidence that distant galaxies appear to be rushing away from us, and that the more distant the galaxy, the faster is its retreat.

In an 1842 article, "On the colored light of the binary stars and some other stars of the heavens," the Salzburg physicist Christian Doppler postulated that the wavelength of light emitted by a receding source is stretched out, so the color of light on arrival at Earth is shifted toward the red and infrared; the color of light emitted by an approaching source is shifted toward the violet and ultraviolet. The Doppler effect for sound is familiar in the change in pitch of an ambulance's siren as it passes. Doppler's insight is the basis for radar guns to measure speed, wind circulation in storms, and many other everyday applications. In Doppler's birthplace, the Café-Konditorei Fürst offers a Doppler Kon(Ef)fekt made of champagne truffle and nougat in a dark chocolate coating. Chris has convinced himself that it tastes different coming and going.

The wavelength of light emitted by a source traveling through an expanding spacetime is likewise stretched out, or redshifted. The greater the redshift—the fractional change in wavelength—the faster the motion of the source. Dividing the distance to a galaxy by its speed of recession gives an apparent age of the universe, which Hubble estimated at about two billion years, suggesting an extent of billions of light-years. Taken together, these discoveries about the galaxies imply that the current universe is old and vast, and that the universe has a past very different from its condition today. Modern measurements confirm the pattern that Hubble inferred, but yield a Hubble ratio of distance to speed about seven times greater.

This is a lot to swallow! In fact, Hubble hesitated to embrace the expanding-universe interpretation of his observations. Describing his studies in 1929, he commented that "It is difficult to believe that the velocities are real; that all matter is actually scattering away from our region of space." In his Rhodes Memorial Lectures, delivered at Oxford University in the autumn of 1936, under the title, "The Observational Approach to Cosmology," he characterizes the significance of the redshifts as "still uncertain. Alternative interpretations are possible, and . . . they lead to totally different conceptions of the universe itself. One conception, at the moment, seems less plausible than the other, but this dubious world, the expanding universe of relativistic cosmology, is derived from the more likely of the two interpretations of redshifts. Thus the discussion ends in a dilemma, and the resolution must await improved observations or improved theory or both."

If redshifts arise from motion, according to the prevailing scientific wisdom, then the universe is expanding according to what is now called the Hubble-Lemaître Law. If, instead, light gradually loses energy through some exotic "tired light" mechanism while traversing long distances, then the redshift does not imply velocity, and the universe may be static. The conjecture that light might experience some sort of gravitational friction as it encountered matter along its flight path first appeared in a 1929 article by a member of Hubble's scientific circle, the Swiss astronomer, Fritz Zwicky.

Hubble's hesitation persisted. He reported to the American Association for the Advancement of Science on December 30, 1941, that a six-year survey with the world's largest telescope showed a uniform distribution of nebulae that—in his view—was incompatible with an expanding universe, and that Earth seemed to exist long before the primordial explosion. The *Los Angeles*

Times headline (page 10) read SAVANT REFUTES THEORY OF EXPLODING UNIVERSE. Today's estimate for the age of the universe, 13.7 billion years, is comfortably greater than the 4.5-billion-year age of Earth, and we are confident that the universe is expanding.

∞

We trace the notion that the universe is changing to Tycho Brahe's report in 1572 of a new star, visible to the naked eye, in the constellation Cassiopeia. Schooled in the Aristotelian tradition that "in the ethereal region of the celestial world no change in the way either of generation or corruption takes place," Tycho expected a reassuring regularity in the starry vault. Since boyhood, he had known all the stars of the heavens perfectly. Now, utterly astonished by the evidence before his eyes, he announced "a miracle indeed," this new star "not previously seen within the memory of any age since the beginning of the world." Forensic astronomers have argued that Tycho's was the omen-star of *Hamlet*'s opening scene.

What Tycho witnessed was the cataclysmic death of a previously faint massive star in a spectacular explosion we call a supernova. He was not the first to record the overnight apparition of a brilliant light in the deep sky. In 1054, Chinese astronomers documented a "guest star" in the constellation Taurus, visible in daylight for three weeks, whose remains are the Crab Nebula. Some scholars go so far as to assert that Sumerian scribes passed on stories of the Vela supernova that occurred some eleven millennia ago. But Tycho's capacity for meticulous observation and the temper of his times, when Tradition and Authority came under challenge, gave special force to his conclusion that the heavens were not perfect and immutable, but a work in progress.

For most of human history, astronomy was practiced by detecting visible light, giving us a rich but highly incomplete impression of the cosmos. Technology developed for practical applications in the 20th century uncovered phenomena to which traditional astronomers had been blind. Commercial transatlantic radiotelephone service was inaugurated between New York and London in January of 1927. Although the first calls were sprinkled with hisses and crackles, when not interrupted by static, the service immediately attracted a following among those who could pay the premium. The following year, Karl Jansky, a recent graduate of the University of Wisconsin, was hired by Bell Telephone Laboratories and assigned to study sources of static, with the aim of improving the reliability and quality of radiotelephone calls. By the fall of 1930, he had installed at Bell's Holmdel, New Jersey, field station a rotatable antenna 14.6 meters (48 feet) across, and begun to study the radio waves it detected, charting the characteristics and intensity of static over time and direction.

In 1932, Jansky had distinguished three sources of noise: static from local thunderstorms, static from distant thunderstorms, and "a very steady hiss static the origin of which is not yet known." In April 1933, after a year of fastidious data-taking and analysis, he announced to the Washington meeting of the International Scientific Radio Union that the most likely source of the radiation was in the constellation Sagittarius, at the center of our Milky Way galaxy, where the signal was strongest. We now identify this source as Sagittarius A*, the apparent location of a supermassive black hole. The May 5, 1933, *New York Times* reported the "discovery of mysterious radio waves which appear to come from the centre of the Milky Way galaxy" and went on to reassure readers: "There is no indication of any kind . . . that these galactic radio waves constitute some kind of interstellar signaling, or that they are the result of some form of intelligence striving for intra-galactic communication." Ten days later, the NBC Blue Radio Network broadcast a live stream of Jansky's "star noise." With the faint condescension it reserved for developments in the provinces west of the Hudson, *The New Yorker* noted on June 17, "It has been demonstrated that a receiving set of great delicacy in New Jersey will get a new kind of static from the Milky Way. This is believed to be the longest distance anyone ever went to look for trouble."

This favorable publicity was not enough to generate support for further studies of the interstellar waves. The extraterrestrial hiss did not compromise transatlantic communications, so Jansky's supervisors—under the financial burden of the Great Depression—declared his project complete and assigned him work that promised immediate practical applications.

Jansky's accounts of his discoveries in the *Proceedings of the Institute of Radio Engineers* captivated a 1933 electrical engineering graduate of Chicago's Armour Institute of Technology (now the Illinois Institute of Technology), Grote Reber. How, Reber wanted to know, did the intensity of cosmic static change with position in the sky and with wavelength? Disappointed to learn that Bell Labs planned no follow-up, and rebuffed when he presented his ideas to astronomers, Reber resolved to create his own purpose-built radio telescope. This he did on his own time, with his own funds, on a vacant lot next to his mother's house in Wheaton, Illinois, then a leafy suburban town of 7,300 residents—about a mile from where Chris has lived since joining Fermilab.

Reber's mother, the former Harriet Grote, who called her firstborn son by her maiden name, encouraged his interest in astronomy. She had taught young Edwin Hubble in seventh and eighth grade at Longfellow School, when his family lived in Wheaton around 1900. She picked Edwin out as a bright boy and took great pride in his accomplishments. Harriet presented Grote with a copy of Hubble's *The Observational Approach to Cosmology* to encourage his pursuit of Jansky's discovery.

The twenty-one-year-old amateur radio operator understood the limitations of wire antennas such as Jansky had used, so he designed a parabolic dish thirty-one feet, five inches in diameter covered in sheet metal to focus signals twenty feet from the surface. Construction was accomplished, largely by Reber's own hands, during the summer of 1937.

The first year of observation yielded no positive results but Reber, a self-professed "rather stubborn Dutchman," carried on—exploring different wavelengths and applying knowledge gained from his professional work in the radio receiver industry to improve his equipment. By the spring of 1939, he had observed the Milky Way at 1.87-meter wavelength and confirmed Jansky's discovery. That was, of course, only the beginning. The goal of an all-sky survey would require both improved sensitivity and a much larger suite of observations.

Reber's day job as a radio engineer gave him access to state-of-the-art test equipment and the latest high-frequency vacuum tubes, with which he could build increasingly capable receivers. He invested in a chart recorder to log data unattended during his workdays in Chicago, an hour distant by train. He published his first sky map in 1944 and switched to a shorter wavelength of 62.5 centimeters to improve the telescope's resolving power. His maps of celestial radio noise revealed what was later confirmed as the first known radio galaxy, Cygnus A, plus the first radio signals from supernova remnants, Cassiopeia A and the Crab Nebula. Not a bad return on Reber's investment of "approximately $11,473" and a decade's passionate commitment!

Reber developed plans for a steerable dish two hundred feet (61 meters) in diameter, only to see his request for funds turned down by several institutions and government agencies. Five decades would pass before such a grand device was built in the United States, but radio observatories—including the celebrated Jodrell Bank Observatory near Manchester—were founded in many places, and radio astronomy took off. The *Third Cambridge Catalogue of Radio Sources*, published in 1959, contained 471 entries.

In 1947, Reber took a position with the National Bureau of Standards in Washington. He joined with collaborators from the Naval Research Laboratory to mount an expedition of eleven men and four radio telescopes to Attu Island, at the far western extremity of the Aleutian chain, to observe the September 12, 1950, annular eclipse of the sun. Remnants of a typhoon that had struck Japan two days earlier brought driving rain to Attu during the eclipse, washing out optical observations. The radio telescopes recorded the progress of the occultation in great detail, making Reber's party the first to learn about the solar atmosphere while observing a total eclipse in the rain. Three years into his Washington sojourn, Grote Reber decamped without notice first to Hawaii, then to Tasmania, where for many years he continued to follow his own unconventional wisdom.

∞

The ageless question, "How did the universe arise?" has stimulated peoples the world over to try to make sense of their origins. David Adams Leeming's *Creation Myths of the World* describes more than 200 traditional accounts, none quite as snappy as these lines from e. e. cummings (*1 X 1*, 1944): "when god decided to invent/everything he took one/breath bigger than a circustent/ and everything began." Science does not yet have an answer. A second cosmological query, "How has the universe evolved?" has become a fully scientific question—but it was not ever thus.

Chris recalls: I first noticed cosmology when I was an undergraduate at Yale. Our physics professors had the charming habit of providing unassigned reading lists of some of the classics of 20th-century physics, many of which were available in Dover reprint editions that cost a couple of dollars—the price of a pig in a blanket at the Yankee Doodle, the local late-night diner. While preparing a term paper on the classic tests of Albert Einstein's General Theory of Relativity, I was reading through the popular works of Sir Arthur Stanley Eddington, the most distinguished astrophysicist of his time and the most accomplished expositor of Einstein's theory. General Relativity reinterprets gravity as a manifestation of curved spacetime, and vice versa. As the revered Princeton theorist, John Archibald Wheeler, put it, "Spacetime tells matter how to move. Matter tells spacetime how to curve." In most circumstances, GR's predictions are indistinguishable—to very high precision—from those of Newton's well-tested law of universal gravitation. In some special circumstances in which the two theories differ noticeably in their outcomes, nature favors General Relativity.

It was Eddington who measured the deflection of starlight by the sun during the 1919 solar eclipse, confirming a key prediction of GR and making Einstein a cultural icon. The opening sentence of chapter 11 of Eddington's 1939 *Philosophy of Physical Science* brought me up short: "I believe there are 15 747 724 136 275 002 577 605 653 961 181 555 468 044 717 914 527 116 709 366 231 425 076 185 631 031 296 protons in the universe, and the same number of electrons." You may not think this is quite as catchy as "Call me Ishmael," but these eighty digits got my attention with just the same intensity. "How could he know that?" I asked myself.

When I discovered that Eddington's number, which equals 136 times 2 to the power 256, issued from his fascination with numerology, and that he had done the arithmetic during an Atlantic crossing, I wondered whether old Sir Arthur had gone 'round the bend. (In this, I joined many of his contemporaries.) Thus, I learned from Eddington that cosmology was faintly disreputable, the realm of speculation rather than prediction and verification. Cosmology was in desperate need of data! But I also learned from Eddington that maybe, just maybe, you *could* know, in a crude sense, about everything in the universe.

That was astonishing; cosmology might be fascinating.

∞

Until Hubble's 20th-century discovery, cosmology could count only one fact—the night sky is dark—plus Tycho's insight that the universe is changing and Nikolaus Copernicus's inference that the Earth is not in a favored location. On large scales, the universe observed from Earth appears similar in all directions, or isotropic. If Earth's location is typical, a good working hypothesis is that the universe is the same everywhere, averaged over large enough scales. We call this idea the Cosmological Principle.

After Albert Einstein formulated his General Theory of Relativity, he turned his attention in 1917 to a model of the universe. For an isotropic universe in which matter is distributed uniformly, the simplest solutions to the equations of general relativity threaten to collapse under the mutual gravitational attraction of all the matter. The puzzle was not new; Isaac Newton, founder of gravitational theory, had fretted two and a half centuries before that if every bit of matter in the universe attracted every other, what prevented the universe from shrinking without limit? Einstein found a way to save his model universe from collapse, at the price of changing his gravitational law by adding a "cosmological constant" called Λ (capital Lambda), a pressure that would offset the gravitational attraction on cosmological scales. Though put in by hand, the cosmological constant did not destroy the underlying principle of general relativity.

Einstein's static, homogeneous, and isotropic model universe tries to enforce a precarious balance between a collapsing universe and one that expands without limit. But it also fails the single cosmological test available, circa 1917—the darkness of the night sky.

Starlight would circulate endlessly around the closed spherical universe, so the longer stars shine, the brighter the sky would become. Hubble's evidence for an expanding universe removed the motivation for a static solution, although the notion of a "steady-state" cosmology would persist for decades.

In Petrograd (Russia), Alexander Alexandrovich Friedmann noticed the latent instability in Einstein's original model and in 1922 published models with uniformly distributed matter that expanded from an extraordinarily compressed state. Working independently in Louvain, Belgium, Abbé Georges Lemaître published similar expanding-universe solutions to Einstein's equations in 1927. When the young man showed his work to Einstein at the famous Solvay Congress of 1927, the master responded that the mathematical manipulations seemed technically correct, but that the physical consequences were "abominable."

In Lemaître's cosmology, the recession velocity of distant galaxies from any location, not just the Earth, is proportional to distance—just as Hubble would observe in 1929.

Friedmann succumbed to complications of typhoid fever in 1925, so he never learned of the evidence for an expanding universe. Arthur Eddington, who had introduced Lemaître to the study of the universe when the Belgian priest spent a year as a research associate in Cambridge, quickly endorsed the expanding-universe cosmology. Einstein's conversion was slower, inhibited in part by the same apparent conflict between the age of Earth and the age of the universe that vexed Hubble. But convert he did; on a visit to Pasadena in 1932, Einstein joined with Willem de Sitter to advocate a spatially flat universe that expands forever at a steadily diminishing rate—and no cosmological constant. Their picture—built on the ideas of Friedmann and Lemaître—became the standard cosmological model for six decades.

Gravitational attraction slows the speeds of the outrushing galaxies, just as Earth's gravity slows the speed of a ball thrown up in the air. The faster the ball is thrown, the higher it rises before falling back to Earth under gravity's pull. For an initial speed a little greater than 25,000 miles per hour, the ball can escape Earth's gravity entirely. In the same way, depending on the initial expansion rate and the amount of matter in the universe, the cosmos might swell forever—or at least for a very long time—before the gravitational attraction won out.

On the hypothesis that nature is dependable, we can run the universe's life story backward in time to help us imagine how things must have been long ago. The pattern Hubble observed suggests that the universe once was very hot and dense—a primordial fireball, a seething state of dizzying hyperactivity. In 1931, Lemaître took the hint to an extravagant limit, arguing that the universe evolved from what he called the primeval atom. We do not know whether we can reliably extrapolate all the way back to a beginning of the universe, as in Friedmann and Lemaître's equations, but the evocative picture of a universe expanding and cooling from a hot and dense fireball has the makings of a good story—the Big Bang.

∞

George Gamow, whose picture of alpha-particle emission and absorption by a nucleus we encountered in chapter 3, moved to the United States in 1934, taking a position at George Washington University. After World War II, he began to apply the newly won knowledge of nuclear physics to the study of the universe, influenced by the thinking of his Leningrad teacher, Friedmann. Gamow's student, Ralph Alpher, worked out a network of nuclear reactions by which the elements might have been formed in the early universe: as the hot and dense universe cooled, protons and neutrons formed, then combined to produce the light elements up to carbon. Alpher estimated at his 1948 thesis defense—an uncommonly public event—that all the primordial nucleo-synthesis had occurred in about 300 seconds. An account in the *Washington Post* bore the headline, WORLD BEGAN IN FIVE MINUTES, NEW THEORY. On April 1, *The Physical Review* published "The Origin of Chemical Elements" under the names of Alpher, Gamow, and the future Nobel laureate Hans Bethe. The frisky Gamow added Bethe's name for the amusement of referring to the work by the first three letters of the Greek alphabet: *alpha, beta, gamma:* αβγ.

Within a year, Alpher and Robert Herman published a remarkable analysis of the evolution of the expanding universe, elaborating the Big Bang picture. They realized that the relative composition of the universe changes with time, so that an era in which most energy is present in the form of radiation would give way to one in which matter predominates.

Matter in the hot early universe is highly ionized. Free electrons, positrons, and heavier particles absorb and scatter the ambient radiation, preventing its propagation over long distances. At an age that we now estimate at 380,000 years, when the expanding universe has cooled to a temperature of a few thousand kelvins, electrons and protons settle into neutral atoms, and the universe becomes largely transparent to quanta of light.

The newly transparent universe would have been teeming with photons, with an intensity and spectrum determined by the temperature of the medium alone, characteristic of an idealized object called a blackbody—a perfect absorber and emitter of electromagnetic radiation.

Although the notion of a blackbody was put forward in 1860, it was not until the very end of the 19th century that physicists in Berlin realized a practical example—a cylindrical cavity (*Hohlraum*) that could be held at a uniform temperature, pierced at one end by a tiny hole through which blackbody radiation passes outward. They also refined the art of measuring the intensity of light emitted, not only in the visible range, but from the infrared into the ultraviolet. In a time of competition between gas and electrical illu-mination, the new tools enabled scientists to define standards to characterize the quantity and quality of light.

Like the heating coil of an electric range, whose glow changes from dull red to bright orange as the temperature and heat output rise, the hotter a blackbody becomes, the more energy it radiates. A blackbody spectrum follows a smooth curve that peaks at a specific wavelength that moves to shorter wavelengths for hotter objects. Astronomers have learned that the color of a star gives an indication of its surface temperature, from about 3,500 kelvins for reddish stars such as Antares to about 5,800 kelvins for our yellowish sun and 11,000 kelvins for the bluish Rigel.

While some scientists turned their knowledge of the blackbody spectrum to the stars and others worked to perfect filaments for incandescent lamps that would supply artificial daylight, still others sought to understand the blackbody spectrum itself. They found that the established laws of 19th century physics could not explain blackbody radiation.

Classical reasoning reproduced the observed spectra at long wavelengths, but failed catastrophically at short wavelengths, predicting an overproduction of radiation that both overshot the data and defied reason. The old physics predicted that hot objects would radiate infinite amounts of energy at short wavelengths.

The solution came in the form of a most radical proposal from a most disciplined thinker—Max Planck in Berlin, of whom Max Born, a pioneer of quantum mechanics and fellow Nobel laureate, wrote, "His was, by nature, a conservative mind; he had nothing of the revolutionary and was thoroughly sceptical about speculations. . . . He did not flinch from announcing the most revolutionary idea which has ever shaken physics." In the traditional picture, the atoms and molecules jiggling in thermal motion could absorb and emit radiant energy in any amount. In what he characterized as an act of desperation, Planck broke with the past, conjecturing in December 1900 that energy could only be transferred in specific discrete amounts—quanta of energy. Planck's new formula matched the observed spectra in every particular. His hypothesis, which spurred physicists to think about nature in new ways, would blossom into a novel vision of the microworld, quantum mechanics.

A half-century after Planck's insight, Alpher and Herman realized that the blackbody radiation at early times would appear as a relic in today's universe.

As the universe expands and cools, it should be awash in a highly uniform bath of electromagnetic radiation, with a temperature in our era that they estimated as about five kelvins—5 degrees above absolute zero. The wavelength of this fossil radiation would be a thousand times longer than the wavelength of visible light, imperceptible to our eyes, in what radio engineers call the microwave band.

If the expanding universe was once hot and dense, it should be possible for radio telescopes to detect the faint afterglow of the fireball. Nevertheless, Alpher and Herman's prediction did not set off a gold rush to find the fossil radiation. Reading their article today, we find it highly perceptive on the essential features of the hot Big Bang scenario, albeit with an imperfect understanding of how the elements are produced. So why did it recede into the scientific underbrush? Perhaps the early universe seemed too remote for serious consideration, or perhaps too many seemingly untestable deductions were presented in one place. Perhaps the article was simply too far ahead of its time, or perhaps cosmology hadn't yet shaken its metaphysical roots. In any event, the insight that a search for cosmic microwave radiation could deliver evidence for Big Bang cosmology was destined for a long season of neglect.

∞

Years later, on May 21, 1965, a three-column spread on the front page of the *New York Times* proclaimed, SIGNALS IMPLY A 'BIG BANG' UNIVERSE. Walter Sullivan's report began, "Scientists at the Bell Telephone Laboratories have observed what a group at Princeton University believes may be remnants of an explosion that gave birth to the universe." When Arno Penzias and Robert W. Wilson joined Bell Labs as young PhDs in 1963, they began charting a campaign of radio astronomy using the Holmdel Horn Antenna, which resembled a giant ear trumpet for microwaves. Originally used to detect microwave signals reflected from Echo balloon satellites, then to capture signals relayed by Telstar, the first active communications satellite, the twenty-foot horn reflector was no longer required for satellite studies. The young radio astronomers set out first to make absolute intensity measurements of several known radio sources and then to prospect across the Milky Way, at a wavelength of 7.35 centimeters. From their first observations, they were bedeviled by an excess of noise that they could not account for.

Atmospheric sources, interference caused by human activity, galactic hiss, and discrete extraterrestrial sources all were examined and ruled out in turn. Might the antenna itself be generating the annoying hum? Desperate to track down the noise that was compromising their measurements, the intrepid investigators noted that a pair of pigeons roosting in the horn had coated the interior with "a white material familiar to all city dwellers," as Wilson so delicately put it in his Nobel lecture. They evicted the avian invaders and cleansed the antenna of guano, but the unexplained hum persisted—arriving with equal strength from every direction and free from circadian or seasonal variations over a nine-month observing period. If interpreted as blackbody radiation despite having been observed at only one wavelength, the intensity of the mystery radiation implied a temperature of 3.5 kelvins (later revised to 3.1 kelvins), with a 1-kelvin uncertainty. Penzias and Wilson could make a convincing case that the universe is filled uniformly with microwave radiation, but what did it mean?

Almost by chance, the Bell researchers learned that a group from nearby Princeton University—Robert H. Dicke, Jim Peebles, Peter Roll, and David Wilkinson—might have some interesting ideas about radiation in the universe. Dicke was exploring the idea that a hot, dense early phase of an expanding universe should have left behind a mist of fossil radiation spread uniformly across the sky. He had tasked Roll and Wilkinson to build a receiver capable of observing the radiation and suggested that Peebles consider the theoretical implications of detecting or not detecting it. Once the two groups met, it became clear that although the Princetonians had been scooped, they could offer a plausible interpretation of Bell's signal that gave it a truly cosmic significance. The July 1, 1965, issue of *The Astrophysical Journal* contains back-to-back letters to the editor: "Cosmic Black-body Radiation" from Princeton, setting out the expectations, and two technical pages from Penzias and Wilson that document "A Measurement of Excess Antenna Temperature," but eschew any cosmological interpretation. Neither paper mentioned Alpher and Herman's prediction of the microwave background.

It was now urgent to find out whether the cosmic microwave background did follow the blackbody spectrum derived by Max Planck. In March 1966, Roll and Wilkinson published their measurement of 3 kelvins with an uncertainty of one-half kelvin at 3.2 centimeters. Within a few months, radio astronomers showed that the temperature of the microwave background radiation, measured by intensity, was close to 3 kelvins at several wavelengths between 21 centimeters and 2.6 millimeters. A blackbody spectrum began to take shape, supporting the view that Penzias and Wilson had indeed discovered the fossil remains of the Big Bang. Further investigation showed that the radiation observed on Earth is slightly skewed by our motion within the universe at large. Once that motion, at a speed of 371 kilometers (230 miles) per second, is taken into account, the microwave background radiation exhibits no preferred direction. Arno Penzias and Robert Wilson shared in the 1978 Nobel Prize "for their discovery of cosmic microwave background radiation." Their observations showed the possibility of gathering information about cosmic phenomena that took place long ago and made Big Bang cosmology the prevailing narrative for the creation and evolution of the universe. Grote Reber, that "rather stubborn Dutchman," was not persuaded: in a 1982 article, he decried what he called the religious dogma of Big Bang creationism. Cosmology would remain a domain of strong opinions!

Measurements made from the ground, mountaintops, airplanes, balloons, and rockets reinforced the evidence for a uniform blackbody distribution corresponding to a temperature of about 3 kelvins. When considered in detail, this turns out to be too much of a good thing for the traditional hot Big Bang picture. Two regions of the visible universe so distant from one another that the light travel time between them exceeds the age of the universe were never in communication or physical contact in the early universe. Why should they have the same temperature?

In 1980, Alan Guth at the Stanford Linear Accelerator Center proposed an audacious path toward a solution. He imagined that, in spite of appearances, all of the visible universe—and more—*had* been in contact in the early universe. A tiny fraction of a second after the primordial event—before Hubble expansion became the rule—the universe underwent stupendously rapid growth that stretched lengths by a hundred trillion trillion times in far less than the blink of an eye. If this were so, everything we see in the universe—and much more—originated from an infinitesimal morsel of matter and energy that existed long enough to come to a uniform temperature. Now refined by many workers, the inflation mechanism has become a standard—though not unchallenged—extension of the Big Bang paradigm.

∞

What makes the universe so uniform and yet so clumpy? This one paradoxical question expresses the tension that has goaded cosmology throughout the past century. The inflation hypothesis shows how we might account for the uniformity. We suspect that the galaxies, clusters, filaments, and voids that make up the universe arose from quantum fluctuations at early times—small patches of above-average or below-average matter densities that grew, under gravitational attraction, to larger structures. Inflation would have stretched these small regions into far larger ones. Rainer Sachs and Arthur Wolfe noticed in 1967 that the gravitational influence of uneven distributions of matter in the early universe would cause small differences in the temperature of the cosmic microwave background radiation. How far back in the history of the universe can we find evidence for the origins of structure?

Bob recalls: In April 1992, I oversaw the particle physics program at the Lawrence Berkeley Laboratory. A couple of our projects were actually outside the scope the Department of Energy had for us. These fringe activities didn't figure much in my concerns and so I really didn't know what to make of the picture I was about to see. George Smoot had come into my office to show me results from the team he led, which was measuring the cosmic microwave background using instruments on the COsmic Background Explorer satellite, COBE (rhymes with Moby). George is a big guy and rather imposing. You might take him for a retired pro football player whose good looks and ability to speak coherently had landed him a lucrative job as a television sportscaster.

"Is this going to be big, George?" I asked.

"Yes, it's going to be big," George replied.

We both knew that COBE's first report had made quite a splash—at least among astronomers. The COBE satellite, proposed to the National Aeronautics and Space Administration in 1974, was launched on November 19, 1989, with Ralph Alpher and Robert Herman in attendance. COBE carried sensitive instruments to detect infrared and microwave radiation from Earth orbit—beyond the distorting filter of our moist atmosphere. Within a few weeks of launch, the Far InfraRed Absolute Spectrophotometer (FIRAS) had measured the intensity of the background radiation at thirty-four wavelengths. In January 1990, John Mather of the NASA/Goddard Space Flight Center, leader of the FIRAS team, presented a verdict based on nine minutes of data-taking to a meeting of the American Astronomical Society in Crystal City, Virginia: the temperature of the background radiation was 2.735 kelvins, with an uncertainty of only 0.06 kelvin, and the spectrum showed no departures from Planck's blackbody formula.

The announcement won a standing ovation from the thousand astronomers in attendance, but an account in the *New York Times*, consigned to page 22 of the Sunday paper, emphasized what COBE had not found. Under the headline, SPACECRAFT SEES FEW TRACES OF A TUMULTUOUS CREATION, Warren E. Leary wrote that "First glimpses of the heavens by a new spacecraft looking for evidence of the beginning of existence have revealed an almost too perfect universe, scientists said today." While the spectrum was beautiful to behold and was discouraging to Big Bang doubters, COBE had not presented any evidence for the density fluctuations suspected of seeding structure in the universe.

Now, a little over two years later, George previewed what the COBE mission would make public at the Washington meeting of the American Physical

Society—a baby picture of the universe! To the uninitiated, George's pictures, constructed using observations from COBE's instrument called the differential microwave radiometers, were the worst baby pictures in history. The baby was not a newborn, but a 380,000-year-old toddler, snapped when the universe turned transparent. That was not the problem; there seemed to be nothing to see. If you printed the images in black and white, you'd have nothing but grey.

That uniformity was the first part of George's tale: in patches roughly the amount of sky your fist blocks out when held at arm's length, the microwave radiation is the same all across the heavens, within one part in a 100,000. It was a refined view of the sameness that cosmic inflation had been invoked to explain.

But if you looked closely enough you could pick out brighter and darker spots representing tiny variations in temperature. To make a dramatic picture you can replace the black-and-white scheme with a color palette that runs from cyan to magenta; where the temperature is thirty millionths of a degree below average, paint it a cool cyan and where it is thirty millionths of a degree above average, paint it a warm magenta. Slight as these variations were, they constituted dramatic evidence of subtle stippling on the spacetime canvas, from which the universe we observe—homogeneous and isotropic on large scales, but rich in structure on smaller scales—might evolve. Here was part two of the story: the early universe was not too perfect, after all. George Smoot was right. This was big. How big would be apparent in a matter of days.

George's talk in Washington was an instant sensation, even beyond the community of physicists and astronomers. "NASA satellite discovery is 'like looking at God,'" announced the April 24, 1992, *Chicago Tribune* in its lead article splashed across the front page. (George had told the press, "If you're religious, it's like looking at God.")

In its lead slot, the *New York Times* was only slightly more circumspect. Under the headline, SCIENTISTS REPORT PROFOUND INSIGHT ON HOW TIME BEGAN: 'BIG BANG' THEORY BACKED," the lede read, "In a momentous discovery supporting the Big Bang theory for the birth of the universe, astronomers looking back toward the beginning of time have detected broad wrinkles in the fabric of space, the first evidence revealing how an initially smooth cosmos evolved into today's panorama of stars, galaxies and gigantic clusters of galaxies." A month later came a *People* magazine profile, teased with "Ever since he and his colleagues found wrinkles in the cosmos that support the leading theory of the birth of the universe, campus life has been one great big bang for astrophysicist George Smoot."

George's celebrity did not soon abate. He appeared in the year-end issue devoted to *People*'s "25 Most Intriguing People of 1992," introduced as "An Astrophysicist Completes the Blueprint of Creation." There he was preceded in notoriety by Madonna ("The Movies! The Album! The Naughty Pictures!") and followed by Katie Couric ("*Today*'s savvy Co-Anchor Put Her Show Back on Top"). Of greater import, John Mather and George Smoot shared the 2006 Nobel Prize, "for their discovery of the blackbody form and anisotropy of the cosmic microwave background radiation," a distinction for the entire COBE team.

COBE was succeeded by two advanced satellites that carried more modern instrumentation. NASA's Microwave Astronomy Probe, renamed to honor David Wilkinson after his death, and the European Space Agency's Planck Mission yielded pictures of the early universe that distinguished much finer features than COBE had resolved. From the sizes of the hot and cold spots, scientists can calculate values for a few fundamental bits of information—the shape, size, age, composition, and rate of expansion of our universe. The exquisite measurements from these missions would provide the most precise quantitative description of the early universe we have, incorporating a new ingredient unknown at the time of the COBE discovery.

20

WHAT IS THE WORLD MADE OF?

T he quest for the substance of substance stretches back in human history as far as we have recorded evidence—to the pronouncements of Zarathustra (Zoroaster) and on to the Babylonian creation myths a millennium earlier. The Greek master Empedocles, born circa 494 B.C.E. in what is now Agrigento, Sicily, transmitted a vision of the classical elements: earth, air, water, and fire, under the influence of the contending forces, love and strife. What became the Aristotelian doctrine held sway well into the 17th century. As science moved from assertion to evidence, mathematical refinement, and experimental tests, the 19th-century chemists regarded the elements of the Periodic Table as the constituents of all matter. By 1961, in the first of his storied *Lectures on Physics*, Richard Feynman proclaimed:

"If, in some cataclysm, all of scientific knowledge were to be destroyed, and only one sentence passed on to the next generations of creatures, what statement would contain the most information in the fewest words? I believe it is the *atomic hypothesis . . . that all things are made of atoms—little particles that move around in perpetual motion, attracting each other when they are a little distance apart, but repelling upon being squeezed into one another.* In that one sentence, you will see, there is an *enormous* amount of information about the world, if just a little imagination and thinking are applied."

Between the rise of quantitative chemistry and Feynman's opening lecture, the existence of atoms had been proved, the atom had been revealed as a tiny nucleus at the center of a cloud of electrons, and the chemical properties of the elements had been explained—at least in principle—by the new worldview of quantum mechanics. The nucleus was composed of protons and neutrons, and a deepening, if still semi-empirical, understanding of nuclear forces could explain which nuclei should exist and how they should behave. Since Feynman first mesmerized his Caltech undergraduates, we have learned that the proton and neutron are composed of quarks bound together by strong-force agents called gluons and that the electron is one member of a family of leptons; and we have developed new laws of nature governing the strong, weak, and electromagnetic interactions.

Might there be more to the world than tangible matter? Already a century after Empedocles, Aristotle advanced the notion of an incorruptible fifth essence, or quintessence, to account for the perceived unchanging perfection of the heavens. With that addition of what he called the æther, Aristotle's Earth-centered universe seemed to leave no room for any new forms of matter. That sort of dogmatic certainty began to fade when telescopes, in the hands of Galileo, Simon Marius, and others, revealed new objects in the sky that had been imperceptible to the naked eye. By the 20th century, the realm of the nebulae—to borrow Edwin Hubble's phrase—yielded hints of something new in the heavens.

In 1933, Fritz Zwicky published a critical examination of interpretations of the redshift of extragalactic nebulae, including his own conjecture about "gravitational friction," in the Swiss Physical Society's journal, *Helvetica Physica Acta*, in the German language of his student days. The paper's abstract gave no hint of the fuse he would light in Section 5: "Remarks concerning the dispersion of velocities in the Coma nebular cluster." In the space of less than two pages, Zwicky argued that galaxies in the Coma cluster—a seemingly stable association of more than a thousand members—moved relative to one another with velocities much higher than one should expect. To account for this motion, the average density of matter in the Coma system "would have to be at least 400 times larger than that derived on the grounds of observations of luminous matter. If this would be confirmed we would get the surprising result that dark matter is present in much greater amount than luminous matter." So did Zwicky challenge the presumption that the bright spots signal the distribution of mass within a cluster.

We're carried back to the words of Lord Rayleigh, on encountering the idle element argon: "[I]n accepting this explanation, even provisionally, we had to face the improbability that a gas surrounding us on all sides, and present in enormous quantities, could have remained so long unsuspected." Are we to believe that the universe at large is filled with enormous quantities of a massive substance, effectively inert like the noble gases, and so long unsuspected? Would not "noble matter" be a fitting name for the mysterious material?

Studying the movement of the planets around the sun early in the 17th century, Johannes Kepler found that the more distant a planet was from the sun, the more slowly it moved. For example, Saturn is on average about nine times as distant from the sun as Earth is, and its mean velocity is about one-third of Earth's. Newton's universal gravitational theory explained Kepler's laws of planetary motion, with the implicit assumption that the sun is overwhelmingly the dominant source of mass in our solar system. Similarly, if the mass in a galaxy is concentrated in the bright regions, the further a star is from the center of a galaxy the more slowly it should move.

As astronomers refined their instruments, they gained the ability to analyze the motions of luminous regions within galaxies, not only the motions of galaxies within clusters. Through the 1970s, both optical and radio studies established that velocities did not decrease with distance from the center of a galaxy, as Keplerian reasoning seemed to suggest. Among the decisive investigations was a comprehensive examination of motions within twenty-one large and small spiral galaxies published in 1980 by Vera Rubin of the Carnegie Institution of Washington and her collaborators. The rotation curves, which display velocity versus distance from the center, implied—if standard gravitational theory holds—that the mass of a galaxy is not concentrated in the center; mass continues to increase even beyond the edge of the optical image. "The conclusion is inescapable," wrote Rubin and company, "that nonluminous matter exists beyond the optical galaxy." Computer simulations suggest that celestial structures such as galaxies would deform and fall apart unless they contain much more mass than we can see.

To many prominent astronomers, Rubin's work clinched the case for dark matter such as Fritz Zwicky had divined in the Coma cluster, and for that contribution she is justly celebrated today. In a *New York Times* obituary, Dennis Overbye wrote that she had "transformed modern physics and astronomy with her observations showing that galaxies and stars are immersed in the gravitational grip of vast clouds of dark matter."

Concluding a 2001 review article written with radio-astronomer Yoshiaki Sofue, Vera Rubin evenhandedly anticipated future developments: "We will ultimately know what dark matter is, the major constituent of the Universe when measured in mass. Elementary particle physics will teach us its origin and physical properties. Perhaps we will be able to put to rest the last doubt about the applicability of Newtonian gravitational theory on a cosmic scale, or enthusiastically embrace its successor." In a less formal setting, she owned up to her preferred outcome: "If I could have my pick, I would like to learn that Newton's laws must be modified in order to correctly describe gravitational interactions at large distances. That's more appealing than a universe filled with a new kind of sub-nuclear particle."

As much fun as that would be for physicists and astronomers alike, it's not likely to happen. So compelling is the evidence for Newton's Laws and the Law of Gravitation in particular, that the odds are very great that particles unknown populate the galaxies and contribute much of their mass, indeed most of the mass of the universe. Physicists are working hard to identify the dark matter (or dark matters!) and to understand its role in the history of the universe.

Astronomers have begun to tell us more about where the dark matter is. Albert Einstein in 1936 published a short note observing that the gravitational attraction of a star upon light that is responsible for the deflection of starlight at the limb of the sun could, under the right circumstances, cause a foreground star to act as a gravitational lens, focusing light emitted by background sources. Zwicky followed up in short order with the suggestion that the dark matter in extragalactic nebulae might focus the light emitted by background galaxies and enable astronomers to trace the total distribution of matter in the clusters. It took seven decades and the Hubble Space Telescope's large-format cameras before international teams could create three-dimensional maps that depict the weblike large-scale distribution of dark matter throughout the universe.

What is the dark matter? Could it be ordinary matter that is not luminous enough to be detected directly by optical telescopes, such as black holes or brown dwarfs intermediate in mass between planets and stars? If heavenly bodies of this kind, which astronomers name MAssive Compact Halo Objects—MACHOs—are very numerous, they might be observed as gravitational lenses when they pass between a distant star and Earth, causing a momentary brightening of the starlight. MACHO searches have shown that they are too rare to explain all the dark matter.

The nuclear reactions that synthesized the light nuclei (deuterium, helium, and lithium) depend on the population of protons and neutrons during the first few minutes of the current universe. Modern accounts of primordial nucleosynthesis place the current abundance of protons and neutrons—the stuff of stars—at about 4 percent of the critical mass-energy density that corresponds to a spatially flat universe; this is equivalent, on average, to a single hydrogen atom in five cubic meters of space. These arguments were first made in the paper by Robert Dicke's group that identified the cosmic microwave background radiation.

What about neutrinos—those fundamental particles that carry no electric charge and very little mass, that fly about at nearly the speed of light and hardly interact at all, but seem to be the most numerous bits of ordinary matter in the universe? Although dark matter in the form of fast-moving ("hot") light neutrinos would help initiate clumping of matter in the early universe, what we have learned of the history of structure formation appears to rule it out as a dominant component of the dark matter.

If it is none of the above, might dark matter be new forms of matter, such as the conjectured weakly interacting massive particles called WIMPS, relics of the hot young universe that we might hope to produce in particle-physics experiments? As the standard model of particle physics came together in the 1970s, theoretical proposals that would extend the standard model to an even more comprehensive and predictive theory yielded a number of candidates. Princeton cosmologist Jim Peebles considered in 1982 the influence of nonrelativistic "cold" dark matter on the evolution of the universe, exhibiting a link between anisotropies in the cosmic microwave background radiation and the development of large-scale structures. He foresaw temperature fluctuations at the level of five parts per million, anticipating the COBE results that George Smoot would announce a decade later. This work focused attention on weakly interacting massive particles as dark matter and was a key element of the work that earned Peebles a share of the 2019 Nobel Prize in Physics.

The inflationary universe paradigm holds that the density of the universe must be very close to the critical value at which the universe expands forever, but at a decreasing rate. Otherwise, spacetime would become highly curved during the inflationary instant, contrary to evidence that it is very nearly flat today. A natural inference would be that dark matter makes up the 96 percent not yet accounted for.

We can't put the universe on a scale to weigh it, and if we're impatient to have hints about The Fate of the Universe (to capitalize as cosmologists compulsively do) we won't want to simply wait for the future to arrive, so what observations can we make? Because the speed of light is finite, looking out to large distances means looking back in time—seeing the universe as once it was. To determine whether the universe is going to continue expanding, or stop and collapse back on itself, we can ask: How fast is it expanding today?

How fast was it expanding a while back? In other words, how is the pull of gravity changing the rate of expansion?

The standard inflationary cosmology of the 1980s was unambiguous: the universe would expand eternally, but at an ever-diminishing rate. A quantity called the deceleration parameter would equal one-half. To abbreviate the technical argument, let us say that *inflationistas* were adamant that the matter density was equal to the critical value, and it is generally foolish to bet against anything even remotely associated with Einstein. A measurement would be a test of the scientists and their techniques, more than a test of nature; the result was foreordained.

Nevertheless, the historical imperative was clear. Already in 1970, Allan Sandage, heir to Edwin Hubble's observational program and the first to determine reasonably accurate values of the Hubble parameter and the age of the universe, characterized cosmology as a search for two numbers: the rates of expansion and deceleration of the universe. Such was the setting for Bob's second brush with cosmological history.

By the 1980s, astronomy was being transformed by charge-coupled devices or CCDs, light sensors that register signals from many pixels in a short time, enabling digital photography. Working at the Lawrence Berkeley Lab, University of California professor Rich Muller and two junior colleagues, Carl Pennypacker and Saul Perlmutter, integrated the new sensors with a computer-controlled telescope and a dedicated minicomputer to automate a search for exploding stars—supernovae. To find a supernova candidate with the time-honored photographic technique, an astronomer could compare before-and-after images and look for a spot that didn't match. With digital technology, CCDs captured the images electronically and computers subtracted *before* images from *after* images, pixel by pixel. A spot on the subtracted image potentially signaled a new light in the heavens—a likely supernova. The digital assistants made comparisons much faster than any human scanner could, so the team could examine hundreds of nearby galaxies in the course of one night's observations. From their first detection in 1986 through June 1991, they found about twenty supernovae—enough to confirm the promise of the automated search approach. Their first CCD was an array of 320 by 512 pixels, for a total of 163,840. For comparison, each camera in an iPhone 14 records twelve million pixels.

Now it was time for the main event. Astronomers had identified distinct classes of supernovae as it became evident that different kinds of stars explode for different reasons. Supernovae of the class known as Type Ia (pronounced one-A), which includes Tycho's supernova of 1572, develop from binary systems in which at least one of the orbiting stars is a white dwarf. They are among the brightest objects in the universe, and they are amazingly uniform. Their spectral features match, when adjusted for redshift, and their light curves—plots of waxing and waning brightness in the weeks following the explosion—match as well, once corrected for the distance from us.

If you can find them, Ia supernovae serve as calibrated light sources, "standard candles" scattered throughout the universe. A supernova's observed brightness over time tells how far away it was at the time of the explosion and—because light travels at a fixed speed—how far back in time the explosion occurred. The redshift of a supernova's spectrum measures how much spacetime has stretched since the light was emitted. These characteristics make Type Ia supernovae precious guides to the history of the universe. By comparing brightness to redshift for numerous supernovae, from nearby to very distant, an observer can infer how the rate of expansion of the universe has changed over time.

Several complications weigh against these positive features. In a typical galaxy, Type Ia supernovae occur only a few times per millennium. They are transitory phenomena that must be measured repeatedly and in spectroscopic detail within a few weeks of exploding, in order to locate the peak brightness that is essential for calibration. They occur at random, but telescope time at the great observatories is a limited commodity, for which astronomers must apply many months in advance. It is too late to request telescope time after a supernova is discovered.

To measure the history of the expanding universe would require observing not just nearby supernovae, but distant ones, whose light waves had been stretched by 20, or 30, or even 100 percent. The idea had been circulating within the astronomical community for years, but experienced people thought measuring expansion with supernovae would be too hard. A big challenge was finding the distant supernovae. Only one had ever been seen. Using the CCD technique in a two-year campaign at the La Silla Observatory in Chile, Hans Urik Nøregaard-Nielsen's team discovered the most distant Type Ia supernova then observed, SN1988U. The wavelengths of its photons had been stretched by 31 percent in the three-and-a-half billion years since they were emitted. In the lingo of the astronomers, its redshift was 0.31. (The redshifts of the historically significant supernovae of 1054 and 1572 would amount to less than one part in a million.)

The LBL team, soon to be known as the Supernova Cosmology Project, grew with the addition of Gerson Goldhaber, a distinguished experimental particle physicist, and Heidi Marvin Newberg, a graduate student working with Muller. Altogether, the team boasted precisely zero certified astronomers. It was, moreover, a group bubbling with ideas. Rich Muller, a font of clever ideas, increasingly turned his attention to research on ice ages and global weather patterns. Carl Pennypacker founded the Hands-On Universe project to make research telescopes and astronomical images available to high school teachers and students.

"One day," Bob recalls, "Gerson came into the office I occupied as head of the Physics Division at Lawrence Berkeley Laboratory and told me that I ought to make Saul the leader of the supernova program although he was younger than Rich, Carl, and himself. In contrast to George Smoot, Saul is not an imposing physical presence. You might take him for a member of a string quartet. In fact, as a violinist whose undergraduate dorm's music tutor was Yo-Yo Ma, Saul would fit the part perfectly. I took Gerson's excellent advice."

To begin their search for distant supernovae with redshifts exceeding 0.3, the Berkeley team obtained scheduled time on the 3.9-meter Anglo-Australian Telescope at the Siding Spring Observatory near Coonabarabran, New South Wales. Three years of observing, with many of the scheduled nights lost to bad weather, yielded not a single identified supernova, but did allow them to develop systematic techniques that led to the ultimate success of the project.

In 1992, the Berkeley team found its first distant supernova signal at an impressive redshift, 0.458, after its 4.8-billion-year journey. Next, a strategy breakthrough made it possible to get batches of supernovae: search the same group of distant galaxies just after a new moon and just before the next, capturing the supernovae that had appeared in the interim. In time, they could be sure to find four to twelve supernovae during each observing period, all of them growing brighter instead of fading. The spectrum of each one would have to be analyzed to determine its redshift and the brightness monitored over time to determine the peak brightness, in order to infer the distance from our location. Having a batch of fresh distant supernovae on demand made it possible to request telescope time in advance and be confident of putting it to good use.

Also in the early 1990s, the Calán/Tololo Supernova Search based at the Cerro Tololo Inter-American Observatory (CTIO) located in the foothills of the Chilean Andes contributed thirty high-quality SN Ia light curves for redshifts between 0.01 and 0.1 that helped to establish Ia supernovae as reliable distance markers. In late 1994, astronomers Brian Schmidt, a Harvard postdoctoral fellow, and CTIO staff member Nicholas Suntzeff founded the High-Z Supernova Search Team (z being the conventional symbol for redshift), convinced that it was now possible to discover SN Ia in large numbers and to use them as reliable distance indicators. The respected astronomers of the High-Z team had well-established connections to the powers who distributed time on the large telescopes. By this time, the Berkeley group included "real astronomers" as well.

The first seven supernovae analyzed by the Supernova Cosmology Project, published in 1997, hinted at a decelerating universe. But adding a more distant supernova, they soon reported that ordinary matter in the universe amounts to less than the critical density, defying the inflationary cosmologists' confident pronouncements. Then the first four supernovae reported by the High-Z team gave agreement. Almost before particle physicists could trot out the old saw that cosmologists are often in error but never in doubt, distant supernovae provided a truly stunning surprise. As 1997 turned to 1998, 34 more from the Supernova Cosmology Project and then ten more from the High-Z team independently revealed that supernovae at a given distance appear fainter than would be expected from their redshifts, even if the universe were free of any matter. In our current epoch, the expansion of the universe is accelerating!

The spectacular agreement between the two independent teams was reassuring, but many potential pitfalls had to be considered and ruled out. Perlmutter recalled years later, "The chain of analysis was long, and the Universe can be devious, so at first we were reluctant to believe our result. But the more we analyzed it, the more it wouldn't go away." For their discovery of the accelerated expansion of the universe, Saul Perlmutter as leader of the Supernova Cosmology Project and Brian Schmidt and Adam Riess as leaders of the High-Z Supernova Search Team shared the 2011 Nobel Prize for Physics.

Subsequent studies, including refined probes of the cosmic microwave background, have confirmed this astonishing conclusion and taught us that the total mass-energy content of the universe is very close to the critical density. Ordinary matter plus dark matter makes up roughly 30 percent of the total; the remainder is a cryptic "dark energy" associated with empty space, that is driving the accelerated expansion of the universe.

The answer to "what is the world made of?" depends on what we mean by "the world." A generation ago, we might have said that but for the fact of our existence, ours is a typical location in the universe. In fact, it is splendidly atypical! A sphere centered on the sun that encloses Earth's orbit contains almost exclusively ordinary matter—the stuff we are made of. Expand the sphere to enclose the Milky Way galaxy, and the dominant component is dark matter, with a good sprinkling of ordinary matter and a negligible dose of dark energy. And the composition of the universe at large—now 4 percent ordinary matter, 27 percent dark matter, and 69 percent dark energy—is changing over time. According to our current understanding, most of the energy of the universe was in the form of radiation for the first 47,000 years. As the universe expands, the density of radiation decreases as the volume increases, and the radiation energy density decreases by a further factor of the radius or scale because the light waves stretch and lose energy through redshift. For the next ten billion years, more or less, the dominant component is matter, which itself diminishes in density with the increasing volume. If dark energy represents a form of energy that is not diluted by expansion—so that its energy density is the same at all times and scales—it will in time account for essentially all the energy density of the universe, as radiation and matter are diluted to negligible components. New observations indicate that dark energy indeed has these miraculous properties.

The discovery that the universe is expanding at an accelerating rate reminds us that pronouncements about the ultimate fate of the universe are subject to dramatic revision in the light of new information. If the laws of nature are as they seem today, and if no new components of the universe show themselves, then we can expect the universe to expand forever, at an ever-increasing rate, becoming very cold and very nearly empty in the far future. It is remarkable that we can rationally make even such a provisional projection. How many more surprises are to be found as we probe the history of the universe?

Astronomy astonishes, with unexpected phenomena such as the expanding universe, dark matter, and dark energy. These are utterly outside everyday experience. But something that seems to us normal—that we inhabit a universe in which matter is common while antimatter is very nearly absent—also requires explanation. That, we shall see, is a task for particle physics.

21

MATTER VS. ANTIMATTER

In May of 1968, Andrei Dmitrievich Sakharov, thrice Hero of Socialist Labor, winner of the Stalin Prize and the Lenin Prize, the father of the Soviet Union's hydrogen bomb, completed his essay "Progress, Coexistence, and Intellectual Freedom." Circulated first in samizdat, but then published by the *New York Times*, the tract laid out the case for peaceful coexistence of the nuclear powers and much more: it demanded intellectual freedom and the release of political prisoners. Sakharov's increasingly strong stands over the next ten years led ultimately to his becoming, himself, a political prisoner and a moral force contributing greatly to the dissolution of the Soviet Union.

At the time he was taking his first steps against the repressive Soviet regime, in September 1966, Sakharov took on a daunting challenge: Why is there any tangible stuff at all? Why didn't the Big Bang produce equal amounts of matter and antimatter? Why do we see only matter and no antimatter? Why didn't it all annihilate into nothingness?

The discovery of the antiproton in 1955 encouraged scientists to contemplate what particle-antiparticle symmetry might mean for the composition of the universe. If somewhere in our observable universe—say on planets other than Earth, in other solar systems, or in other galaxies—antimatter were present in significant amounts, the mutual annihilation of matter and antimatter should make quite a spectacular display. Astronomers Geoffrey Burbidge and Fred Hoyle reported from the Mount Wilson and Palomar Observatories that astronomical observations could tolerate at most one anti-proton for every ten million protons in our galaxy.

Might the antimatter simply be out of sight? At Brookhaven National Laboratory on Long Island, Maurice Goldhaber wondered whether the birth of the universe might somehow have given rise to our cosmos, in which matter predominates, and a hitherto unobserved anticosmos that is rich in antimatter, so that matter and antimatter are balanced overall. The light emitted by the anticosmos would be the same in every respect as the light emitted by the normal cosmos, so—if we could observe that light at all—it would not reveal that it came from an antiworld.

For more than six decades, astronomers have explored in vain for large con-centrations of antimatter. Everywhere they have learned to look, they have found at least a million times more matter than antimatter; in the interstellar medium, where scrutiny has been most intensive, matter is at least a million-billion times more common than antimatter. Here on Earth, CERN and Fermilab have manufactured antiprotons in high-energy reactions since the early 1980s, enabling proton-antiproton collider experiments—including the discoveries of the top quark and the weak force particles W and Z—and comparisons of the antiworld with our own. Together, the two laboratories have produced and accumulated perhaps ten million billion antiprotons, about twenty nanograms worth. Nuclear antimatter is a rare commodity indeed!

Unless you have experienced a PET scan, you are unlikely to recall any personal encounters with antimatter. Positron-emission tomography is an imaging technique used in clinical studies and research to assess how tissues and organs are functioning. It determines in real time the location of short-lived positron-emitting radioactive nuclei that label tracer molecules, which may be injected, swallowed, or inhaled. Because our bodies are primarily made of low-mass atoms in the top rows of the periodic table, these make the best tracers. What is needed are isotopes that have more protons than they should, protons that would rather be neutrons. They can achieve this transformation by emitting a positron, which carries away the unwanted charge, together with a neutrino. The isotopes that can fulfill this function are carbon-11 (six protons, five neutrons), nitrogen-13 (seven protons, six neutrons), oxygen-15 (eight protons, seven neutrons), and fluorine-18 (nine protons, nine neutrons). The tracer collects in areas of the body that have high levels of metabolic activity. A positron typically travels a few millimeters in tissue before annihilating with an electron to release two gamma rays nearly back-to-back. The gamma rays point back to the annihilation site and so reveal where the radioactive tracer decayed. The amount of antimatter introduced into the body is so tiny that the two million patients scanned each year in the United States need not fear the fanciful fate of the Teller-Anti-Teller twins we mentioned in chapter 18, "Trading Places."

In the years that followed the antiproton discovery, physicists assumed that baryon number—a quantum number that equals plus one for protons and neutrons, minus one for antiprotons and antineutrons—was absolutely conserved, like electric charge. No one had ever witnessed the creation or destruction of a single proton or neutron without an accompanying antiproton or antineutron; nor have we to this day. What kept a proton from decaying into a positron and a neutral pion, for example? To be sure, no principle seemed to forbid such a process. As Maurice Goldhaber put it in the 1960s, "People always said, 'The proton is stable,' and I said, 'How do you know? All you know is from experiment. You feel it in your bones it is stable because if it lived less than [ten quadrillion] years it would kill you.'" Your radioactive body would doom itself! Goldhaber's lower bound on the proton lifetime was 100,000 times the age of the universe. Experiments pushed the lifetime to nearly a trillion-trillion years, but proton stability remained a statement of experience, not a law with a secure foundation.

How might we understand the composition of the universe if baryon numbers were absolutely conserved? If the amounts of matter and antimatter had been equal at early times, either everything would have annihilated, leaving a spare "radiation-dominated" nothing-universe with neither protons nor antiprotons, or matter and antimatter must somehow have been segregated into widely separated domains. The first is ruled out by our existence, the second severely constrained by the absence of annihilation gamma-ray signals from the frontiers where matter and antimatter domains must collide.

The alternative to a symmetric universe is the possibility that—in the face of laws of nature dictating that baryons are always created and destroyed in pairs—that a primordial excess of matter has persisted since the universe began. That just-so story is hardly persuasive. Moreover, a satisfying cosmogony must account not only for the dominance of matter over antimatter, but for the considerable amount of matter the universe contains.

That is a tall order in an expanding universe and is hard to imagine if, as most cosmologists now think likely, any initial matter density was diluted in an epoch of rapid inflation.

When he asked himself how our universe of matter came to be, Sakharov had in mind two striking recent discoveries. The observation of the cosmic microwave background by Penzias and Wilson indicated that the universe began with a Big Bang thirteen billion years ago, or so. More obscure was a small but striking deviation from the established picture of the neutral K mesons. As we described in chapter eighteen, the Jekyll-and-Hyde kaons were understood to have taken on two guises, each one equal mixtures of matter and antimatter: the K-short, whose tenth-of-a-nanosecond lifetime usually ended with a decay into two pions, and the K-long, which lived 500 times longer because it was forbidden to decay into two pions. The absence of the K-long decay into two pions was consistent with a principle called CP invariance that had survived the discovery of parity violation. CP stands for charge-parity, the imaginary process that changes a particle into its antiparticle and reflects spatial directions. CP invariance guaranteed that for the decay to two pions, the Jekyll-and-Hyde pieces of the K-long decay into two pions exactly canceled. A neat story—except that when Val Fitch, Jim Cronin, and their team looked very carefully in 1964 they found that about one K-long in 500 did—against expectations—decay to two pions. CP invariance was overthrown—quite a surprise for aficionados, but so what? It seemed a small effect only observed in this arcane system of neutral kaons. It was Sakharov's genius to recognize that this one-in-five-hundred effect and the notion of an expanding universe that was at early times incredibly hot and teeming with energy were clues to an existential puzzle: why the universe can be made of real stuff rather than just "radiation," cosmologists' term for particles moving near the speed of light, such as photons and neutrinos. Sakharov hypothesized that the universe began with a balance between matter and antimatter—equal numbers of baryons and antibaryons. From that beginning, he argued that three requirements must be met to explain why we are here, or indeed why there is anything much at all. First, there must be processes in which baryons and antibaryons are not created or destroyed in equal numbers, even though this had never been seen and still has never been detected.

To create an imbalance wasn't enough though. While in principle now the number of baryons could exceed the number of antibaryons, matter and antimatter might be inclined to equilibrate in a hot environment into equal numbers of baryons and antibaryons. In fact, this would be the case if they could come into thermal equilibrium, cooked long enough so to speak, because then the number of each particle would depend only on the particle's mass.

Since the proton and antiproton have exactly the same mass, there would be equal numbers of them. Sakharov concluded that there must have been a time when there wasn't thermal equilibrium, when the particle soup wasn't just slowly simmering. The expanding universe supplied this second condition. Instead of simmering, the soup didn't just boil over, it exploded so that the ingredients were no longer in touch with each other and thus didn't make a nice uniform broth.

Finally, Sakharov realized if every process worked the same for particles and antiparticles, baryon-number-violating reactions might produce extra protons, but they would also yield equal numbers of antiprotons. Things must not work the same for particles and antiparticles. This meant that CP invariance must fail. The Fitch-Cronin experiment showed that CP violation was fair game.

Sakharov's three-page note was not widely recognized, in part because the Soviet literature was only fitfully read in the West in the 1960s. Moreover, even if there was no good reason to think that baryon number was absolutely conserved, no one had advanced a plausible mechanism for baryon-number violation. And Sakharov had other things on his mind: having challenged the universe, he would challenge the Soviet colossus.

In the mid-1970s, speculative theories about the nature of forces led physicists to turn their attention to the matter excess in our universe—and to Sakharov's three conditions. Encouraged by the first successes of quantum chromodynamics and the electroweak theory, theorists began to imagine a "grand unification" of the strong, weak, and electromagnetic interactions. Such a fabulous theory would have two features: the separate families of quarks and of leptons would be joined together in extended quark-lepton families, while the strong and electroweak forces would emerge from some single force that followed from a comprehensive symmetry principle. Of the known forces, only gravity is left aside. Once quarks and leptons are cousins, it is likely that new forces might transform quarks into leptons (or even antiquarks), and vice versa, violating baryon number and raising the prospect—or specter—of proton decay.

Since we have never seen such interactions, they must be very feeble. Taking a cue from the electroweak theory—in which the observed weakness of the weak interactions is owed to the considerable 100-GeV masses of the W and Z force particles—we anticipate that the extended gauge symmetry of the grand unified theory must be hidden in such a way as to make the new force particles extremely massive. In our human experience—at the energies we explore in our laboratories—the strong, weak, and electromagnetic interactions act with very different strengths.

How can all these forces be unified if their observed strengths are so different? In collisions at extraordinarily high energies, energies we can't reach in any existing or projected machine, the masses of the W and Z would seem small: we could pretend their masses and those of the quarks and leptons are zero. Then the strength of the various interactions would depend only on the coupling strengths as discussed in chapter 8, "Function Follows Form." At the energy of the Z boson, the parameter that measures the strength of electromagnetism is approximately 1/129; the analogous measure of the strong interaction is about fifteen times greater. We saw that quantum effects tend to make the strong interaction progressively less potent at high energies—through the phenomenon of asymptotic freedom—in contrast to the electromagnetic interaction, which grows in strength. The precise details depend on the ins and outs of a particular unified theory, but a very long extrapolation in energy indicates that the strong and electroweak couplings converge in the neighborhood of a million billion GeV to a value near 1/40.

This is good news for a grand unified theory. The new gauge bosons that change quarks into leptons and make protons decay will have masses near the unification energy, so the interactions they mediate are indeed too weak for us to have observed them. Within a very minimalist unification model presented by Howard Georgi and Sheldon Glashow at Harvard University, protons will decay, but with a lifetime estimated by Georgi, Helen Quinn, and Steven Weinberg to exceed a million-trillion-trillion years—far surpassing then-existing limits. Not even the most devoted graduate student can monitor a single proton for a million-trillion-trillion years—the age of the universe is only 13.8 billion years—but a team could watch a million-trillion-trillion protons for one year, or a few. That many protons can be found in three cubic meters, or three tonnes, of water. A telling signature would be the display caused by a proton decaying into a positron and a neutral pion, which itself decays into two photons.

In one of those happy twists that punctuate the history of science, the series of Kamioka experiments that has yielded such dramatic discoveries in the realm of neutrino oscillations was conceived to search for proton decay. The first version, called Kamiokande, for Nucleon Decay Experiment, monitored 880 tonnes of ultrapure water beginning in 1983. After 201 days of surveillance, the team could set lifetime limits about ten times longer than the grand-unified-theory predictions. How odd it is that in seeking evidence for a most elegant, if esoteric, theory of the Laws of Nature, the experiment is surpassingly prosaic: simply staring attentively at a vat of water. Perhaps it is a good complement to the discovery of the Law of Gravity, inspired, at least in folklore, by a falling apple. The Kamioka saga has the virtue of being true! Enhanced versions of the experiment, culminating in Super-Kamiokande, at 50,000 tonnes, have made the bounds a hundred times stronger. Other versions of grand unified theories push the expected proton lifetime just beyond today's experimental constraints. The hunt goes on; one does not easily give up the chance to find a new force of nature and establish a key to the universe. We are still waiting for the proton-decay apple to fall.

During the late 2020s, two Brobdingnagian detectors sited deep underground to filter the annoying flux of cosmic-ray muons will extend the search for proton decay. In Japan's Kamioka mine, the Hyper-Kamiokande experiment will push the well-established water-Cherenkov technique to new levels, with a sensitive volume of 260,000 tonnes of water in a tank sixty meters high and seventy-four meters in diameter, viewed by 40,000 photosensors—approaching a dragonfly's 60,000 ommatidia. In the Black Hills of South Dakota, a vast cavern 1,500 meters deep in the Homestake Mine will contain the Deep Underground Neutrino Experiment (DUNE). Over a century and a quarter, the mine yielded more than forty-one million ounces of gold and nine million ounces of silver; now it will be on the receiving end of 68,000 tonnes of a very different element: argon. The DUNE experiment foresees four super-boxes, each fifteen meters wide, fourteen meters high, and sixty-two meters long, containing liquid argon Time Projection Chambers. Think of a gymnasium as tall as a four-story building with two basketball courts laid out end-to-end inside. While awaiting a sign of proton decay, Hyper-K and DUNE will exploit their large volumes to explore the physics of neutrinos with unprecedented sensitivity.

Our grandest dreams for particle accelerators reach toward one million GeV, a billion times short of the unification scale, so we will not soon—not ever—carry out direct experiments there. But we can look for insights from grand unified theories and test the predictions they make for our circumstances. A long-enough proton lifetime is one success. A more quantitative check is that at the energies we can explore, the first grand unified theories predict the relative strengths of the neutral-current and charged-current weak interactions closely enough for us to take them seriously.

The idea of grand unified theories is so seductive that even without experimental evidence for baryon-number violation, physicists hastened to carry out Sakharov's program, which drives an important area of contemporary research in particle physics. At the moment the verdict is mixed. We have an independent motivation for baryon-number violation from grand unified theories and we have the observation of CP violation. Since no proton decay or neutron decay has been observed, Sakharov's requirement of CP violation seems at first to be in better shape. After all, Sakharov was motivated by the observation of CP violation in kaon decays. The possibility of greatly extending observations of CP violation arose with the discovery of the *b* quark. Bottom quarks, with their electric charge of minus ⅓, are similar to strange quarks whose electric charge is the same. Particles made from a bottom quark and an everyday up or down quark, or rather antiquark, are called *B* mesons. They turned out to be an ideal stage for observing CP violation and this was pursued both with all-purpose accelerators like Fermilab's Tevatron Collider and with specially designed electron-positron colliders at Stanford and at KEK, the Japanese Laboratory for High-Energy Physics.

The standard model of particle physics makes very specific predictions relating the lifetimes, decay characteristics, mixing properties, and degree of CP violation for *B* particles and their strange counterparts, the *K* mesons. The amount of CP violation found in the neutral kaons and predicted for *B* particles by the standard model does lead to a matter excess, but the baryon density it would produce is a billion times smaller than that in our universe. Under Sakharov's hypothesis there must be other sources of CP violation. Would the *B* mesons conform to the standard-model expectations, or would their study identify a break in the standard model that pointed to the CP violation Sakharov's program required?

Bob: I was fortunate to be part of the BaBar collaboration at SLAC. A team assembled to measure *B* mesons and anti-*B* mesons—written $\bar{B}$ and pronounced *B*-bar—couldn't resist naming their experiment BaBar after the famous elephant in Jean de Brunhoff's marvelous children's books. However, the owners of the rights to the de Brunhoff books could resist. Only after a prolonged negotiation was permission obtained to use the name and branding. Designing and building a novel detector suitable for the CP measurements consumed several years. As the time drew close for our team to make its first measurements, which would give an early glimpse of agreement or disagreement with the expected range, special steps were taken to avoid any conscious or unconscious bias that might push our results near the expectation.

There had been instances in the history of particle physics when results in agreement with anticipations had been announced only to be shown later to be incorrect. How did this happen? Of course, care was always taken before presenting final results. The scrutiny of the data and their analysis might take longer than the experiment itself. When should this analysis stop? It was a natural temptation to stop when the results agreed with expectations: "We must have gotten all the bugs out!" We needed to take steps to prevent falling into this trap. The collaboration adopted a policy of "blinding the results," hiding the final result while looking for flaws. Only after we had checked everything would we allow ourselves to see the result and we had agreed that no matter what, that is the result we would report. To reveal the result, we "opened the box," and out popped an answer pretty far away from what we expected, and that's what we reported.

If only that first result had been verified by much larger sets of data later on! We might have had a great discovery: the standard model for CP was wrong; it was missing something. Perhaps we would have had evidence of a new source of CP violation, perhaps even what Sakharov's theory needed. Alas, when data poured in from our experiment and from the competing Belle experiment in Japan, the final results agreed with the theory—another triumph for the standard model, but a failure to find something new. Years later, with many more results from many more experiments, this part of the standard model remains frustratingly secure.

To date we have no direct evidence for baryon-number violation or for the sort of CP violation that could lead to the observed imbalance between baryons and antibaryons. On the other hand, there is an observation that requires explanation: we are indeed here. Perhaps these phenomena reside only in particles outside the standard model. Speculations of this sort clearly provide great scope for theorists, but what if we insist that no new particles be added? Are there some particles already attached to the standard model that we haven't considered sufficiently?

In part because the simple story that followed the invention of grand unified theories did not work out, theorists have turned their attention to an indirect mechanism called leptogenesis. Processes in the early universe might have induced an asymmetry between leptons and antileptons, which then propagated into the baryon-antibaryon asymmetry we witness. The most studied possibility is that very heavy relatives of the electron, muon, tau, and neutrinos were produced out of matter-antimatter balance and subsequently decayed into baryons and antibaryons out of balance.

The leptons are the particles that do not experience the strong interaction; they carry no color charge. Physicists conferred the name—from the ancient Greek *leptós*, meaning slight or slender—in the late 1940s, when the known examples were the electron and the hypothetical neutrino, both far lighter than a proton. We may blush with etymological shame when we refer to heavy leptons, but the name has outlived its origins.

In any event, the notion of very heavy leptons, specifically very heavy neutrinos, predates the need for a new paradigm for baryogenesis, the creation of baryons.

Paradoxically, the simplest model that explains why the neutrinos we know are very, very light—hundreds of thousands of times lighter than the electron—relies on the existence of some very, very heavy neutrinos, so massive that we haven't seen them and never will. This balancing of heavy and light neutrinos is called the see-saw mechanism; as the mass of one goes up, the mass of the other goes down.

It is these still hypothetical very, very heavy neutrinos that may provide the CP violation to help explain why there is something rather than nothing. How can we test the idea when these heavy neutrinos are so massive that we have no chance of observing them? The most we can do, it seems, is to test the neutrinos we can observe and hope that they will provide hints about their unobservable cousins. The experiments that study neutrino oscillations by directing intense beams of neutrinos through the Earth to distant detectors can test for CP violation in the ordinary neutrinos. The pattern of oscillations from one neutrino flavor at production to another at detection is influenced by the feeble interactions of neutrinos or antineutrinos with the matter they encounter on the way.

Two experiments are searching for matter-antimatter differences, and two more are on the way. In the Tokai-to-Kamioka (T2K) experiment, neutrinos are sent 295 kilometers across the island of Honshu from the Japan Proton Accelerator Research Complex to the Super-K detector. The NOvA experiment, sited in Ash River, Minnesota, detects the interactions of neutrinos sent from Fermilab, 800 kilometers distant. These will be succeeded by the DUNE (1,300 kilometers from Fermilab) and Tokai-to-HyperK experiments in the coming years.

Particle physics continues to probe the workings of nature at the smallest scales imaginable. Thanks to the insight of Andrei Sakharov we now understand that the secrets still hiding from us in the depths of the nanonanoworld may also hold the explanation for the existence of all the tangible matter we see before us.

22

THE BEST OF ALL POSSIBLE WORLDS?

Gottfried Wilhelm Leibniz is celebrated as a "universal genius" of the 17th and 18th centuries for his contributions ranging from history and philosophy to mathematics and the sciences. Physicists know him best as the first to publish the calculus—for which he conceived some of our standard notations—and as combatant, late in life, in a bitter priority dispute with Isaac Newton. Less familiar is Leibniz the Christian apologist, preoccupied with the problem of evil: How can an infinitely wise, benevolent, and infallible deity permit evil to exist in the world? The divine wisdom of Leibniz's *Theodicy* (1710), like an inconceivably powerful, massively parallel, chess-playing supercomputer, "weighs [all the possibles] one against the other, to estimate their degrees of perfection or imperfection, the strong and the weak, the good and the evil. It goes even beyond the finite combinations, it makes of them an infinity of infinites, that is to say, an infinity of possible sequences of the universe . . . The result of all these comparisons and deliberations is the choice of the best from among all these possible systems . . ."

This justification earned the delicious lampoonery that Voltaire delivered in his 1759 philosophical tale, *Candide*, a road novella about the tribulations of a credulous and incurably optimistic youth. According to the great fictional buffoon Dr. Pangloss, Candide's tutor in matters of metaphysico-theologico-cosmolonigology, ours is incontestably the best of all possible worlds. Neither the Iberian Inquisition nor the Lisbon earthquake of 1755 could shake Pangloss's faith in the perfection of his world, but the good doctor did not have a scientist's respect for doubt and uncertainty. Here is Pangloss's confident chorus in Leonard Bernstein's comic operetta version of *Candide*:

> *Once one dismisses*
> *The rest of all possible worlds*
> *One finds that this is*
> *The best of all possible worlds.*

Conscious as we moderns are of our world's shortcomings, "the best of all possible worlds" strikes us as an eccentric and ironical assessment of the human condition. All the same, an important and radical idea lurks in that Panglossian—Leibnizian—phrase: the proposition that other worlds could be possible. Might other worlds be possible, or do the laws of nature, properly understood, lead ineluctably to a world with the character of the world we know? It is a question that fascinated Albert Einstein, who wondered "whether God could have made the world in a different way; that is, whether the necessity of logical simplicity leaves any freedom at all." Arthur Schopenhauer, the "philosopher of pessimism," concluded that this is the *worst* of all possible worlds! No fan of Leibniz, he argued in an 1844 essay, "On the vanity and suffering and life," that a world just a little worse than ours could not maintain itself, and is thus impossible.

If our world is one among many conceivable possibilities, what selected it as the outcome? Are such questions metaphysical or theological? Might they be scientific questions—even, as the mountebank's immodest honorific suggests, cosmological?

We don't have all the answers. We cannot say, for now, in what measure the character of our world is determined by chance and in what measure by necessity.

But scientific progress is measured by the questions that once seemed outside the purview of science but have come, over time, within reach. Answering those questions—sometimes merely being able to formulate them as scientific questions—draws still other questions to the domain of science.

We can begin to explore what determines the character of our world by changing, in our imagination, some of the ground rules. This practice has a long tradition as an instrument for the initiation—some might say, hazing—of physics graduate students. "What would happen," the professors ask, "if the mass of the proton or the mass of the electron changed a little bit? How would physics change if the strength of electromagnetism changed this way or that?" When we were on the receiving end of those questions, we had little patience for them. To tell the truth, we really hated them, because the world wasn't that way, so why waste time thinking about it? Andy Warhol's ten variations on Mao Zedong may be sly sociopolitical commentary, but it seems pointless to consider a world in which Mao's face is blue, because everyone knows it never was. Now that we have shaken off some of the comfortable certainty of youth, we have come to understand that the "what ifs" were much better questions than our teachers realized. They just didn't go far enough.

Poets speak of attaining some distance from the world—finding some strangeness—to express the truths that make a rock stony or a love sublime. Like the poet, the physicist seeks not merely to name the properties of nature, but to plumb their essence. To truly understand the world around us, to fully delight in its marvels, we must step outside ordinary experience and look at the world through new eyes.

Particle physicists look at the world quite literally through new eyes when we construct accelerators and detectors to create and observe fresh experience. In a more figurative sense, we expand our vision by dreaming other worlds, the better to understand our own. To conduct a thought experiment that transports us to an alternative world, we might simply ask what would happen if we could change a certain parameter, a quantity on which the consequences of physical laws depend, like the mass of a particle or the strength of an interaction. Or we might idealize our own world, trying to get at its essence by sweeping aside the chaff of distracting detail. Or we might forge a more radical vision by changing the rules that shape the world.

What distinguishes thought experiments about alternative worlds from idle daydreaming is the reality check at the very core of science, Galileo's *cimenti*, our reliance on observation and experiment. Nature is our test of whether something is so or not. And—if we are attentive—we can learn something even from unconventional ideas that fail the reality check.

The point of dreaming new worlds—and the goal of our science—is not merely to describe our world, but to understand why the world is the way it is. A physicist who asks, "Why?" is not necessarily trying to discover motive or intent. "Why?" is a token for "How did it come to be that . . . ?" In other words, how does an outcome follow from a certain set of principles? Is the outcome inevitable, given the assumptions, or is it one of many possibilities? How narrowly do the laws of nature prescribe the outcome? And what happens when we change the assumptions—either a little or a lot?

∞

The quotidian world is built from the atom, a structure a billion times smaller than the human scale. Most of the mass of the atom resides in a tiny, positively charged core called the nucleus, which occupies only one million-billionth of the atom's volume. The nuclear charge is balanced by a diffuse cloud of lightweight electrons. The electron is one of the most common objects in the universe. Your little finger alone contains more than ten trillion trillion of them! Electrons conduct electricity in metals, reflect light from silvery surfaces, define chemistry and the structure of matter, and write these words on our computer screen. The nucleus harbors protons and neutrons, elementary particles for an earlier generation of scientists.

In 1926, the Viennese theorist Erwin Schrödinger found a compact equation that describes the behavior of the electrons that orbit the nucleus and many other aspects of the microworld. In principle—and to a large degree in practice—Schrödinger's equation makes the properties of matter computable.

Put the masses and charges of the electron and the nuclei into the equation and out pops everything—from the hydrogen atom to diamonds to DNA.

To Schrödinger and the other inventors of quantum theory, where those masses and charges might come from was not an issue—and certainly not a scientific question. No experiment anyone had contemplated could shed light on the origin of masses and interaction strengths, and no theory provided a clue. Four decades later, when we were students, the masses of the proton and neutron and electron, the strengths of the fundamental interactions, still seemed just to be arbitrary numbers, written down capriciously in the back of the *Book of Nature* at the moment the universe was born. To ask why the electron mass is about 1,836 times smaller than the proton mass seemed like an exercise in metaphysics, and not a particularly productive one, at that.

The inventor of the neutrino, Wolfgang Pauli, was possessed of a frightening intelligence backed up by a self-confidence that was rare even for a theoretical physicist. Pauli's colleagues revered his deep intuition, and trembled before his ferociously candid critiques. Our physics grandfather Viki Weisskopf told us how wonderful it was to be Pauli's assistant in Zürich: "You could ask him absolutely anything without worrying that he would think your question was stupid. Pauli thought *all* questions were stupid—except his own!"

According to legend, when Pauli died in 1958, he was greeted at the gates of heaven by God himself, who asked if he had any questions. "Oh, yes, Lord," said Pauli. "The proton's mass is 1,836 times the electron's. You should have a reason." God turned to a blackboard and started to explain, but Pauli cut him off. "*Quatsch!* Rubbish!"

It seems perfectly natural to us that the apocryphal Pauli didn't ask about the meaning of life or the theory of everything, but about a single observation that he hadn't been able to understand—about a small thing of nature. The impulse of particle physics—of all science—is the search for laws of ever-increasing universality, of ever widening applicability, but our goal is to understand nature in every particular. The individual facts we wonder about, the daily mysteries of our experience, drive our curiosity.

Ever since Galileo, science has thrived on the minute particular—finding answers to limited questions, and weaving those partial truths into a tapestry of broad understanding, instead of reaching directly for cosmic truths and coming up with empty platitudes. As Galileo wrote, "I would rather find one small fact about the slightest thing than endlessly debate the greatest questions without reaching any truth at all."

What we understand of particle physics suggests alternative worlds that might have been—worlds governed, perhaps, by the same fundamental laws as ours, but utterly different in appearance and conduct. Nature appears to have chosen values for certain fundamental quantities—the masses of particles and the strengths of interactions. Other values—other choices, perhaps—would make for a universe very different from the one we know.

When we contemplate alternative worlds, it is fun to imagine ourselves seated before a console of dials. We suspect that the values displayed were determined early in the history of the universe by physical laws that we may hope to discover. To physicists and other curious children, a dial invites playful twiddling—experimentation, if you like. No real experiment can reset the dials, so we can't test-drive different universes, but thought experiments aren't constrained by what is practical in the lab. Turn the dials of imagination, and behold another world for Doctor Pangloss!

The proton and neutron are not-quite-identical twins, one carrying a positive charge, the other electrically neutral. It's a curious fact about our world that the neutron is just a tiny bit heavier than the proton: about one-seventh of a percent, or a little less than three electron masses. Without experiment to guide us, we might have expected the proton to be a little heavier than the neutron, because its electric charge ought to give it some extra energy, that is, some extra mass. But the proton is the lighter twin.

In our world, neutrons have long life expectancies only within nuclei. A free neutron survives a little under fifteen minutes on average, before it undergoes radioactive decay, leaving a proton, an electron, and an antineutrino in its place. Within common nuclei, the binding among neutrons and protons inhibits or excludes neutron decay. If we twisted our dials to interchange the proton mass and the neutron mass, a free proton would decay into a neutron—now stable, because it is the lightest particle of its kind—plus an antielectron, and a neutrino. How symmetrical a turnabout! But how disastrous for hydrogen atoms, and for us!

Once the proton and neutron masses were interchanged, protons—the nuclei of hydrogen atoms—would begin to disappear. With no electric charge on its nucleus to restrain it, the hydrogen atom's electron would simply wander off. In a quarter of an hour, only about 37 percent of the world's hydrogen would be left. About a day later, every hydrogen atom on Earth would be gone, leaving no water, no organic molecules, no hydrogen economy.

Score one for Pangloss!

Best not to fiddle with the dials that control the proton and neutron masses. What about the dial for the electron's mass? Though it contributes little to the masses of atoms and molecules, the electron mass is very important: it determines the radius of the electron cloud that defines the size of an atom. If we could slowly turn the dial marked "electron mass," increasing its value gradually, everything around us would start to shrink. Dial the electron mass 5 percent larger, and Schrödinger's equation tells us atoms become 5 percent smaller, that you become 5 percent shorter!

Keep turning that dial, raising the electron's mass by 20 percent, 30 percent, and zap! At 31 percent, catastrophe: the common form of nitrogen, nitrogen-14, self-destructs. As the electron cloud crowds closer around the nucleus, one of nitrogen's seven protons gobbles up an electron, spits out a neutrino, and turns into a neutron. A nucleus containing six protons is carbon, not nitrogen; when nitrogen turns to carbon, all amino acids become pure hydrocarbons. Life—at least in the form we know—ceases. To add insult to mortal injury, the atmosphere, 78 percent nitrogen in the normal world, rains carbon-14 until the Earth is twelve feet deep in soot.

Score two for Pangloss—or maybe Schopenhauer! "Don't touch that dial!" is beginning to seem like sound advice.

The first subatomic physicists viewed the proton as an elementary particle—one of a few constituents, with the electron and neutron, of all matter. And why not? It is as ubiquitous as the electron, and different combinations of protons and neutrons make the nuclei of all the elements. Since our student days, physicists have looked ever more closely at the proton—inside the proton, it is fair to say—and have found it to be a complex structure made of simpler units, quarks. We have learned a lot about how the proton is put together, and how nature might have set the dials for the proton and neutron masses.

It is fun to think about turning the dials for the proton and neutron masses, but now we see that they are linked, behind the imaginary console, to dials that set the quark masses and the intensity of the strong interaction. Hidden from view, gears and sprockets connect the dials, and we are beginning to puzzle out how all the dials may be hooked together.

Knowing that the proton mass comes from somewhere—mostly from the strong interaction among quarks—impels us to ask how it might have been set to a different value. If the strong interaction had been just a little stronger or a little weaker, for example, the proton and neutron would both weigh a little more or less than they do in our world. Now we can see how, but not yet why, the neutron outweighs the proton. The up quark is a little bit lighter than the down quark. The proton has two of the lighter up quarks and one of the heavier down quarks, while the neutron contains two of the heavier down quarks and one of the lighter up quarks: about 9 MeV of quark masses for the proton, but 12 MeV for the neutron. This small difference more than makes up for the proton's extra electromagnetic energy, as lattice QCD studies now make precise. This is a very delicate matter. A slight change in the quark masses, and the proton would come out heavier than the neutron, with the dramatic consequences we have explained.

At the current limits of our experimental resolution, we have not discerned any internal structure for the quarks or leptons, so we take them, provisionally, to be elementary particles. There is not much to say about an elementary particle like an electron or a quark. You can't really describe it—but then, there isn't much to describe. An elementary particle has no size and no shape; it has no hue or odor. It does have a definite mass and a definite electric charge; it may act as a magnet—a tiny compass needle, and if so its magnetic strength can be measured. That's about it. Precisely because they are elementary, elementary particles have very few properties. By the way, if electrons are indeed elementary, then no force is required to hold them together—unlike atoms, nuclei, protons, and the macroscopic matter of everyday experience.

The apparent immutability of our world is deceptive. The moon's orbit around the Earth—which bespeaks stability, order, and reliability—is regular because the moon is continuously falling toward the Earth, accelerated by the gravitational attraction between the two bodies, its direction of motion changed from the straight path it would follow in the absence of any force to the elliptical orbit it follows, day in and day out. Newton had it right: a body in motion continues in a straight line at constant speed unless a force acts upon it.

Beneath the surface of the chunks of matter around us, the constituents are always in motion. Some agent—a force—has to keep them tied together, tugging them back when they begin to stray. A proton may always be a proton, but when we look deep inside, we find that the constituents are not only moving, but changing all the time. The quarks' continual change of colors is the fingerprint of the exchange of gluons, which is what keeps the quarks attracted to one other.

We find it evocative to speak of interactions as agents of change because interactions are mediated by force particles—literal agents—in quantum field theory. The photon of electromagnetism, the W and Z bosons of the weak interactions, and the gluon of the strong interaction are force particles. So, too, is the hypothetical graviton. The observed force particles merit a section of their own in the massive *Review of Particle Physics*. Not only do the force particles act on the quarks and leptons that we call fundamental constituents, but one agent of change can act on another.

We have already met the weak interaction in the guise of neutron decay, an example of radioactivity. The weak interaction might seem an insignificant element of everyday life. With the proton and neutron masses set to their proper values, the hydrogen atom does exist, and the atmosphere is indeed mostly nitrogen, and not soot! But in fact, the weak interaction plays a central role in our lives: it powers the sun.

Inside the sun, in a series of steps made possible by the weak interaction, four protons combine to yield one alpha particle (made of two protons and two neutrons) plus two positrons, plus two neutrinos, plus 25 MeV of energy in the form of light and heat and neutrinos. The first step in the solar furnace is the fusion of two protons into a deuteron (proton plus neutron). Normally, in our world, a proton does not decay into its heavier twin, the neutron. But if two protons are very close together, one can change into a neutron because the resulting neutron can combine with the spectator proton to make a deuteron, a stable nucleus. The binding of the deuteron provides enough energy savings to allow the decay of the proton into neutron plus positron plus neutrino. The problem is to bring two protons close enough to make the deuteron in the end.

The two protons in a hydrogen molecule (H_2), like those in a water molecule (H_2O) are, on average, separated by roughly a tenth of a nanometer—about 100,000 times too distant to form deuterons. That is why there is no spontaneous "cold fusion." At the center of the sun, the temperature is about fifteen million kelvins, so hot that the electrons are stripped from hydrogen atoms. Still, the protons resist being brought together because their like electric charges repel. Occasionally, the stupendous temperature of the solar interior gives one proton enough energy to approach near enough to another and fuse into deuterium. Thus the weak interaction, which produces deuterium at the center of the sun, is ultimately responsible for all the energy that sustains us on Earth.

At the heart of the weak interaction is an agent of change, the W boson. Neutron decay is a two-step process. First, a down quark turns into an up quark plus a negatively charged W boson. Then the W materializes into an electron and an antineutrino. There is an impediment: the W weighs 80,377 MeV, about eighty-five times the mass of the neutron, so this is a transaction that could never occur by the rules of our macroscopic world. The neutron has only 1.3 MeV to give, to become a proton.

Fortunately, the uncertainty principle of quantum mechanics makes generous loans available on variable terms. The more energy you borrow, the sooner you must repay it. The W boson will negotiate an energy loan as long as it can come out ahead in the end, with energy to spare. The W is created deeply in debt in neutron decay—or in proton-proton fusion—and must cash in very rapidly as an electron and an antineutrino. The heavier the W boson, the more rapidly it must disappear.

To make deuterium in the sun, one proton must turn into a neutron by temporarily borrowing enough energy to make a positively charged W boson. If the W were heavier than it is in our world, the energy loan would fall due sooner and the production of deuterium would be slowed. The slower heating of the sun's core would cause the sun to collapse a little, in turn increasing the temperature enough to restore the thermonuclear burning to nearly its original rate. If we set the dial that controls the W mass to twice its normal value, the sun's thermal output would drop only by about 5 percent. That sounds like a subtle effect, but we would soon learn firsthand what the ice ages were like!

Chastened by the devastation we've wrought by twiddling the dials, we concede that Dr. Pangloss might have had a point, and we resolve to leave the masses and coupling strengths just where we found them. But then we notice at the top of the console, almost out of reach, a simple on-off switch labeled "exclusion principle." It's set to on, and held in that position by a strip of duct tape, to discourage casual changes—because this switch doesn't simply alter values, it changes the rules. The duct tape only encourages our curiosity, of course!

When we remarked that Schrödinger's equation holds the key to understanding the structure of matter, we took for granted one very important—and very weird—fact. The flight patterns that electrons occupy in an atom, called orbitals, have limited capacity. Each can accept only two electrons. Despite having no perceptible size, an electron spins like an inexhaustible top. If an orbital contains an electron spinning one way, that orbital can only accept another electron if it spins the other way. This limitation on an orbital's capacity is the exclusion principle, conjectured in 1925 by Wolfgang Pauli to understand the chemical regularities depicted in the periodic table of the elements. As the number of electrons in an atom increases—from one in hydrogen to two in helium, three in lithium, and so on—the electrons are accommodated in new orbitals increasingly distant from the nucleus.

Pauli's exclusion principle seems like an esoteric rule that should only concern atomic scientists who devote themselves to atomic spectra or the distribution of electrons inside atoms. In reality, the Pauli principle shapes the everyday world. Thanks to the exclusion principle, the atoms of the periodic table are a rich chemical palette from which the remarkable diversity of matter can be painted. When Paul Ehrenfest presented the 1931 Lorentz Medal to Pauli at the Royal Netherlands Academy of Arts and Sciences, he described the consequences of the exclusion principle this way:

"Pick up a piece of metal, or a stone. When we think about it, we are astonished that this quantity of matter should occupy so large a volume. Admittedly, the molecules are packed close together, and likewise the atoms within each molecule. But why are the atoms themselves so big?

"Consider Bohr's model of an atom of lead. Why do so few of the eighty-two electrons run in the orbits close to the nucleus? The attraction of the eighty-two positive charges in the nucleus is so strong. Many more of the eighty-two electrons could be concentrated into the inner orbits, before their mutual repulsion became too large. What prevents the atom from collapsing in this way? Answer: only the Pauli principle, 'No two electrons in the same quantum state.' That is why atoms are so needlessly big, and why metals and stones are so bulky.

"You must admit, Herr Pauli, that if you would only partially repeal your prohibition, you could relieve many of the practical worries of everyday life, for instance the traffic problem on our streets."

It wasn't in Pauli's power to repeal the exclusion principle, even when Ehrenfest tweaked his prosperously pear-shaped friend by remarking that Pauli might have considerably reduced what he paid his tailor for the beautiful, black formal suit he wore to the ceremony. It's not in our power either; the exclusion principle isn't just an empirical law that applies to electrons, but a deep and general statement of how the world works, called the spin and statistics connection. Particles are divided into two great classes: bosons (after the

Indian physicist Satyendra Nath Bose), and fermions (after Enrico Fermi). The photon and the weak force particles W and Z—and all the particles whose spin is a whole number—are bosons that welcome being in the same quantum state. The electron, proton, neutron, and all the quarks—and all the particles whose spin is 1/2, 3/2 . . .—are fermions that shun being crowded into the same quantum state as their peers. Fermions and bosons are odds and evens: combine a neutron and a proton—two fermions—and you get a deuteron—a boson.

We can recast Ehrenfest's little fantasy of a world without the exclusion principle: could the electron have been a boson? Next to the dial that controls the electron mass on our master console of the universe should there be a toggle switch labeled "fermion or boson"? Almost. An audacious grand symmetry principle called supersymmetry asserts that every one of the fundamental particles we have mentioned should have a fraternal twin that shares all its attributes—charge, interactions, mass, and such—except for spin, which is different by a half unit. Every boson we know would have a fermion superpartner, and every fermion we know would have a bosonic superpartner. The spin-one photon, W, and Z would be paired with the spin-one-half photino, wino, and zino; the spin-one-half quarks would be paired with spin-zero squarks; and the spin-one-half electron would be paired with the spin-zero selectron. Not one of these exotic superpartners has been observed despite prodigious efforts over four decades, but superpartners remain a prime quarry for the Large Hadron Collider at CERN.

Supersymmetry can't be an exact symmetry of our world; the superpartner masses must be much larger than the normal-particle masses, or we would already have discovered them. But what if it had turned out the other way around?

If we suppose that the world is supersymmetric, then we are invited to ponder what would have happened if the selectron had been lighter than the electron and the squarks lighter than the quarks. In other words—to put Ehrenfest's challenge in modern terms—what would the universe be like if the tangible world were made of bosons? If selectrons replaced electrons as the lightest charged particles, every selectron would seek the lowest quantum state, and bosonic atoms would indeed be far smaller than atoms made of electrons, just as in Ehrenfest's fantasy.

Thinking about a single atom doesn't tell us the whole story. In the normal world, the forces among atoms are simply electric forces, but with the extra discipline imposed by the exclusion principle. The latticework of crystals and the intricate structures of proteins arise because when atoms join together, each electron must find its own solitary quantum state. "More is different!" is rightfully the rallying cry of condensed-matter physicists.

In the superpartner world, more is different in a very dramatic way: superpartner matter—dare we call it *smatter?*—would be unlike anything we've seen in the everyday world, for in the superpartner world, bulk matter is amorphous. In our normal world, should you join two handfuls of clay together, you hold in your hands one lump of clay twice as big.

If one of our supersymmetric colleagues should ever hand you two lumps of bosonic clay, please resist the temptation to join them together. With no exclusion principle to ensure the integrity of individual molecules, the two lumps would fuse into a *smaller* lump, liberating more energy than a thermonuclear explosion. Far more devastating than the voracious ice-nine that Kurt Vonnegut imagined in *Cat's Cradle*, bosonic matter would be an insatiable, shrinking, undifferentiated blob, menacing the superworld with an inevitable, all-consuming, amorphous death!

Maybe Pangloss had a point, and we shouldn't be tempted to flip the electron's toggle switch from fermion to boson. But in a supersymmetric world, the question isn't idle, for in supersymmetry lurks the potential for catastrophe. We must be grateful that the breaking of supersymmetry—had it been present—left us with fermions lighter than bosons, and we will need to understand what made it so.

∞

Particle physics opens possibilities that science fiction can riff on: a world without amino acids, a sooty atmosphere, and the menace of the incredible shrinking bosonic blob. These alternative worlds—and many others—are conceivable, given our current understanding of the fundamental particles and their interactions. Many alternatives lack the potential for the rich chemistry and structure of our world. We may not inhabit "the best of all possible worlds," as Dr. Pangloss would have had it, but it is a lot richer and better for the likes of us than it might have been.

It is amusing and instructive to ask how today's world would change if we could indulge our fantasy of giving the dials a spin, but particle physics doesn't just deal with the world as it is today. The laws of nature that we are learning are those that prevailed early in the history of the universe, so we can also ask how the universe would have evolved had the dials been set to different values.

Examining alternative worlds encourages us to ask what actually determines the nature of the world around us. What sets the parameters that dictate that hydrogen exists, stabilized by the delicate balance that makes a proton lighter than a neutron? What fixes the electron mass? What regulates the mass of the weak boson? We build our particle accelerators many kilometers long and our detectors the size of basketball pavilions in search of answers.

Within the past two decades, visionary theorists have explored a radical, and potentially revolutionary schema inspired by string theory and the cosmic inflation. Leibniz's supreme being considered all possible hypothetical universes and created only the best. Proponents of the string landscape hold that ours is but one world in an immense cosmological ensemble—the *multiverse*—and that different laws of nature—of stunning variety—prevail in different universes.

This is a wild idea, and it departs from the evidence we have that, in our experience, the laws of nature are the same everywhere and at all times. But is it more extraordinary to assert that the universe we can see is all there is, or that other possibilities might be realized? Scientists must be disciplined by evidence, and we must be ready to open our minds when evidence is lacking.

As our Panglossian fable illustrates, only a tiny fraction of the universes that the string landscape conjecture admits would support the evolution of intelligent life. Given the observation that we exist, the thinking goes, the laws of nature that we have uncovered are not so improbable. For the construction of a world view, this new anthropic picture seems a rupture with the practice of half a millennium of seeking the cause behind experimental observations. Perhaps the laws of nature cannot be derived once and for all, but are environmentally determined, influenced at least in part by circumstance.

We have seen that, over time, new questions come within our scientific reach. It also happens that questions that preoccupied our scientific ancestors now seem to us the wrong questions. A famous example is contained in Johannes Kepler's 1596 *Mysterium Cosmographicum (The Cosmographic Mystery)*, the first published defense of the heliocentric solar system advocated by Nicolaus Copernicus. Kepler further tried to explain why the sun should have exactly the six planetary companions he knew—Mercury, Venus, Earth, Mars, Jupiter, and Saturn—in their observed orbits. He arrived at a cunning construction, in which the five platonic solids, nested together in the order octahedron (eight faces), icosahedron (twenty), dodecahedron (twelve), tetrahedron (four), and cube (six), define six inscribed or circumscribed spheres whose diameters correspond to the sizes of planetary orbits. We now know that the sun holds in its thrall more than six planets, not to mention the asteroids, periodic comets, and dwarf planets, nor all the moons around Kepler's planets. Moreover, as Kepler himself showed in his 1609 treatise, *Astronomia Nova (The New Astronomy)*, planets move not in circles, but in ellipses with the sun at one focus.

But improved observational knowledge is not why Kepler's project seems ill-conceived to us; we just do not believe that it should have a simple answer. Neither symmetry principles nor stability criteria make it inevitable that those six planets, formed in the maelstrom of a protoplanetary disc of dust, should orbit our sun precisely as they do, or that another solar system should look the same as ours.

This example holds two lessons for us: First, it is hard to know in advance which aspects of the physical world will have simple, beautiful, informative explanations, and which we shall have to accept as "complicated." Second, and here Kepler's quest to explain quantized circular orbits is particularly inspiring, we may learn very great lessons indeed while pursuing challenging questions that—in the end—do not have illuminating answers.

While chasing his wild goose, Kepler uncovered the three great regularities of planetary motion—elliptical orbits with the sun at a focus, equal areas swept out in equal times, and the relation between orbital period and mean distance from the sun—that led Isaac Newton to his theory of universal gravitation. If we are as serious, perhaps we too will be led to fundamental insights about our world, even if it should turn out to be just a speck in the multiverse.

And what if we humans are forced to contemplate not just our place in the universe, but our place in a multiverse? Confronting in 1864 a seemingly limitless and eternal universe, an infinitude of space alongside an infinitude of duration, Victor Hugo asked, "Mere microbes upon our imperceptible globe during the fleeting instant of our existence, are we not, in the presence of this overwhelming Infinite, terribly lowly and insignificant?" His response offers courage and hope: "No, because we comprehend it."

Humanity's place in the universe is even less central than Victor Hugo and his astronomical muse, Camille Flammarion, could conceive. We are, as they knew, exponentially smaller than the reaches of their 19th-century cosmos, but that is not all. There is much more to the natural world than we can see and touch. The extraordinarily ancient universe that we inhabit is expanding, and it evolved from conditions that do not at all resemble its current state. Even as our instruments and our imagination take us back billions of years in time and out farther than we can ever hope to travel, we are also exploring nature's inner workings at sizes a billion-billion times smaller than our own. We are studying phenomena so ephemeral that they transpire in a trillionth of a trillionth of the blink of an eye.

Titans on the scale of electrons or quarks, we humans do not have dominion over that world of the infinitesimal; instead, what takes place there, so remote from everyday experience, makes the rules that shape our existence. The more closely we examine the submicroscopic realms, the better we understand the human scale and beyond. The uncertainties that come into play at very short distances and fleeting time intervals, expressed in probabilities and quantum fluctuations, seem increasingly likely to have seeded the large-scale structure of the universe. Perhaps we are subjects rather than monarchs!

And yet . . . though nature elicits in us a sense of awe and humility, the understanding we have earned instills in us a sense of purpose that lifts us above languishing in our insignificance. What an understanding it is! Building on the working hypothesis that nature is reliable, not capricious, and the expectation that cause must precede effect, we humans have come to comprehend a great deal about the two infinities—surpassingly large and small—and about our own in-between existence.

We celebrate the elegance and economy of today's picture of particle physics—the standard model—in our book's title, *Grace in All Simplicity*. We need only identify a few basic constituents plus a few symmetries that dictate the fundamental interactions, recognize that symmetries that are concealed foster richness and diversity, and we can derive the essential features of the physical world. We can read some outcomes from the patterns themselves, expressed in a few acts of the particle play. Others—such as the confinement of quarks and gluons, the collisions of heavy nuclei at high energies, the properties of an ensemble of atoms or molecules acting collectively—are emergent phenomena that require analysis of a different sort.

We delight in the phenomena understood, the mysteries resolved, the hints of a more complete understanding, but what animates our work is all that we don't know. We and our colleagues live for the questions still unanswered, the puzzles that remain, the mysteries not yet encountered, the questions we don't know enough to ask.

We two have been lucky to come of age as physicists when the standard model came together, when the two infinities met and merged as never before, when technological innovations made possible new instruments to help physicists explore the boundless horizons of our science. As theorists, we have had the good fortune to work with gifted experimenters, instrument makers, accelerator builders, computational scientists, and technologists of many sorts. It has been our pleasure to share something of those interactions, along with the work of our scientific forebears, in this book.

Particle physics is a planetary undertaking. It brings together people of different backgrounds and diverse styles who are curious about nature and driven to explore. Professional and personal lives are enriched by teachers and students, colleagues and friends, from many countries and cultures. Together, we elevate experimental test over attachment to received truths. We welcome uncertainty not as motivation to be cynical, but as incentive to open our minds and investigate. As individuals and as teams, we compete and collaborate. Our science thrives in a culture in which standing is measured by knowledge shared, not secrets withheld. For that tradition of openness, we thank the citizens of the world, who provide material support and active encouragement for the explorations we undertake together. What we discover is for all to share.

SUGGESTED READING

We would be delighted if a reader of *Grace in All Simplicity* developed a desire to learn more about some of the discoveries or personalities we describe. Among many excellent books that treat topics featured in *Grace*, we offer:

Close, Frank. *Half-Life: The Divided Life of Bruno Pontecorvo, Physicist or Spy.* New York: Basic Books, 2015.

Farmelo, Graham. *The Strangest Man: The Hidden Life of Paul Dirac, Quantum Genius.* London: Faber and Faber, 2009.

Gamow, George. *Thirty Years that Shook Physics.* Mineola, N.Y.: Dover, 1985.

Heilbron, John L. and Robert W. Seidel. *Lawrence and His Laboratory: A History of the Lawrence Berkeley Laboratory.* Berkeley: University of California Press, 1981.

Panek, Richard. *The 4% Solution: Dark Matter, Dark Energy, and the Race to Discover the Rest of Reality.* Boston: Mariner Books, 2011.

Redniss, Lauren. *Radioactive: Marie & Pierre Curie: A Tale of Love and Fallout.* New York: Dey Street Books, 2015.

Rowe, David E. and Mechthild Koreuber. *Proving It Her Way: Emmy Noether, A Life in Mathematics.* Cham, Switzerland: Springer, 2002.

Segrè, Gino. *The Pope of Physics: Enrico Fermi and the Birth of the Atomic Age.* New York: Henry Holt and Co., 2016.

Smoot, George and Keay Davidson. *Wrinkles in Time: Witness to the Birth of the Universe.* New York: William Morrow and Co., 1993.

At a more technical level, we call attention to:

Close, Frank. *Elusive: How Peter Higgs Solved the Mystery of Mass.* New York: Basic Books, 2022.
Pais, Abraham. *Inward Bound.: Of Matter and Forces in the Physical World.* Oxford, UK, and New York: Clarendon Press, 1986.
Segrè, Emilio. *From X-rays to Quarks: Modern Physicists and Their Discoveries.* New York: W.H. Freeman, 1980.

Both historical and contemporary information is available on the World Wide Web.

Wikipedia, the free online encyclopedia, has grown in authority as a source for biographical information about individual scientists and for introductions to ideas and discoveries, https://www.wikipedia.org.

Since the year 2000, the American Physical Society has published *This Month in Physics History,* capsule summaries of signal events, https://www.aps.org/publications/apsnews/features/history.cfm.

Among science periodicals, three freely available online publications are of special note: The CERN *Courier* (https://cerncourier.com), which reports on international high-energy physics; *Symmetry* magazine (https://www.symmetrymagazine.org), a joint publication of Fermilab and SLAC in the United States; and *Quanta Magazine* (https://www.quantamagazine.org), an editorially independent publication supported by the Simons Foundation.

Only a fraction of notable research has been honored with a Nobel Prize, but most Nobel Lectures, which are freely available on the World Wide Web, offer descriptions of the research that serve as bridges to the scientific literature. A complete listing of laureates, with short descriptions of the work being rewarded and links to supporting material is at https://www.nobelprize.org/prizes/lists/all-nobel-prizes-in-physics/.

ACKNOWLEDGMENTS

Grace in All Simplicity is a reflection on how we have come to comprehend the world around us at its most fundamental level. We are indebted to a multitude of men and women from the past and to our colleagues and friends for the discoveries and insights that shape today's marvelous, but tantalizingly incomplete, understanding. *Grace* is our homage to all of them. Very special thanks are due to Kate Metropolis, who contributed directly to the early stages of the work and whose sage literary counsel we have tried to keep ever in our minds. It has been a joy to be represented and nurtured over many years by Michael Carlisle, along with Mike Mungiello and their colleagues at InkWell Management. Jessica Case and the Pegasus team have guided us from a manuscript to the book you have before you. We thank the Rockefeller Foundation Bellagio Center for the grant of a stimulating residency, and we salute the institutions and governments that enable education and research to be shared for the benefit of all. Throughout our adventures as scientists and authors, our wives, Fran and Liz, have provided constant inspiration and support.

INDEX